Active Volcanoes of the World

Series Editors

Corrado Cimarelli, München, Germany
Sebastian Müller, Bristol, UK

For further volumes:
http://www.springer.com/series/10081

Juan Carlos Carracedo
Valentin R. Troll
Editors

Teide Volcano

Geology and Eruptions of a Highly Differentiated Oceanic Stratovolcano

Springer

Editors
Juan Carlos Carracedo
Dpt. Física-Geología
University of Las Palmas de Gran Canaria
Las Palmas de Gran Canaria
Spain

Valentin R. Troll
Department of Earth Sciences
Uppsala University
Uppsala
Sweden

Managing Editor
Sebastian Wiesmaier

Editorial Assistants
Lara S. Blythe
Christine Karsten

Additional material to this book can be downloaded from http://extras.springer.com.

ISBN 978-3-642-25892-3 ISBN 978-3-642-25893-0 (eBook)
DOI 10.1007/978-3-642-25893-0
Springer Heidelberg New York Dordrecht London

Library of Congress Control Number: 2012953115

Printed on acid-free paper

Springer is part of Springer Science+Business Media (www.springer.com)

Preface

The Canary Islands Archipelago, offshore of the northwestern coast of Africa, originated from ocean-island volcanism over a span of 20 million years. This 600-km-long chain of islands (total population ~2 million), with their beautiful volcanic landscapes, beaches, and year-round mild climate, receives more than 12 million visitors each year. The prime tourist destination is Teide Volcano on the Island of Tenerife, the centerpiece of Teide National Park and the focus of this scientific volume. In 2010, Teide National Park was the most heavily visited national park of any European country and the second most visited worldwide. Teide is a huge volcano that towers 3,718 m (a.s.l.) above the central part of Tenerife, reaching the highest elevation in the Canaries and Spain. Moreover, if its height is measured relative to the seafloor, Teide is the third tallest (~7,718 m) volcanic edifice on Earth after the Hawaiian shield volcanoes Mauna Kea and Mauna Loa. In 2007, the United Nations Educational, Scientific, and Cultural Organization (UNESCO) inscribed Teide National Park as a World Heritage Site, in recognition of its diverse, abundant evidence of the geological processes that underpin the evolution of volcanic islands, complementing other volcanic properties such as Hawaii Volcanoes National Park (USA) and Galápagos National Park (Ecuador).

Because of its imposing physical visage, Teide naturally has long attracted scientific attention following the colonization of the Canaries, but especially during the eighteenth and nineteenth centuries when the emerging "science" of geology began to develop. Beginning in the latter part of the twentieth century, many geoscience and related studies—including the systematic geologic mapping and dating of volcano-related deposits—have been conducted at Teide as well as other Canarian volcanoes, resulting in a substantial scientific literature. For example, during the past 6 years, one of the editors (Carracedo) has published and edited three major books (in Spanish) summarizing the volcanic geology and associated hazards of Canarian volcanoes in general, and of Teide in particular. Unfortunately, to date no comparably comprehensive works in English about Canarian volcanism exist. Thus, this volume marks a milestone in remedying this long-standing deficiency. It provides a wide-ranging summary of the geologic evolution of Teide—the emblematic volcano of the Canaries. In 14 chapters, this volume addresses a wide diversity of topics and disciplines, including: the prehistoric to present-day scientific understanding of Teide, its geodynamic setting within the context of plate tectonics (i.e., "hotspot" model), development of rift zones and other volcanic structural elements, radiometric and paleomagnetic dating studies, petrologic-geochemical-isotopic evolution of Teide's magmatic system, island-wide geophysical investigations, eruptive history and styles, and volcanic and other geological hazards.

It is noteworthy that the book's last chapter emphasizes the volcanic hazards of the Teide Volcanic Complex (TVC). While the TVC has erupted five times during recorded history (most recently in 1909), such activity has been relatively weak, causing minimal damage and no fatalities. However, larger prehistoric eruptions and flank collapses along the volcano's rift zones testify to much more hazardous activity in Teide's recent geologic past. The episode of volcanic unrest at Teide during mid-2004, together with the related, highly controversial specific "prediction" of an eruption in October 2004 that did not materialize, has greatly enhanced public awareness of volcanic hazards in Tenerife. The 2004 Teide volcanic "crisis" adversely affected Tenerife's tourism economy and disrupted the daily lives of many of its residents. In addition, the submarine eruption near La Restinga (Island of El Hierro) during 2011–2012—the first since 1971 in the Canaries—has further increased public anxiety regarding hazards posed by future volcanic eruptions. On the positive side, however, these recent developments also have prompted the expansion of real-time monitoring studies of Canarian volcanoes.

Carracedo and Troll are perfectly suited to coedit this volume, because of their own extensive experience in working at Teide and other Canarian volcanoes. This fact is immediately apparent from a quick glance at the Table of Contents, which shows that they are authors or coauthors of many of the book's chapters. With its comprehensive discussion and broad spectrum of topical coverage—well illustrated by photographs, diagrams, and tables—this volume should prove to be highly useful to nonSpanish speaking practitioners within the global volcanologic community, especially those specializing in ocean-island volcanism. Given its scope and breadth, the Carracedo-Troll book is destined to have a long shelf life, serving as a valuable reference work for decades to come. Moreover, this book sets a benchmark for the production of similar summaries of the other historically and potentially active volcanoes of the Canary Islands. The lessons that can be learned from the existing data, and new data to be accrued from future studies, are critical for the preparation of effective emergency-response plans when the next episode of volcanic unrest at Teide, or at some other Canarian volcano, culminates in significant and possibly hazardous eruptive activity.

10 October 2012

Robert I. Tilling
Senior Research Volcanologist, Emeritus
Volcano Science Center
U.S. Geological Survey
Menlo Park CA USA

Acknowledgments

The editors would like to thank Robert I. Tilling for writing the preface and revising Chap. 14 and Sebastian Wiesmaier, Ph.D., for taking over the responsibilities as managing editor of this book. Lara S. Blythe, Ph.D., (chief lector) and Pauline Agnew provided countless hours of meticulous effort and are greatly thanked for improving our English. Christine Karsten, M.A., is thanked for masterly assisting the editor team, and for being there whenever help was needed.

Those researchers who have inspired our thinking on the geology of the Canary Islands, and Teide in particular, shall be acknowledged here also, especially J. M. Fúster, H.-U. Schmincke, J. Martí, R. Cas, C. Stillman, K. Hoernle, T. Hansteen, J. Geldmacher, A. Klügel, A. Gurenko, T. R. Walter, S. Krastel, E. Ibarrola, H. Hausen, P. Rothe, N. D. Watkins, M. Canals, G. Ablay, J. M. Navarro, V. Araña, A. B. Watts, and F. Logopito.

Corrado Cimarelli, Ph.D., Johanna Schwarz, Ph.D., and Sebastian Müller, Ph.D. are thanked for outstanding support from the publisher's side.

Institutional support for the editors and authors is greatly acknowledged, in particular from the University of Las Palmas de Gran Canaria (Spain), Uppsala University (Sweden), Ludwig-Maximilians-Universität München (Germany), IPSL CEA-CNRS Paris (France), Blaise Pascal Université Clermont-Ferrand (France), and Trinity College Dublin (Ireland).

Contents

From Myth to Science: The Contribution of Mount Teide to the Advancement of Volcanology

1

Juan Carlos Carracedo and Valentin R. Troll

Abstract

This chapter outlines the progress of geological research into the origin and evolution of the Teide Volcanic Complex within the framework of Tenerife Island, the Canary Islands, and oceanic volcanism in general. Initially considered to relate to either the entrance to 'Hell' or to mythical Atlantis, for von Buch, von Humboldt, Lyell and the other great eighteenth and nineteenth century naturalists Teide eventually helped to shape a new, and at that time revolutionary concept; the origin of volcanic rocks from solidified magma. This school of thought slowly cast aside Neptunism and removed some of the last barriers for the development of modern Geology and Volcanology as the sciences we know today. Despite the volcanic nature of the Canaries having been already recognised by the twentieth century, modern geological understanding of the archipelago progressed most significantly with the advent of plate tectonics. While some authors still maintain a link between the Canaries and the Atlas tectonic regime (see also Chap. 2), geological research truly advanced in the Canaries through comparison with hotspot-derived archipelagos, particularly the Hawaiian Islands. This approach, initiated in the 1970s, provided a breakthrough in the understanding of Canary volcanism, demonstrating Tenerife and Teide to be one of the world's most interesting, complex and to many, one of the most iconic of oceanic volcanoes.

1.1 Introduction

European volcanoes such as Etna and Vesuvius have been constant references in Volcanology since Greek and Roman times. Detailed and accurate accounts, most notably the description by Pliny the Younger of the 79 A.D. eruption of Vesuvius that destroyed Pompeii and Herculaneum, laid the foundations of modern Volcanology. Volcanic terminology as common as

J. C. Carracedo (✉)
Departamento de Física (GEOVOL), Universidad de Las Palmas de Gran Canaria, Las Palmas de Gran Canaria, Canary Islands, Spain
e-mail: jcarracedo@proyinves.ulpgc.es

V. R. Troll
Department of Earth Sciences CEMPEG, Uppsala University, Uppsala 75236, Sweden
e-mail: valentin.troll@geo.uu.se

J. C. Carracedo and V. R. Troll (eds.), *Teide Volcano*, Active Volcanoes of the World,
DOI: 10.1007/978-3-642-25893-0_1, © Springer-Verlag Berlin Heidelberg 2013

"volcano" and "basalt" were first used in accounts penned by Pliny the Elder, as was "crater" by Aristotle. Etna and Vesuvius became historically relevant because of their frequent catastrophic eruptions that destroyed entire cities, such as Catania, in 1669, or Naples, in 1631, both causing many thousands of victims.

In contrast, the only aspect of interest of Mt. Teide until the eighteenth century was its exaggerated height (Figs. 1.1, 1.2). Teide was considered the highest mountain on Earth until Mont Blanc and the Andean volcanoes were measured and observed to be higher. It is interesting to note, however, that present-day Volcanology has reinstated Teide amongst the highest volcanic structures on the planet (only surpassed by Mauna Loa and Mauna Kea, on the island of Hawaii). If the base level is taken to be the ocean floor and not sea level, Mt Teide rises above 7,000 m (3,718 m a.s.l.).

While Vesuvius and Etna defined important catastrophic episodes in the history of Italy from Roman times to present, Teide volcano only posed a threat to the smaller population of aboriginal inhabitants on the island of Tenerife (the Guanches). The absence of explosive eruptions and victims since the colonisation of Tenerife at the end of the fifteenth century promoted the image of Teide as the main stable element in the landscape of the entire archipelago and as a prime cultural reference, even locally acquiring a protective role in folklore for example as "Father Teide". The eruptions on Tenerife in historical times have had a limited impact on the population and the economic infrastructure of the island, with the exception of the 1706 eruption which partially destroyed the town of Garachico and filled the harbour with lava (the main commercial port in Tenerife at the time). This eruption, however, was not directly related to Teide, its vent being located 17 km away on the NW rift zone.

The role played by the Canaries and Mt. Teide changed lastingly upon the arrival of well established naturalists such as Leopold von Buch, Charles Lyell, Alexander von Humboldt and Georg Hartung, among many others. During the eighteenth century, geology was at the centre of a long-lasting controversy between those who held the view that all rocks, including what we now see as volcanic rocks, were marine deposits formed by chemical precipitation in the ocean (Neptunists, after the god of the sea in Roman mythology) and those who believed that volcanic rocks resulted from the solidification of molten masses from the Earth's interior (Plutonists, after Pluto, Greek god of the underworld).

The former school, led by Abraham Gottlieb Werner (1750–1817), a renowned German professor of Geology (Fig. 1.3), and the latter by the Scot James Hutton (1726–1797), established a lively debate with strong religious overtones that lasted almost an entire century. The neptunistic theories rigorously adapted the teachings of the book of Genesis, contrasting the more "enlightened" ideas of the plutonists. The controversy contributed decisively to the development of Geology as a modern science and was based to quite an extent on the observations made in the Canaries by the now famous eighteenth century naturalists.

The relevant role of the Canaries and Mt. Teide in the resolution of crucial problems in Geology and Volcanology arose from the European continent, particularly from Germany, France and Scotland, due to the fact that the volcanic settings in those countries are much more difficult to interpret than Canarian volcanoes. Fervent neptunists and co-workers of the influential Professor Werner, such as von Buch and initially even von Humboldt himself, who had expressed numerous doubts, gradually became ardent defenders of plutonism after travelling to the Canaries, thereby irreversibly opening the door to the advancement of purely scientific Geology that was largely free from religious restrictions. To von Buch we owe the basic concept that minerals in lava are formed by magmatic crystallisation and to von Humboldt that volcanic alignments are due to tectonic activity at depth.

Regrettably, the essential role of the Canaries and Mt. Teide during this important stage in the development of modern Geology and

Fig. 1.1 The island of Tenerife and a towering Mount Teide in an engraving by Olfert Dapper in 1686

Volcanology did not continue. There were no research groups or centres in Spain or the Canary Islands at that time devoted to the discipline of Geology. Nonetheless the Canary Islands offered a privileged setting in which to study the Geology of oceanic islands, made possible by exceptional conditions: the absence of significant subsidence, allowing observation of all stages of evolution starting with the oldest formations. This is impossible in most similar archipelagos, where subsidence is a relevant factor causing the insular edifices to be submerged during relatively early stages of their evolution and the scant plant cover and low relative meteorisation rate of rocks and formations, being much lower in the Canaries because of the comparatively low rainfall. These favourable circumstances converted the Canaries, and Teide and the surrounding area in particular, into a world-renowned setting for the study of Volcanology, but this was not understood until the second half of the twentieth century.

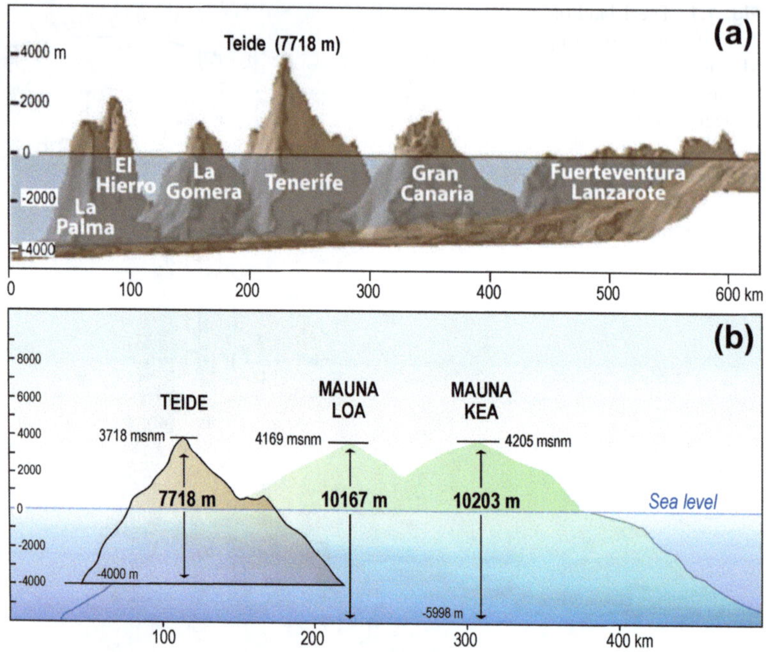

Fig. 1.2 a and b. Teide is by no means earth's highest mountain, as was generally accepted until Mont Blanc was measured (the first recorded ascent of Mont Blanc, 4,810 m, was in August 1786). However, besides having the highest elevation in the Canaries and Spain, it is the third highest volcanic feature on earth, with only Mauna Kea and Mauna Loa being higher

Fig. 1.3 Werner described the basalts of Stolpe (the birthplace of Leopold von Buch) as sediments without traces of melting. He interpreted the columnar features as desiccation cracks, like those found in drying mud

Establishment of the Hawaiian Volcano Observatory (HVO) by the U.S. Geological Survey at the beginning of the twentieth century is acknowledged as the key element in advancing the study of the Hawaiian Islands and leading to the development of modern Volcanology. Although an intense and continuous study of the Canaries began shortly thereafter, during the 1960s, there is a fundamental difference: while in Hawaii the above-mentioned

Observatory was a centre for the great majority of volcanological studies since 1912, a similar centre was never created in the Canaries, but research was led from Madrid, with the corresponding loss of efficiency and the dispersion of efforts, hindering the possibility that the Canary Islands could have become a similar world-famous setting for the development of Volcanology some 100 years ago.

This is exemplified in the development of volcanological terminology employed in the eighteenth and nineteenth centuries derived from Latin (volcano), Greek (crater, pyroclast, phonolite, etc.) and, to some degree Canarian Spanish (caldera, *malpaís*), but American English was the language used, coinciding with the creation of the Hawaiian Volcano Observatory, since the start of the twentieth century (hotspot, pillow lavas, surge, shield volcano, etc.) and especially Hawaiian terms (e.g., pāhoehoe and ʻaʻā lavas) became internationally accepted.

1.2 Teide Volcano in Classical Mythology

There have been some references to Teide, mainly of a mythological nature, in the Classical Era. The best-known and most enduring legend involving the Canaries is the one related to Atlantis, narrated by the Greek philosopher Plato (427–347 BC) in his work Timaeus and Critias. According to this legend, a civilisation, the Atlantean, as advanced and powerful as the Egypt of the Pharaohs, disappeared overnight when the continent sank into the ocean. Only the highest peaks remained above water, to form the archipelagos of Macaronesia: Azores, Madeira, Cape Verde and the Canaries.

It is through Jean Baptiste Bory de Saint Vincent that this legend became scientifically significant in relatively modern times, when he related the Canaries to Atlantis during a visit to the archipelago, described in his work entitled *Essais sur les Îles Fortunées et l'Antique Atlantide* (Kunzli 1911). Acknowledged as a distinguished naturalist, Bory de Saint Vincent conferred scientific credibility on this legend, which was considered to be one of the possible theories of the origin of the Canaries until the mid-twentieth century. It was only when the Canaries were found to overlie oceanic crust, which moreover is more than 180 million years old, that any scientific basis ascribed to this attractive legend was radically dismantled.

However, reality exceeds even the most imaginative legends. Plato would probably have been stunned by a story involving an entire continent (Africa) moving several thousand kilometres away from America over more than 180 million years to form an ocean (the Atlantic) through which, more than 20 million years ago, the volcanic Canarian Archipelago was formed by a magmatic plume originating from the Earth's interior at a depth of almost 3,000 km, and producing at its highest point, Teide, stretching vertically over 7,000 m from the ocean floor.

1.3 Mt. Teide in the Pre-Hispanic World

For the Pre-Hispanic population of Tenerife (the Guanches) Teide was the dwelling place of Guayota, an evil mythical creature, god of the deceased and identified with Hell (von Fritsch and Reiss 1868). The Guanches therefore envisioned Mt. Teide as a demonic spiritual force that brought death and destruction, quite the opposite of the image it adopted later in Hispanic Canarian folklore. The fear and superstition of the Guanches developed as they lived alongside the volcano and may have witnessed at least 6, possibly 8 of its eruptions, mostly around the base of the stratovolcano and on the NW Rift. On the other hand, they learnt how to take advantage of the resources provided by volcanism: the *cañadas* (flat, pumice-covered paths) for the seasonal migration of their goat herds; the volcanic rocks for building their huts, and the caves and volcanic tubes for occasional shelter. They were adept at mining the glassy volcanic obsidian, with which they skilfully fashioned cutting tools.

Similarly to most nomadic tribes, it is very possible that they used fire to clear the land of brush in order to make new pastures for their livestock, thus providing the source of several references to eruptions on Teide reported in ships' logs. As an example, the pre-historical age (1430 AD) for the volcanic cones nested in the La Orotava Valley comes from a Guanche oral tradition, reported by Humboldt on his journey to Tenerife in 1799. However, charcoal underlying lapilli from this eruption yielded a ^{14}C age of 29.090 ± 190 years BP, and the lavas, a $^{39}Ar/^{40}Ar$ age of 27.000 ± 5.900 ky (Carracedo et al. 2010). The Guanche tradition seems to fit better with the calibrated radiocarbon age of 590 ± 66 years BP, most probably related to a forest fire, obtained from charcoal underlying a pumice deposit mantling the Orotava Valley, probably from the Montaña Blanca eruption (Figs. 1.4, 1.5) (Carracedo et al. 2007).

Only a few Guanche words have survived, mostly in geographical and toponymical terms. The very name of Teide has its origin in the Guanche term *Echeide* (Hell). It is surprising, however, that this name was given to Teide and not to the island of La Palma, where volcanoes have been much more active during the Guanche period, causing several victims amongst the local population (Rodríguez Ruiz et al. 2002). Perhaps it was the continuous fumarolic activity at Teide's summit (with temporal emission of hot sulphurous gases forming a plume that may occasionally have been quite voluminous) that contributed to Teide being named after Hell, as eruptive activity on Teide's cone itself was limited to a single eruption during the Guanche Pre-Hispanic period (Carracedo et al. 2007).

1.4 References in the Fourteenth and Fifteenth Centuries

The first references to volcanic eruptions in Tenerife are limited to distant sightings by fifteenth century sailors, who used Teide as a natural landmark during their voyages across the

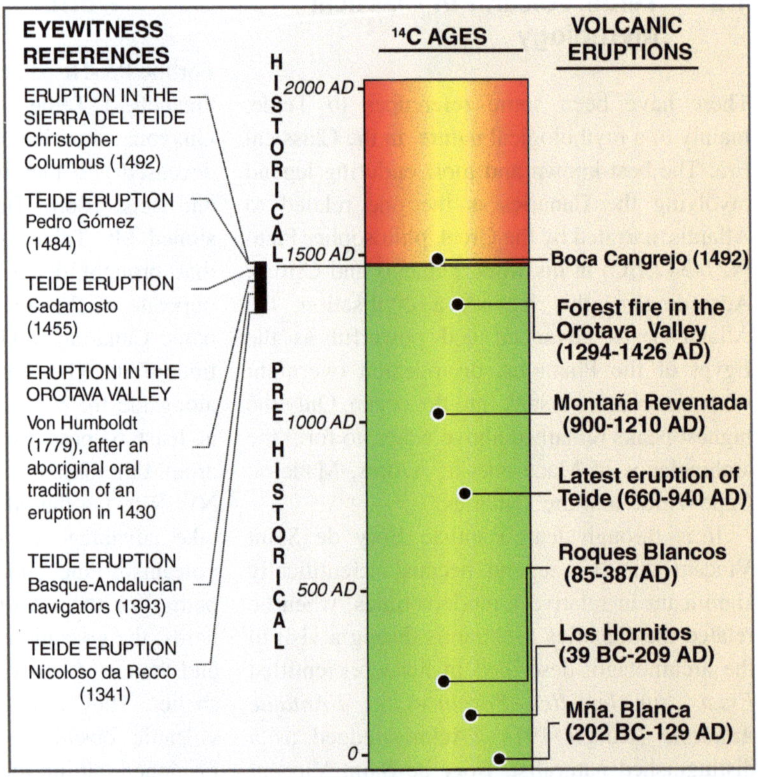

Fig. 1.4 Eyewitness references to supposed volcanic eruptions of Teide volcano (*left*) and actual geological events in the last 2000 years in Tenerife (of historical age or C^{-14} dated; *right*). The alleged date of the eruption in the Orotava Valley (1430 AD) coincides with an extensive forest fire in the valley dated at 1294–1426 AD (modified from (Carracedo et al. 2010)

Fig. 1.5 **a**, **b**. Frequent spectacular 'plumes' in the summit area of Teide, locally known as 'Teide's headdress'. **c**. Small plume at the top of Teide in October 2004, initially interpreted as evidence of volcanic reactivation causing considerable alarm, later confirmed as a meteoric cloud (La nube que quiso ser protagonista. EL DÍA. Santa Cruz de Tenerife, 21-10-2004). **d**. Model of the formation of clouds at the summit of Teide volcano by local orographic convergence (Álvarez and Hernández 2006)

Atlantic because of its great height. Many of those references include descriptions of possible volcanic eruptions.

An account of a possible eruption of Teide contained in the ship's log kept by Nicoloso da Recco, copied by Boccaccio, was put forward by Santiago (1948) as indicating a Teide eruption: "*it must be remembered that, in 1341, the Italians, Castilians and other Spaniards who accompanied Recco observed that smoke issued from the Peak*" (Friedlander 1915).

However, the original source, the account by Giovanni Boccaccio ("About Canaria and other islands newly found in the ocean beyond Spain" 1341) clearly describes a well-known meteorological phenomenon, the so-called "Teide's headdress" (Fig. 1.2), a cloud that forms over the summit area due to an adiabatic process similar to the foehn effect: "*They found an island at which they did not wish to disembark because a certain wonder occurred there. A mountain is said to exist there, which, according to their calculations, is thirty miles high, or even more… at whose summit there is a mast the size of a ship's, from which hangs a large lateen sail, taut as a shield, that swollen with the wind extends over a large area, only to appear to decrease little by little, as in ships, to rise again at once, always in this same manner.*"

It is quite surprising that this accurate description of an atmospheric feature—clouds at the summit of Teide volcano that formed by local orographic convergence—has been interpreted as eruptions of Teide even in very recent scientific articles, using this feature to assess the probability of the eruptive hazard of the volcano. It is equally surprising that the development of one of these clouds over Teide on October 20, 2004 was believed to signal the onset of an eruption and caused great alarm among the residents of the island. Scientific and technical personnel continued relating this cloud to an eruptive column, asserting that this phenomenon was related to an increase in seismic activity observed at the time.

Another description of a possible eruption dates from 1393, the original source being the accounts of Andalusian and Basque seamen included in the chronicles of King Henry III, quoted for the first time in 1839 by Webb and Berthelot (Hausen 1955). That account states that "*on coming closer to the island they saw flames and smoke issuing from the highlands, whereupon they did not dare to disembark and sailed away from what they then began to call Hell Island*". Since the last Teide eruption has been dated at a much earlier date (eighth century), this putative eruption does not fit into the history of the volcano, and probably reflects a meteoric phenomenon.

A further reference to a possible eruption of Teide is that of Ca'da Mosto (*Il libro de la prima navegazione per l'Oceano alle terre de Negre de la Basse Etiopia*, 1455), citing "*Tenerife, the most populated of the islands and one of the highest on Earth…and in clear weather a mountain can be seen from a great distance burning continually in the centre of the island*". The radiometric ages obtained do not allow any leeway for a known eruption of Teide in that period (Fig. 1.4), thus these seafarers were most probably describing fumarolic activity at the peak, forest fires or the spectacular meteorological phenomena above Teide.

1.5 References to Teide Volcano at the Dawn of Science: The Renaissance and Baroque Periods (Sixteenth and Seventeenth centuries)

Rather than studying Teide as an active volcano, a task that would be approached centuries later, during the pre-scientific period in which the magical vision of the mountain was maintained, the most important issue was identifying its position and altitude (for navigational purposes).

Teide was surrounded by a mystical aura and believed to be the highest mountain on Earth until the altitude of Mont Blanc was measured. It was said that the sun seemed to be closer when viewed from the peak of Teide and that the heat was irresistible. In fact, Teide is a mountain that is relatively easy to climb (the custom nowadays is for all Canarians to climb the volcano on foot at least once, but there is also access by cable car). Back then, it seemed an extraordinary challenge, however. It is therefore not surprising that the main objective of the early scientists visiting the Canaries was to make the ascent to the Peak of Teide (Fig. 1.6).

If we examine the altitudes assigned to Teide until the latter part of the seventeenth century (see Fig. 1.1) one notes that they are expressed in miles and even leagues (about 3 miles), while the more suitable method of measuring in toises (190 cm approx.) or fathoms (90 cm approx.) was only introduced in the late seventeenth century.

In 1631, an eruption of Vesuvius that generated "torrens cineris" or torrents of ash—known today as pyroclastic flows—caused more than 4,000 victims. Shortly thereafter, in 1669, Mt. Etna erupted catastrophically, devastating one third of the area of Catania. Those catastrophic events prompted the study of volcanoes. At that time, the newly explored Andean volcanoes were the subject of continual reports, with even greater altitudes and with yet more frequent eruptions.

The scenario was prepared for the crucial visits and observations of the great naturalists of the eighteenth century—von Buch, von Humboldt, Lyell—fully exploiting the possibilities afforded by the industrial and cultural revolution at that time for exploration and scientific progress.

In the mid-seventeenth century, the scientific revolution (Galileo, Descartes, Newton, etc.) established the firm basis of a fundamental tool, the application of scientific method, in contrast with prevailing religious beliefs. Teide would no longer have strong magical connotations and would instead slowly transform to the theme of research it has become today.

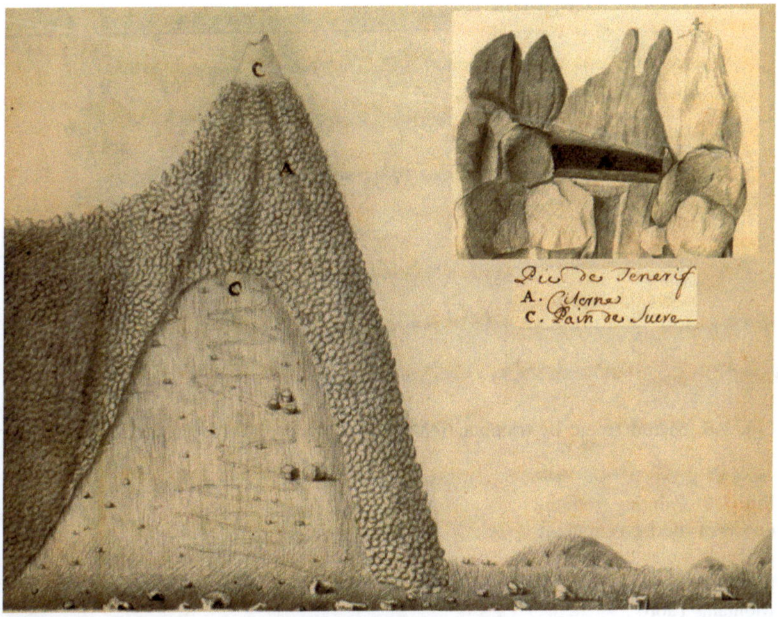

Fig. 1.6 Drawing by Louis Feuillée of the "Pic de Tenerife" (Teide Volcano) and the summit cone (Pain de Sucre or Sugar Loaf, Pan de Azúcar in Spanish) (Feuillée 1724). The path to climb the volcano and a natural reservoir (holding melting ice) are shown as facilities for the ascent

1.6 The Contribution of the Great Eighteenth and Nineteenth Century Naturalists

The main objectives early in this epoch were the ascent of Teide and measuring its altitude. The exact height was crucial for ships to calculate their position by means of simple trigonometric approximations (Figs. 1.7, 1.8, 1.9).

The Royal Academy of Sciences of Paris commissioned the astronomer Louis Feuillée in 1724 to set the precise position of the first meridian (on the island of El Hierro) and the altitude of Teide (see Fig. 1.7). Feuillée's measurement (2193 toises or 4,274 m) was considered incorrect and remained unpublished. In 1776, Jean Charles Borda, sent by the Royal Academy of Sciences of France to Tenerife with the same objective, obtained a value of 1905 toises (3,713 m), very close to the true elevation of 3,718 m (Borda 1776). Even Alexander von Humboldt, who arrived in Tenerife in 1779, was unable to improve Borda's work, which remained the best measurement of Teide's altitude until 1851 (Fig. 1.10).

However, the importance of Humboldt's visit to Tenerife was not only related to the accurate assessment of Teide's altitude, but to his geological and volcanological observations. Despite the fact that von Humboldt was a former student of Abraham Gottlieb Werner, the founder of the school of Neptunism, he completely changed his

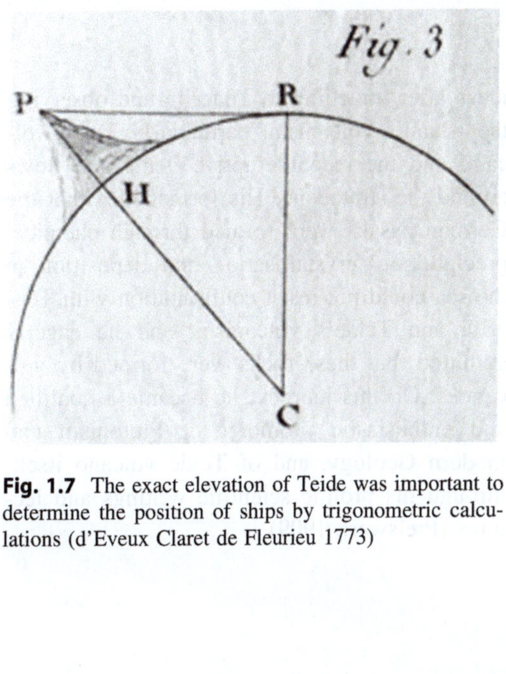

Fig. 1.7 The exact elevation of Teide was important to determine the position of ships by trigonometric calculations (d'Eveux Claret de Fleurieu 1773)

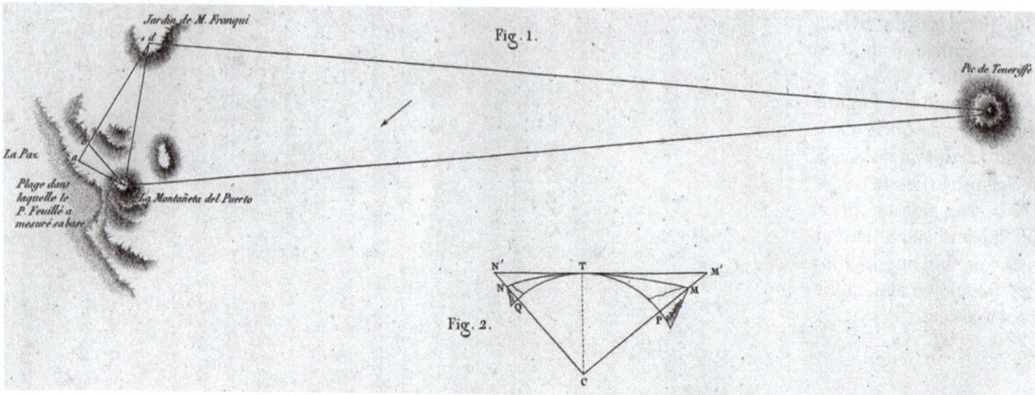

Fig. 1.8 Measurement by triangulation of the altitude of Teide volcano made by Borda in 1776 (Preiswerk 1909)

Fig. 1.9 Painting from the epoch of the measurement of Teide's elevation. Here, a view at the summit of Montaña Taoro (Montañeta del Puerto in the drawing of Fig. 1.8), one of the stations used in the triangulation

ideas after travelling to Tenerife and observing the island's volcanism, particularly Teide volcano and the recent eruptive vents and flows around the stratocone. His former idea that the German basalts were formed through chemical precipitation, crystallisation and deposition in the sea could not resist confrontation with Tenerife and Teide's volcanism, and he eagerly admitted that these rocks were formed by volcanoes. On this journey, he became a qualified and enthusiastic promoter of Plutonism and modern Geology, and of Teide volcano itself, through his prolific scientific writings and lectures (Preiswerk 1909).

Giovanni Bocaccio	1341	30 miles	
Alvise Ca'da Mosto	1455	60 Italian miles	
André Thevet	1555	18 marine leagues	
Thomas Herbert	1624	15 miles	
Bernhardus Varenius	1650	4 miles and 5 furlongs	
Edward Barlow	1668	27 miles	
Allain manesson Mallet	1683	15 miles	
Robert Challe	1690	2730 toises	5320 m
Louis Feuillée	1724	2193 toises	4274 m
Manuel Hernández	1742	2658 toises	5180 m
John Cross	1742	2408 toises	4693 m
Thomas Astley	1744	2,25 miles	4162 m
Michel Adanson	1794	2052 toises	3999 m
Jean Charles Borda	1776	1095 toises	3713 m
Alexander von Humboldt	1799		3736 m
Charles Phillipe de Kerhallet	1851		3715 m
Charles Piazzi Smyth	1856		3717 m
Parque Nacional del Teide	1954		3718 m

Fig. 1.10 Recorded elevation of Teide volcano through history

Leopold von Buch, also a former student of Abraham Gottlieb Werner and an ardent neptunist, visited Tenerife in 1815 following Humboldt's advice. He also was soon persuaded of the volcanic origin of Teide and the surrounding Las Cañadas Caldera; contradicting Werner, he admitted that volcanism is one of the main processes on Earth. However, after taking this

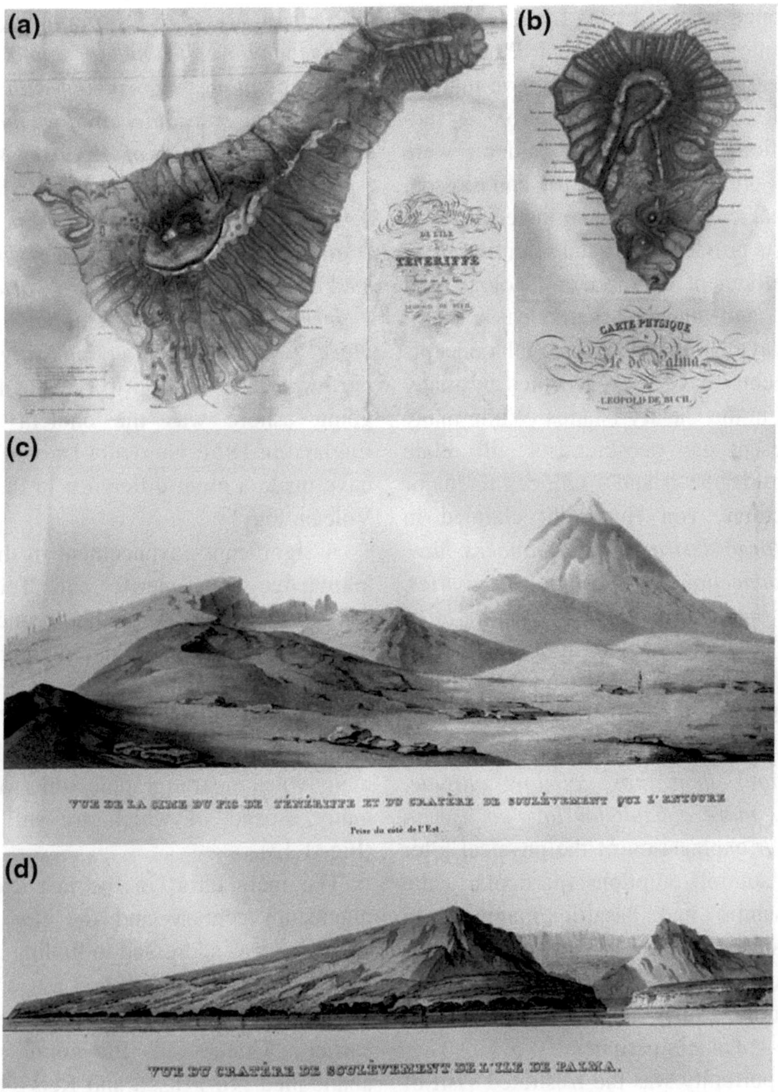

Fig. 1.11 To Leopold von Buch, Tenerife and La Palma were the prototypical examples of uplifted craters (or "craters of elevation", to distinguish them from eruption craters). The islands had been thrust upwards and then collapsed at their centres to form an uplifted crater or "caldera", a term that he took from La Palma. In this theory, that surprisingly had immediate success, the island was not the result of lava accumulation but "emerged ready-made from the interior of the earth". **a** Map of the island of Tenerife by Leopold von Buch (Jeremine 1930). **b** Map of the island of La Palma by Leopold von Buch (Jeremine 1930). **c** The Pic de Tenerife (Mt. Teide) and the encircling "uplifted crater" (the Caldera de Las Cañadas) viewed from the east in a drawing by Leopold von Buch (Jeremine 1930). **d** View from the west of the "uplifted crater" of La Palma in a drawing by Leopold von Buch (Jeremine 1930)

crucial step forward, von Buch took one step backwards with his theory of Craters of Elevation (*Erhebungscrater*), interpreting the Caldera de Las Cañadas and the Caldera de Taburiente (La Palma) as prototypical examples (Dittler and Kohler 1927) (Fig. 1.11a–d).

The Craters of Elevation theory was definitively abandoned when Charles Lyell, a student of James Hutton (the founder of Plutonism), arrived in the Canaries in 1853 to prove that the islands were formed by accumulation of successive eruptions, and that their calderas were not caused by uplift, but by collapse and erosion.

These great eighteenth and nineteenth century naturalists provided a crucial scientific basis for the development of modern Geology and Volcanology, and many of their ideas are still accepted today. Humboldt expressed concepts that only recently have been accepted by many geoscientists in the Canary Islands. While many of these present day geoscientists still relate seismicity inside the island's edifices to major oceanic fractures, von Humboldt claimed in 1800 that "*large destructive earthquakes have no direct connection with volcanic activities, which are the cause only of small local shocks…*", precisely the current distinction between tectonic earthquakes and local seismicity related to volcanism. It was von Humboldt's idea that "*Very high volcanoes have fewer eruptions than those of low altitude, because it is more difficult for lava to ascend them*", a clear explanation of the physical filter imposed on summit eruptions (particularly the heavier basanitic and basaltic magmas) in stratocones such as Mount Teide when magmas reach a critical height, favouring the eruption of lighter, phonolitic ones and eventual focus of vents on the volcano's periphery.

A lost opportunity was the frustrated visit of Charles Darwin to Tenerife. Inspired after reading Alexander von Humboldt's account of his ascent of El Teide, Darwin arrived in Tenerife in 1831 as the expedition naturalist aboard the HMS Beagle. However, as Darwin reports "*After heaving to during the night we came in sight of Tenerife at daybreak… The peak or sugar loaf has just shown itself above the clouds. It towers in the sky twice as high as I should have dreamed of looking for it. Oh misery, misery, we were just preparing to drop our anchor within half a mile of Santa Cruz when a boat came alongside bringing with it our death-warrant. The consul declared we must perform a rigorous quarantine of 12 days. Matters were soon decided by the Captain ordering all sail to be set and make a course for the Cape Verde Islands… And we have left perhaps one of the most interesting places in the world, just at the moment when we were near enough for every object to create, without satisfying, our utmost curiosity*". The reason to prevent Darwin from going ashore was the cholera outbreak in England in 1831. No doubt Darwin's visit could have made a great difference in the progress of Volcanology!

A significant advancement in the geological knowledge of Tenerife and Teide Volcano came with the work in the second half of the nineteenth century of the German geologists Fritsch, Hartung and Reiss (von Fritsch 1867). The first geological map of Tenerife was compiled by W. Reiss, already depicting the main volcano-stratigraphic units of the island, many aspects of which are still valid today (Fig. 1.12).

The main effort in the last decades of the nineteenth century and the first part of the twentieth was addressed to finding a solution for the origin of the Caldera de Las Cañadas and the Orotava and Güímar Valleys, once von Buch's earlier "Craters of Elevation" theory was abandoned. To Fritsch and Reiss, the two morphological depressions forming the Las Cañadas Caldera—divided by the Roques de García large spur—are the headwalls of two main drainage

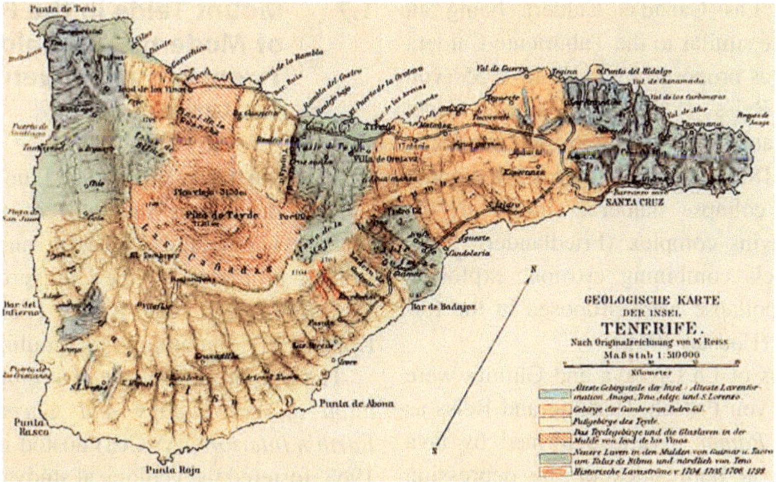

Fig. 1.12 The first geological map of Tenerife (von Fritsch 1867). The main volcano-stratigraphic units of the island are clearly defined: *in blue*, the oldest lavas (the Miocene Shields); *orange*, the Cumbre de Pedro Gil (the NE rift zone); *yellow*, the flanks of Teide (the Las Cañadas volcano); red stripes, Teide lavas filling the Caldera de Las Cañadas and the Icod Valley; *green*, recent lavas; *red*, historic eruptions (Meyer 1896)

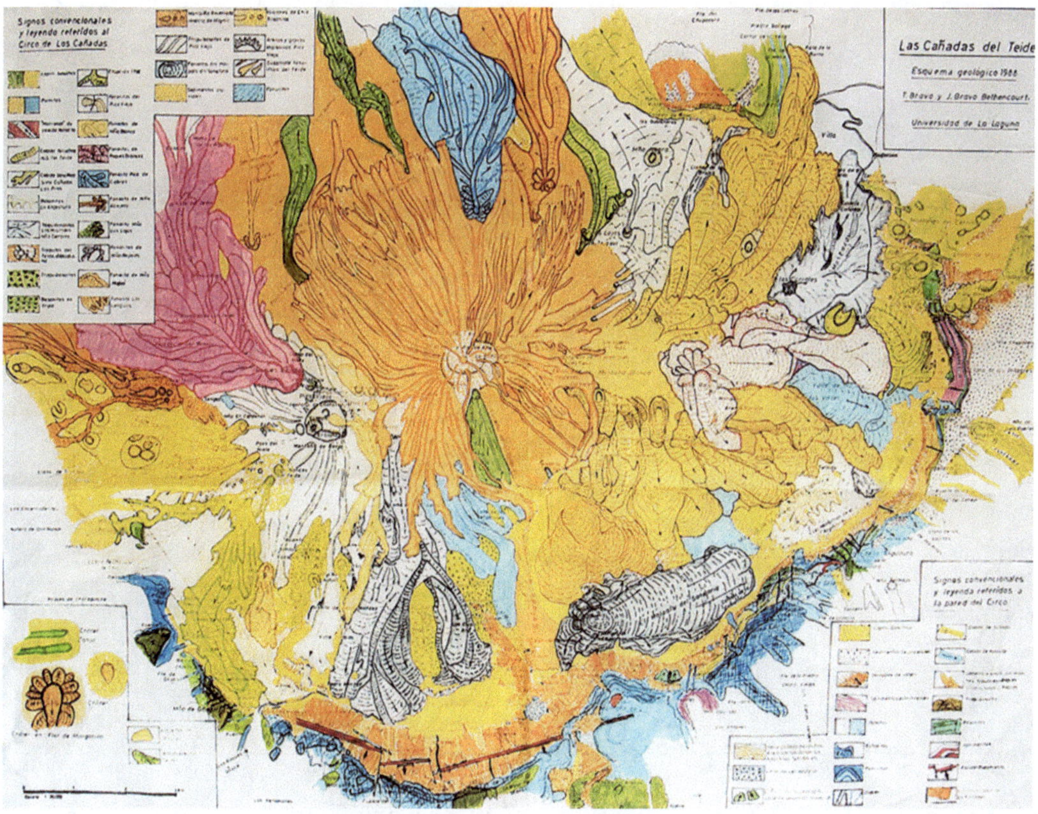

Fig. 1.13 Geological sketch map by Bravo-Bethencourt and Bravo (1989); from (Araña and Coello 1989)

systems, the Las Cañadas Caldera being an erosive feature similar to the Taburiente Caldera in La Palma, as proposed by Lyell in 1835 (von Fritsch and Reiss 1868). In contrast, Gagel (1910) postulated an explosive origin, similar to the Krakatau 1883 eruption, whereas Friedlander suggested a collapse caldera, similar to the Somma-Vesuvius complex (Friedlander 1915). Several models combining erosion, explosion and vertical collapse were proposed in the following years (Hausen 1955).

The valleys of La Orotava and Güímar were explained by von Fritsch, Hartung and Reiss as "*intercolline Räume*", valleys formed by lava accumulation at both sides of the depression (von Fritsch 1867).

1.7 Mount Teide in the Framework of Modern Volcanology: The Twentieth and Twenty-first Centuries

Research on Teide Volcano and the Las Cañadas Caldera during the first half of the twentieth century was mainly focused on petrological studies, prompted by the Chinyero eruption in 1909 (Preiswerk 1909; Kunzli 1911; Dittler and Kohler 1927; Jeremine 1930; Smulikowski 1937).

The Symposium of the *International Association of Volcanology and Chemistry of the Earth's Interior* (IAVCEI) hosted in Tenerife in 1968, fostered the geological study of the Canary Islands, particularly Tenerife and Teide Volcano,

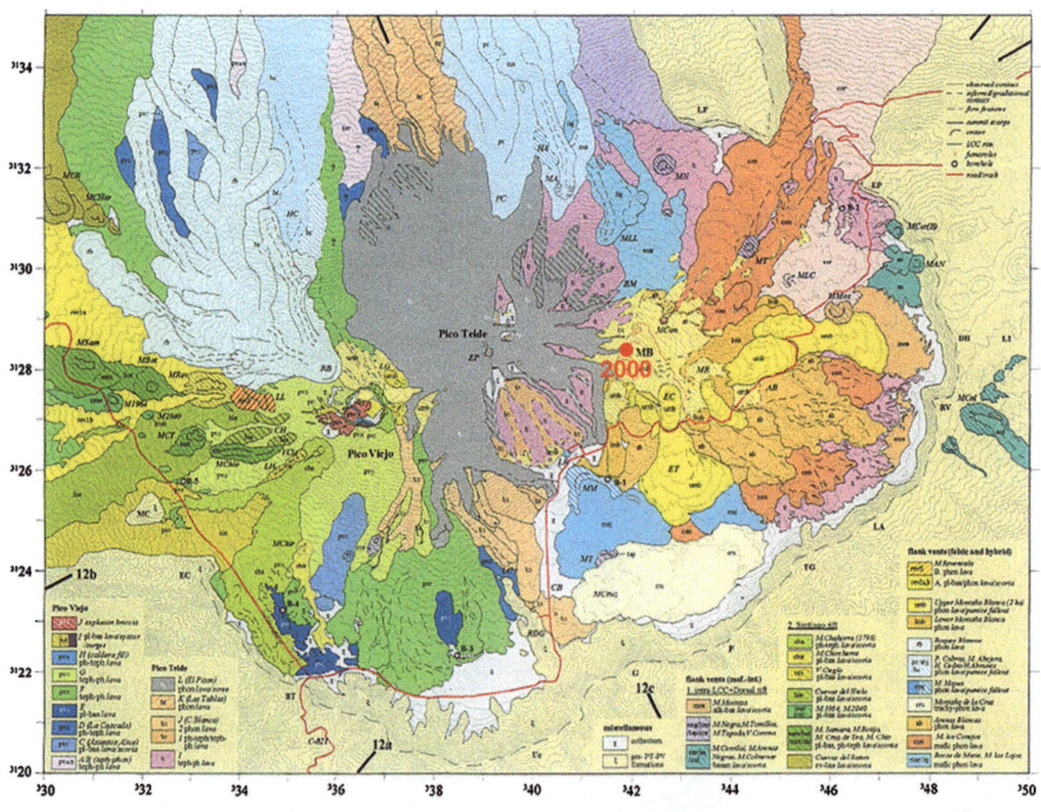

Fig. 1.14 Geological map of the Teide volcanic complex (Ablay and Martí 2000). This map is restricted to the Teide-Pico Viejo stratocones and vents, and the proximal edges of the NW and NE rift zones. Only one radiometric age is provided, which dates the 2 ky eruption of Montaña Blanca (Ablay and Marti 2000)

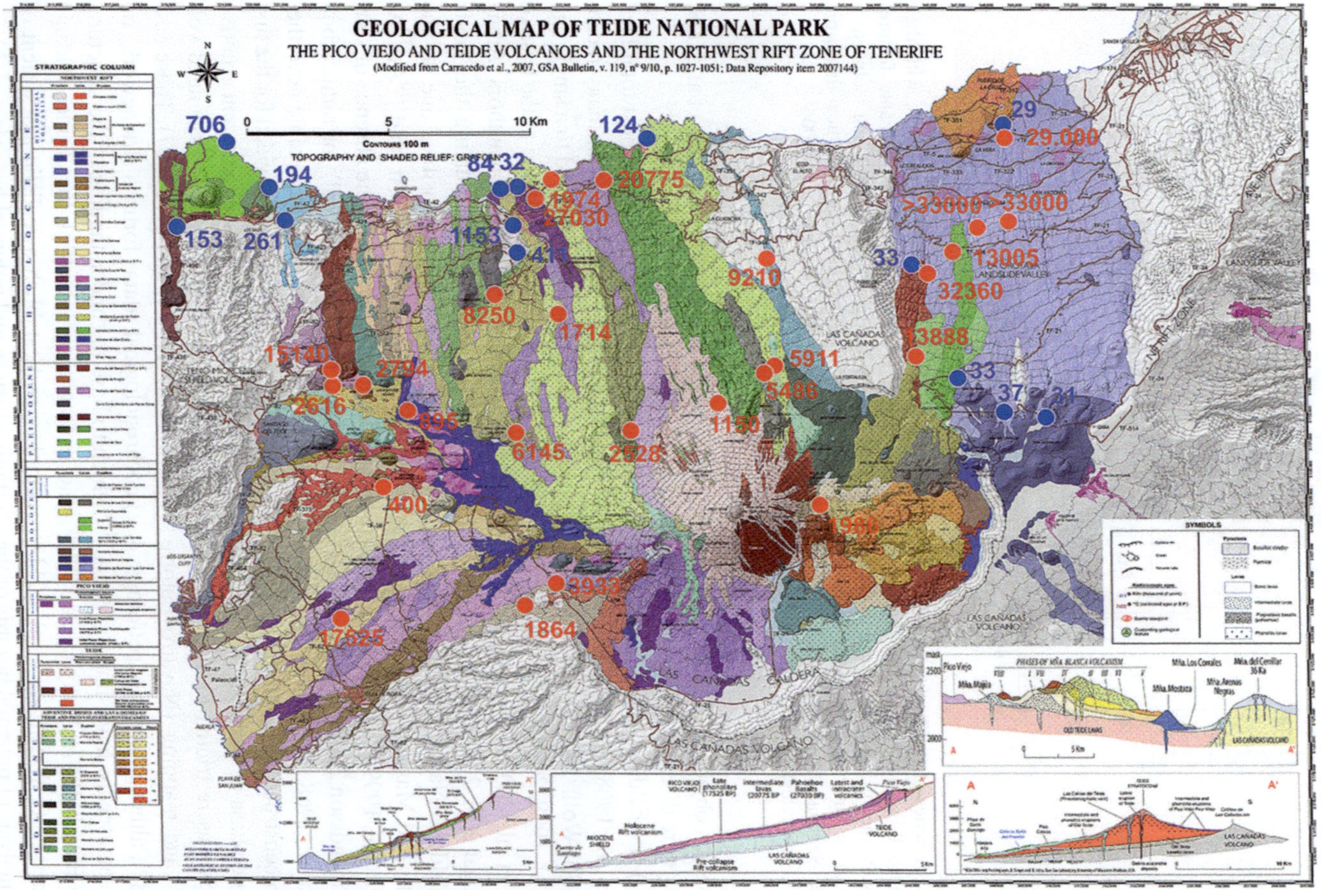

Fig. 1.15 Geological map of the Teide volcanic complex (Carracedo et al. 2007). This map includes the Teide and Pico Viejo stratocones, their vents, and the NW and NE rift zones. The map shows the ^{14}C and new K/Ar ages, on which the reconstruction of the structural and eruptive evolution of the volcanic complex was based. Ages shown in red are ^{14}C in years; in *blue*, K/Ar in ky (modified from Carracedo et al. 2007)

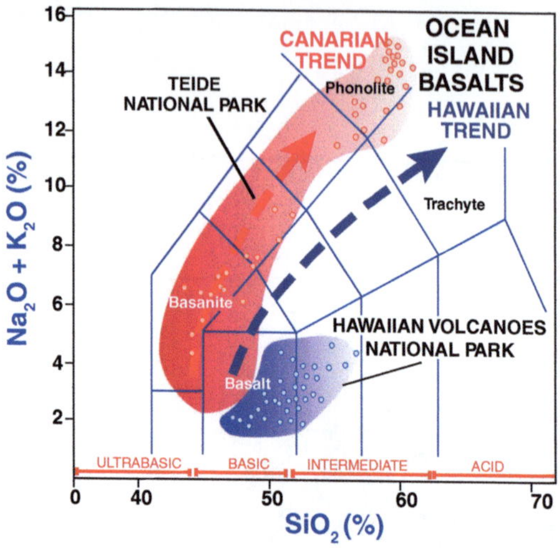

Fig. 1.16 Magmatic series of the Teide volcanic complex and the Mauna Loa and Mauna Kea volcanoes, forming respectively the Teide and Hawaii National Parks. In contrast to the basic magmas of the latter, the eruptions of Teide National Park include more evolved rocks (phonolites, trachytes). Combined, both sites represent the entire series on a large scale, with their corresponding eruptive mechanisms, volcanic features and landforms, justifying both Parks being included in the UNESCO World Heritage list (analytical data from Clague 1987; Rodríguez-Badiola et al. 2006)

since then a research objective of global interest. Research efforts were directed to the study of the older (>200 ky) pre-caldera Las Cañadas Volcano (Fúster et al. 1968; Ridley 1970; Araña 1971; Booth 1973; Wolff 1985, 1987; Martí et al. 1994; Bryan et al. 1998, 2000, 2002; Edgar et al. 2002; Huertas et al. 2002; Pittari et al. 2005; Bryan 2006; Edgar et al. 2007) and the genesis of Las Cañadas Caldera (Navarro Latorre and Coello 1989; Watts and Masson 1995; Martí et al. 1997; Ancochea et al. 1999; Cantagrel et al. 1999; Marti and Gudmundsson 2000).

However, since 1968, limited progress was made on the reconstruction of the latest (post-caldera) volcanic phase of Tenerife (Fúster et al. 1968). Research was restricted to a revision of the early work and mapping (Navarro Latorre and Coello 1989), although recently petrological and geochemical aspects have improved considerably (von Fritsch 1867; Ablay et al. 1998; Ablay and Marti 2000; Wiesmaier et al. 2011), as well as the analysis of potential hazards of the volcano (Araña et al. 2000; Márquez et al. 2008; Martí et al. 2008).

Particularly surprising is the almost total lack of geochronological information in many recent papers, since dating was restricted to a single age for the Montaña Blanca lava dome at the base of Teide (Ablay et al. 1995). Several authors (Araña et al. 2000) even stated that dating the Teide volcanic complex was unfeasible, due to the impossibility of applying K/Ar and $^{40}Ar/^{39}Ar$ techniques to this period and the absence of suitable organic material (charcoal) for radiocarbon dating. Eventually this proved possible nevertheless, and a set of 54 new ages provided for the first time precise age constraints of the recent eruptive history of Teide Volcano and its associated volcanism (Carracedo et al. 2003, 2007). These new geochronological data form a framework on which to base the understanding of the structural and volcanic evolution of the Teide volcanic complex, and establish a realistic assessment of eruptive history and potential hazards.

Besides the geological maps of the Teide volcanic complex (Figs. 1.13, 1.14, 1.15) other newly developed resources facilitate the study of this geological area. Very accurate topographic maps (1:5000 and 1:10000) with shaded relief DTMs, 1:1000 and 1:500 orthophotographs and other thematic maps can be downloaded from http://visor.grafcan.es/visorweb/.

Teide National Park was inscribed on UNESCO's World Heritage List in 2007 (http://whc.unesco.org/en/list/1258) recognised for its natural beauty and its importance in providing

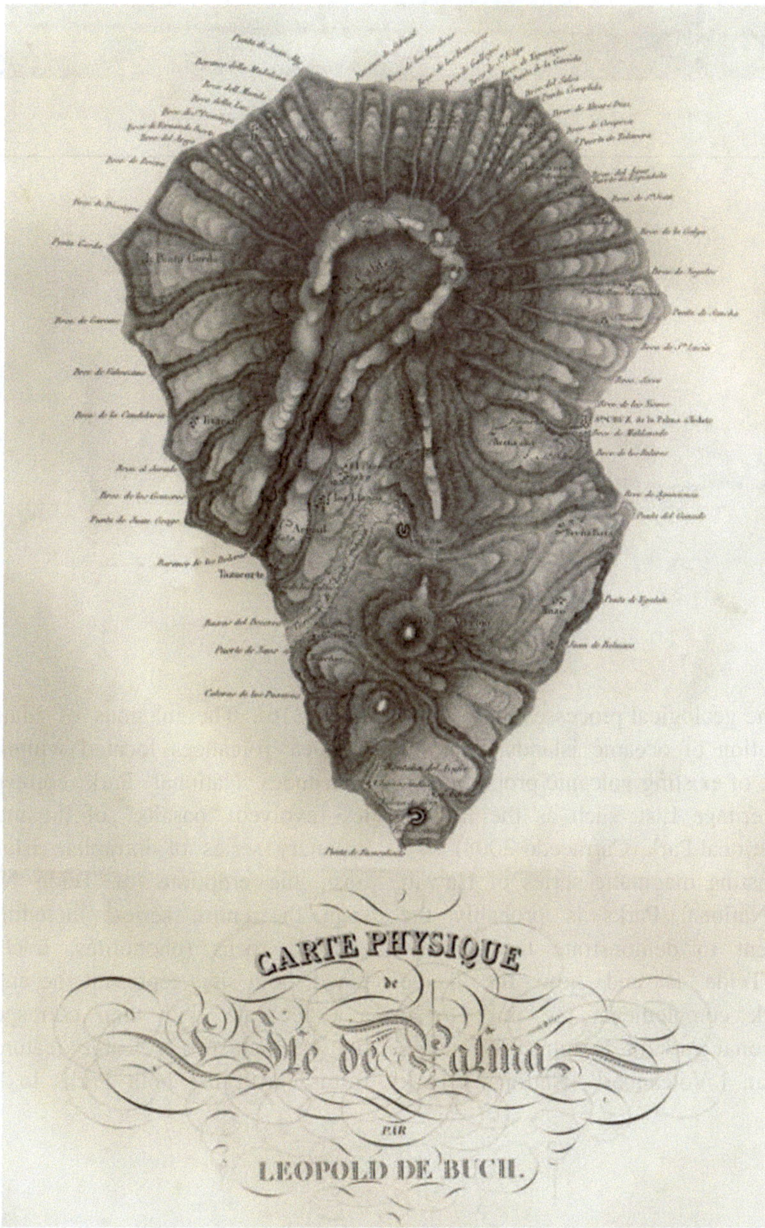

evidence of the geological processes that underpin the evolution of oceanic islands, complementing those of existing volcanic properties on the World Heritage List, such as the Hawaii Volcanoes National Park (Carracedo 2008).

The contrasting magmatic series of Hawaii and Teide National Parks is probably the basic argument to demonstrate how exceptional Mt. Teide is and how the Teide National Park complements the only listed volcanic National Park in an intraplate island, the Hawaiian Volcanoes National Park (Fig. 1.16). The magmas of Mauna Loa and Kilauea volcanoes located within the Hawaii Volcanoes National Park correspond to the less evolved "basalts" of the magmatic evolutionary series of intraplate islands. In contrast, the eruptions of Teide National Park span the entire series, including the more evolved rocks (phonolites, trachytes). Combined, both sites represent the entire series on a large scale, with their corresponding eruptive mechanisms, volcanic features and landforms, justifying both Parks to be registered

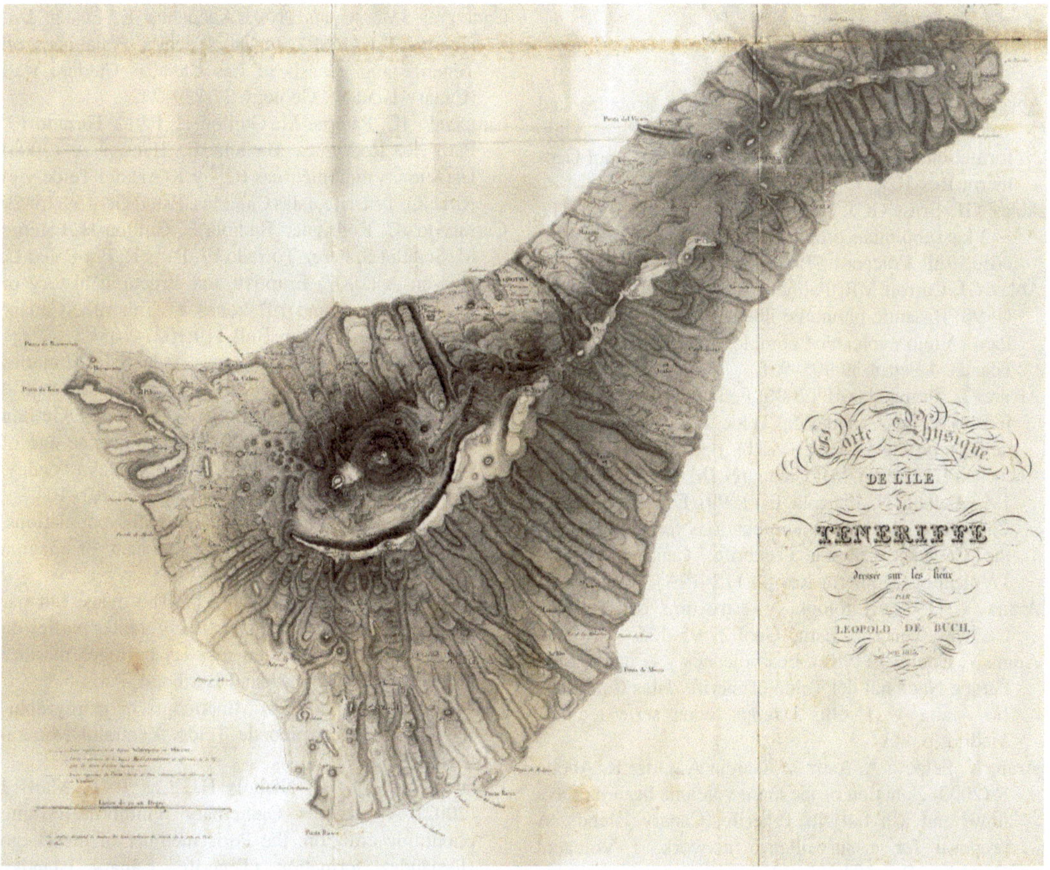

in the UNESCO World Heritage List (in 1987 and 2007, respectively).

The Teide area is a major setting for international research with a long history of influence on Geology and geomorphology, which, as we have seen goes back to the works of von Humboldt, von Buch and Lyell, and which made Mount Teide a significant site in the evolution of Volcanology as a science. Access within the Park is restricted these days, with visitors confined to marked paths and roads. Permission is required for collecting rocks and accessing Teide's summit.

References

Ablay GJ, Martí J (2000) Stratigraphy, structure and volcanic evolution of the Pico Teide-Pico Viejo formation, Tenerife, Canary Islands. J Volcanol Geotherm Res 103:175–208

Ablay GJ, Ernst GGJ, Marti J, Sparks RSJ (1995) The ~2 ka subplinian eruption of montaña blanca, Tenerife. Bull Volcanol 57:337–355

Ablay GJ, Carroll MR, Palmer MR, Martí J, Sparks RSJ (1998) Basanite-phonolite lineages of the Pico Teide-Pico Viejo volcanic complex, Tenerife, Canary Islands. J Petrol 39:905–936

Álvarez L, Hernández JL (2006) Fenómenos tormentosos locales en Las Cañadas del Teide. Asociación Canaria de Meteorología (ACANMET), pp 11

Ancochea E, Huertas MJ, Cantagrel JM, Coello J, Fúster JM, Arnaud N, Ibarrola E (1999) Evolution of the Canadas edifice and its implications for the origin of the Canadas Caldera (Tenerife, Canary Islands). J Volcanol Geotherm Res 88:177–199

Araña V (1971) Litología y estructura del Edificio Cañadas, Tenerife. Estud Geol 27:95–135

Araña V, Coello J (1989). Los volcanes y la caldera del Parque Nacional del Teide (Tenerife, Islas Canarias). In: Araña V, Coello J (eds) Icona serie técnica, Madrid, p 443

Araña V, Felpeto A, Astiz M, Garcia A, Ortiz R, Abella R (2000) Zonation of the main volcanic hazards (lava flows and ash fall) in Tenerife, Canary Islands. A proposal for a surveillance network. J Volcanol Geotherm Res 103:377–391

Bravo T, Bravo-Bethencourt J (1989) Mapa volcanológico de Las Cañadas y Teide-Pico Viejo. In: Araña V, Coello J (eds) Los volcanes y la Caldera del Parque Nacional del Teide. ICONA, Madrid

Booth B (1973) The Granadilla pumice deposits of southern Tenerife, Canary Islands. Proc Geologist Assoc 84:353–370

Borda JC (1776) Journal de voyage fait en 1776 aux Iles Canaries. Bibliothèque Muséum National d'Histoire Naturelle, Paris

Bryan SE (2006) Petrology and geochemistry of the Quaternary caldera-forming, phonolitic Granadilla eruption, Tenerife (Canary Islands). J Petrol 47:1557–1589

Bryan SE, Martí J, Cas RAF (1998) Stratigraphy of the Bandas del Sur formation: an extracaldera record of Quaternary phonolitic explosive eruptions from the Las Canadas Edifice, Tenerife (Canary Islands). Geol Mag 135:605–636

Bryan SE, Cas RAF, Martí J (2000) The 0.57 Ma plinian eruption of the Granadilla member, Tenerife (Canary Islands): an example of complexity in eruption dynamics and evolution. J Volcanol Geotherm Res 103: 209–238

Bryan SE, Martí J, Leosson M (2002) Petrology and geochemistry of the Bandas del sur formation, Las Canadas Edifice, Tenerife (Canary Islands). J Petrol 43:1815–1856

Cantagrel JM, Arnaud NO, Ancochea E, Fúster JM, Huertas MJ (1999) Repeated debris avalanches on Tenerife and genesis of Las Canadas Caldera wall (Canary Islands). Geology 27:739–742

Carracedo JC, Paterne M, Guillou H, Pérez Torrado FJ, Paris R, Rodríguez Badiola E, Hansen A (2003) Dataciones radiométricas (C^{14} y K-Ar) del Teide y el Rift NO, Tenerife, Islas Canarias. Estud Geol 59:15–29

Carracedo JC, Rodríguez Badiola E, Guillou H, Paterne M, Scaillet S, Pérez Torrado FJ, Paris R, Fra-Paleo U, Hansen A (2007) Eruptive and structural history of Teide Volcano and rift zones of Tenerife, Canary Islands. Geol Soc Am Bull 119:1027–1051

Carracedo JC (2008) Los volcanes de las Islas Canarias IV, La Palma, La Gomera, El Hierro. Rueda, Madrid

Carracedo JC, Singer B, Jicha B, Pérez Torrado FJ, Guillou H, Badiola ER, Paris R (2010) Pre-holocene age of Humboldt's 1430 eruption of the Orotava Valley, Tenerife, Canary Islands. Geol Today 26:101–104

Clague DA (1987) Hawaiian xenolith populations, magma supply rates, and development of magma chambers. Bull Volcanol 49:577–587

d'Eveux Claret de Fleurieu CP (1773) Voyage fait par ordre du Roi en 1768 et 1769, à différentes parties du monde, pour éprouver en mer les horloges marines inventées par M. Ferdinand Berthoud, París

Dittler E, Kohler A (1927) Mineralogische-petrographische notizen von pico de Teide. Zentralbl Mineral 4:134–143

Edgar CJ, Wolff JA, Nichols HJ, Cas RAF, Martí J (2002) A complex Quaternary ignimbrite-forming phonolitic eruption: the Poris member of the Diego Hernández formation (Tenerife, Canary Islands). J Volcanol Geotherm Res 118:99–130

Edgar CJ, Wolff JA, Olin PH, Nichols HJ, Pittari A, Cas RAF, Reiners PW, Spell TL, Martí J (2007) The late quaternary diego hernandez formation, Tenerife: volcanology of a complex cycle of voluminous explosive phonolitic eruptions. J Volcanol Geotherm Res 160:59–85

Feuillée L (1724) Voyage aux Iles Canaries ou journal des observations physiques, mathematiques, botaniques e historiques faites par ordre de Sa Majesté. Bibliothèque Centrale du Muséum National d'Histoire Naturelle, París

Friedlander I (1915) Über vulkanische verwerfungtäler. Zeitschrift für Vulkanologie 2

Fúster JM, Araña V, Brandle JL, Navarro JM, Alonso V, Aparicio A (1968) Geology and volcanology of the Canary Islands: Tenerife. Instituto Lucas Mallada, CSIC, Madrid

Gagel C (1910) Die mittelatlantischen Vulkaninseln. Handbuch der regionalen Geologie, vol 7–10. Winter, Heidelberg

Hausen H (1955) Contributions to the geology of Tenerife (Canary Islands). Societas scientiarum fennica, commentationes physico-mathematicae, geologic results of the Finnish expedition to the Canary Islands 1947–1951, vol 18, Issue no 1. Centraltryckeriet, Helsingfors

Huertas MJ, Arnaud NO, Ancochea E, Cantagrel JM, Fúster JM (2002) Ar^{-40}/Ar^{-39} stratigraphy of pyroclastic units from the Cañadas volcanic edifice (Tenerife, Canary Islands) and their bearing on the structural evolution. J Volcanol Geotherm Res 115:351–365

Jeremine E (1930) Composition chimique et mineralogique de la roche du Pico de Teide. Bull Soc Fr Mineral 53:210–215

Kunzli DE (1911) Petrographische resultate von einer Teneriffareise. Mitteilungen der Naturforschenden Gesellschaft Solothurn

Márquez A, López I, Herrera R, Martín-González F, Izquierdo T, Carreño F (2008) Spreading and potential instability of Teide volcano, Tenerife, Canary Islands. Geophys Res Lett 35:L05305. doi: 10.1029/2007GL032625

Marti J, Gudmundsson A (2000) The Las Canadas caldera (Tenerife, Canary Islands): an overlapping collapse caldera generated by magma-chamber migration. J Volcanol Geotherm Res 103:161–173

Martí J, Mitjavila J, Araña V (1994) Stratigraphy, structure and geochronology of the Las Cañadas caldera (Tenerife, Canary Islands). Geol Mag 131:715–727

Martí J, Hurlimann M, Ablay GJ, Gudmundsson A (1997) Vertical and lateral collapses on Tenerife (Canary Islands) and other volcanic ocean islands. Geology 25:879–882

Martí J, Geyer A, Andujar J, Teixidó F, Costa F (2008) Assessing the potential for future explosive activity from Teide-Pico Viejo stratovolcanoes (Tenerife, Canary Islands). J Volcanol Geotherm Res 178:529–542

Meyer H (1896) Die Insel Tenerife. Wanderungen im canarischen Hoch-und Tiefland. In: Hirzel G (ed) Leipzig, p 328

Navarro Latorre JM, Coello J (1989) Depressions originated by landslide processes in Tenerife. In: ESF meeting on Canarian volcanism, Lanzarote, 1989. ESF, Strasbourg, pp 150–152

Pittari A, Cas RAF, Martí J (2005) The occurrence and origin of prominent massive, pumice-rich ignimbrite lobes within the late pleistocene abrigo ignimbrite, Tenerife, Canary Islands. J Volcanol Geotherm Res 139:271–293

Preiswerk H (1909) Sodalittrachyt von Pico de Teide. Zentralbl Mineral 13:393–396

Ridley WI (1970) The abundance of rock types on Tenerife, Canary Islands, and its petrogenetic significance. Bull Volcanol 34:196–204

Rodríguez-Badiola E, Pérez-Torrado FJ, Carracedo JC, Guillou H (2006) Petrografía y geoquímica del edificio volcánico Teide-Pico Viejo y las dorsales noreste y noroeste de Tenerife. In: Carracedo JC (ed) Los volcanes del Parque Nacional del Teide/El Teide. Pico Viejo y las dorsales activas de Tenerife. Naturaleza y parques nacionales-serie técnica. Organismo Autónomo Parques Nacionales Ministerio de Medio Ambiente, Madrid, pp 129–186

Rodríguez Ruiz P, Rodríguez Badiola E, Carracedo JC, Pais Pais FJ, Guillou H, Pérez Torrado FJ (2002) Necrópolis de La Cucaracha: único enterramiento con restos humanos asociados a una erupción prehistórica de La Palma (Islas Canarias). Estud Geol 58:55–69

Santiago M (1948) Edición crítica y estudio bibliográfico y nota de D. Agustín del Castillo "descripción histórica y geográfica de las Islas Canarias". Ed. El gabinete Literario, Tomo I y II, Madrid

Smulikowski K (1937) Sur l'anorthose du Pico de Teide. Archiwum Mineralogiczne 13

von Fritsch K (1867) Tenerife geologisch topographisch dargestellt. Ein beitrag zur kenntniss vulkanischer gebirge. In: von K, Fritsch V, Hartung G, Reiss W (eds) Eine karte und sechs tafeln mit durchschnitten und skizzen nebst erläuterndem text. Wurster, Winterthur

von Fritsch K, Reiss W (1868) Geologische beschreibung der insel tenerife: ein Beitrag zur kenntnis vulkanischer gebirge. Verlag con Wurster & Co, Winterthur

Watts AB, Masson DG (1995) A giant landslide on the north flank of Tenerife, Canary Islands. J Geophys Res 100 (B12):24487–24498

Wiesmaier S, Deegan FM, Troll RV, Carracedo JC, Chadwick JP, Chew D (2011) Magma mixing in the 1100 AD Montaña Reventada composite lava flow, Tenerife, Canary Islands: interaction between rift zone and central volcano plumbing systems. Contrib Mineral Petrol 162:651–669

Wolff JA (1985) Zonation, mixing and eruption of silica-undersaturated alkaline magma: a case study from Tenerife, Canary Islands. Geol Mag 122:623–640

Wolff JA (1987) Crystallisation of nepheline syenite in a subvolcanic magma system: Tenerife, Canary Islands. Lithos 20:207–223

Geological and Geodynamic Context of the Teide Volcanic Complex

2

Juan Carlos Carracedo and Francisco J. Perez-Torrado

Abstract

Long-lived and lively debates commenced in the Canaries several decades ago regarding geological evidence that potentially helps to clarify important features and processes of ocean island volcanism. This included the true nature of the crust underlying the islands, the ultimate cause for the existence of the magmatism in the archipelago, and how large-scale morphological features that shape the islands, such as rift zones and giant landslide scars, have actually formed. The Canaries, once considered to be remnants of an older and larger sunken landmass, are now firmly integrated into the general framework of ocean island volcanism, thus gaining from the abundant geological information published in this field, and in return, providing volcanological data of global significance for ocean islands elsewhere.

2.1 Introduction

As volcanoes develop, they initially go through a constructive phase of evolution in which growth of the edifice through volcanic activity outpaces destruction through mass wasting (Hoernle and Carracedo 2009). During the destructive phase of evolution, mass wasting and erosion exceed volcanic growth and island volcanoes decrease in size until they are eroded to sea level. In this context, Teide Volcano currently represents the peak of development of Canarian volcanoes, the western islands not having yet attained this stage, and the eastern ones being already beyond it.

2.2 The Canary Volcanic Province

Tenerife lies, in time and space, at the centre of the Canary archipelago, the emerged islands forming a 490 km-long chain that increases in age towards the African continent (Fig. 2.1). However, to understand the genesis and evolution of this archipelago we have to take into consideration not only the presently emerged islands (Neocanaries) but the older islands, already submerged (Palaeocanaries). As the African plate moves over the magma source, it cools and subsides, and the older volcanoes of the chain sink beneath sea level forming

J. C. Carracedo (✉) · F. J. Perez-Torrado
Departamento de Física (GEOVOL), Universidad de Las Palmas de Gran Canaria, Las Palmas de Gran Canaria, Canary Islands, Spain
e-mail: jcarracedo@proyinves.ulpgc.es

J. C. Carracedo and V. R. Troll (eds.), *Teide Volcano*, Active Volcanoes of the World,
DOI: 10.1007/978-3-642-25893-0_2, © Springer-Verlag Berlin Heidelberg 2013

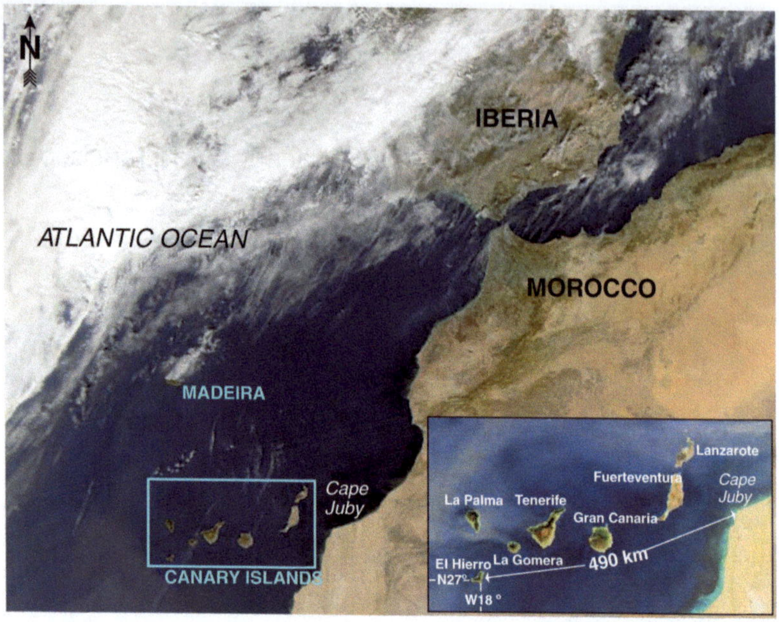

Fig. 2.1 Image (NASA) showing the Canary Islands, in the central east Atlantic off the African coast

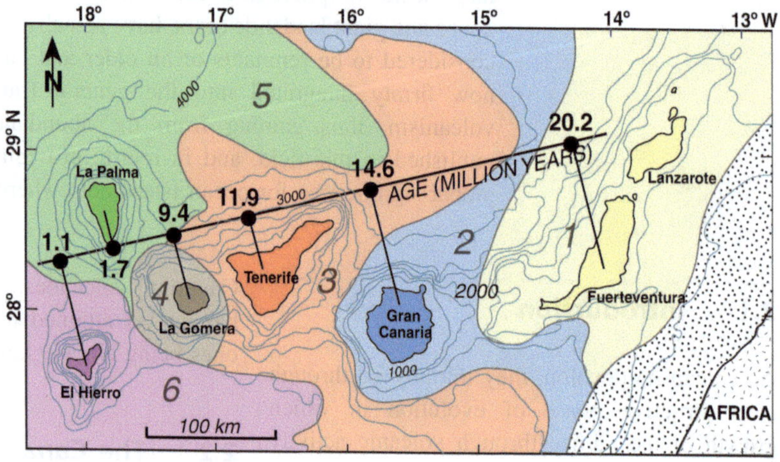

Fig. 2.2 Constant W–E aging of the Canary Islands, consistent with the progressive overlap oceanwards of the islands' aprons (in respective colours), starting at Fuerteventura-Lanzarote. Ages from Guillou et al. (2004). Aprons from Urgeles et al. (1998)

seamounts. Therefore, from a geological point of view, it is crucial to take into account the entire chain of islands and seamounts, summarised as the Canary Volcanic Province (CVP).

The west to east aging of the Canaries is very well documented from abundant radiometric age determinations and from marine geophysical data, indicating that the ages of the oldest rocks of the different islands consistently increase from west to east, whereas their aprons consistently overlap in the opposite direction (Fig. 2.2).

Evidence for age progressive volcanism in the submerged, northern part of the CVP (Fig. 2.3) comes from radiometric dating of seamounts (Geldmacher et al. 2001, 2005). As quoted by these authors, additional evidence for age progressive volcanism in the Palaeocanaries is proven by a widespread and time-transgressive seismic layer, interpreted to reflect volcanic ashes from the Canary hotspot (Holik et al. 1991), present in oceanic sediments marking the Cretaceous/Tertiary boundary near Lars

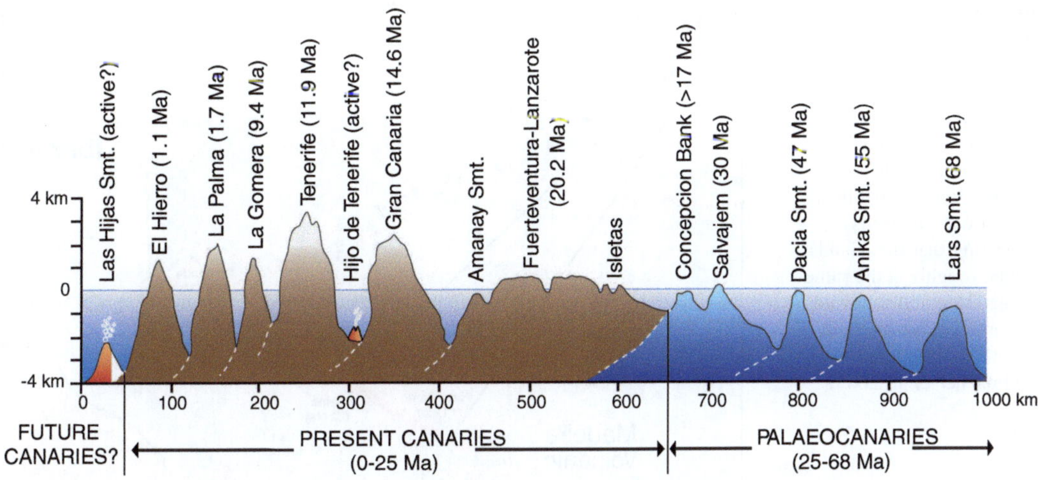

Fig. 2.3 Schematic diagram showing the age progressive chain of islands and seamounts that forms the Canary Volcanic Province (ages from Geldmacher et al. 2001; Guillou et al. 2004)

seamount, but getting younger towards the Canary Islands.

The CVP and the Madeira Volcanic Province (MVP) show some interesting common features. Both volcanic lineations follow parallel curved trends (Geldmacher et al. 2001), suggesting that the islands formed roughly at the same average rate and in the same direction over the last 70 My (Fig. 2.4).

2.3 Genetic Models for the Canaries

Different hypotheses have been published to account for the origin and structural evolution of the Canary Islands. However, two models have been the subject of a lively debate since 1975. Anguita and Hernan (1975) attributed the Canarian magmatism to a propagating fracture from the Atlas mountains, a model based upon structures that cut through the lithosphere to be the cause of, and the control for, the location of the Canary volcanism. Alternatively, Carracedo (1975) postulated an upwelling mantle plume (cf. Morgan 1971), a feature largely independent of the lithosphere.

Although volcanic chains can be formed in relation to transform faults or propagating fracture zones (e.g., Azores), it is not easy to explain how large volcanic chains such as the Canary Islands can be generated within the context of decompression fracturing (McKenzie and Bickle 1988; White and McKenzie 1989). Furthermore, the lithosphere around the Canaries is among the oldest (Jurassic) and thickest on Earth, and therefore lithospheric faults would be problematic to account for the large volumes of magma required to develop the Canary and Madeira Volcanic Provinces. Stress-induced magmatism, reactivation of pre-existing fracture zones (Favela and Anderson 2000) or propagating fractures (Anguita and Hernan 1975), may channel the magma inside the lithosphere and control the geographic arrangement of island volcanoes. However, hotspot trails intersecting fracture zones (e.g., Azores) generally do not show a systematic age progression as is evident in the Canary archipelago (Guillou et al. 2004).

Although local seismicity has been detected around the Canaries, no evidence has been found to prove the existence of any major fault connecting the Atlas mountains with the Canaries in any detailed geophysical studies of the area (Martínez and Buitrago 2002) or in the Atlantic around the Canarian archipelago (Watts 1994; Funck et al. 1996; Watts et al. 1997; Urgeles et al. 1998; Krastel et al. 2001; Krastel and Schmincke 2002). Features interpreted to be crustal fractures that predated and facilitated the

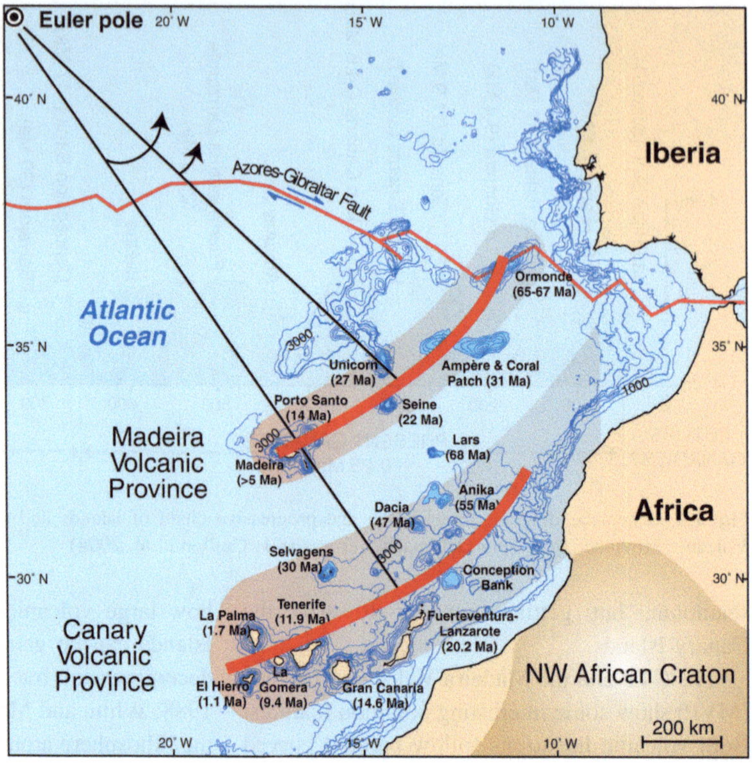

Fig. 2.4 Bathymetric map showing the Canary and Madeira Volcanic Provinces, consisting of islands and associated seamounts, in the central east Atlantic. Both volcanic lineations follow parallel curved trends, suggesting that the islands formed roughly at the same average rate and followed the same course over the last 60 Ma (modified from Geldmacher et al. 2005)

formation of the Canaries, supporting their fracture-related origin (Geyer and Marti 2010), proved to be artifacts associated with ship tracks created during multi-beam data acquisition (Carracedo et al. 2011a).

Conversely, Canary and Madeira Volcanic Provinces age progression and curved synchronous tracks, clearly different from the E–W orientation of fractures or transform zones in the East Atlantic (Geldmacher et al. 2005), can be better explained in the context of a hotspot model (Carracedo et al. 1998).

Several features of the CVP, however, are not easily explained within the context of the classical mantle plume model, particularly the exceptionally long period of volcanic activity of islands in the CVP (e.g., at least 23 My for Fuerteventura). Geldmacher and coworkers (2005) proposed interaction of a Canary plume with edge-driven convection at the margin of the African craton (Fig. 2.5), consistent with further observations by Gurenko et al. (2006).

2.4 Hot Spot Dynamics and Plant Radiation

Macaronesia is a biogeographical region based on the existence of many common elements of flora and fauna. Recent phylogenetic analyses provided evidence of close similarities between species of the Macaronesian flora and the Iberian and Moroccan populations—particularly laurel forest communities, considered to be relicts of the Paleotropical Tethyan flora, which suggests a common origin.

The wet and warm climate in Southern Europe and North Africa during the Paleogene was conditioned by the influence of the warm east-to-west circum-equatorial global marine current, ensuring high temperatures and monsoon summer rains (Uriarte 2003). These conditions changed dramatically, and the tropical flora became extinguished on these continents as a result of the climatic deterioration

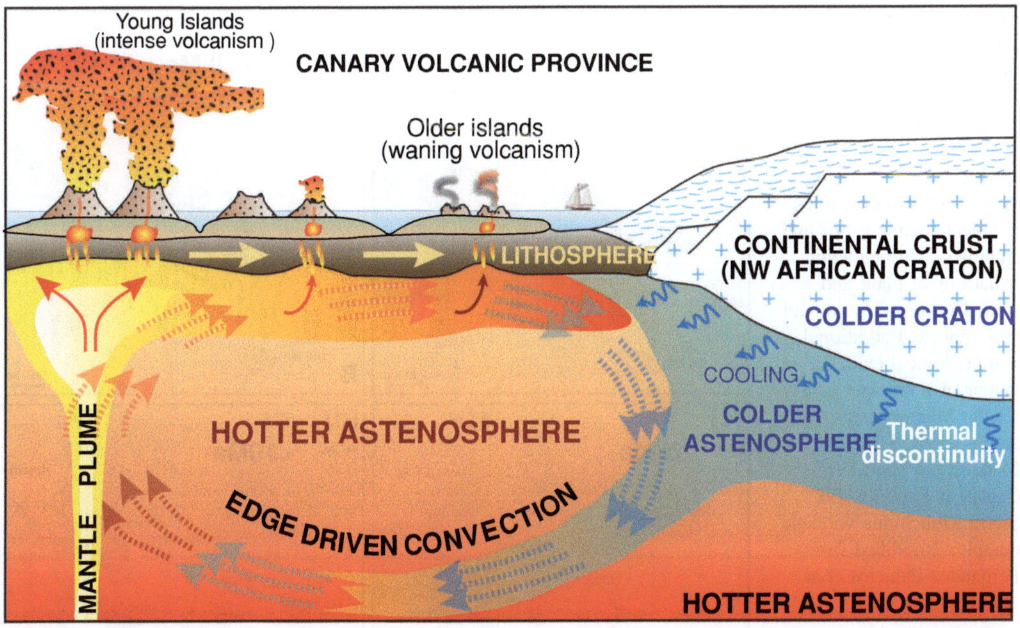

Fig. 2.5 Hotspot or mantle plume model that can adequately explain the linear younging direction along a NE–SW oriented path for the Canary Islands (Carracedo et al. 1998). The conventional hot spot model cannot readily explain the long history of the Canary Islands and the occurrence of historic volcanism in Lanzarote. However, a coherent explanation may be interaction of small-scale upper mantle convection at the edge of the African craton with the Canary mantle plume (modified from Carracedo 1999; Geldmacher et al. 2005)

triggered by the arrival of the glaciations at about 3.2 My (Meco et al. 2006) and the onset of the Canarian marine current. The Iberian and Moroccan regions became a late refugium for these populations until the late Pliocene.

However, the presence of palaeo-endemic floral elements in the laurel forests of contemporary Macaronesia is difficult to explain because of the age differences and the excessive distances from paleotropical sources for the ocean-crossing dispersal abilities of species.

A new approach, linking radiation of paleotropical flora to the Macaronesian archipelagos and the hot spot model has been proposed by (Fernandez-Palacios et al. 2011), suggesting that large and high islands may have been continuously available in the region for as long as 60 million years (Geldmacher et al. 2005), functioning both as stepping stones and as repositories of paleoendemic forms and crucibles for neoendemic radiations of plant and animal groups. In turn, this model (Fig. 2.6) represents additional, non-geological evidence that is consistent with a hot spot origin for the Macaronesian archipelagos.

2.5 Absence of Significant Subsidence as a Crucial Feature in the Canaries

Possibly one of the most relevant differences in the geological evolution of the Hawaiian and the Canarian archipelagos is the absence of high-rates of subsidence characteristic of the majority of mantle plume-related islands in the Canaries. While ocean islands generally rapidly subside below sea level to become guyots, the Canaries remain above sea level for very long periods (e.g., Fuerteventura >23 My; Fig. 2.7). Had the Canaries experienced a subsidence history similar to that of the Hawaiian archipelago, only La Palma and El Hierro would still be above sea level.

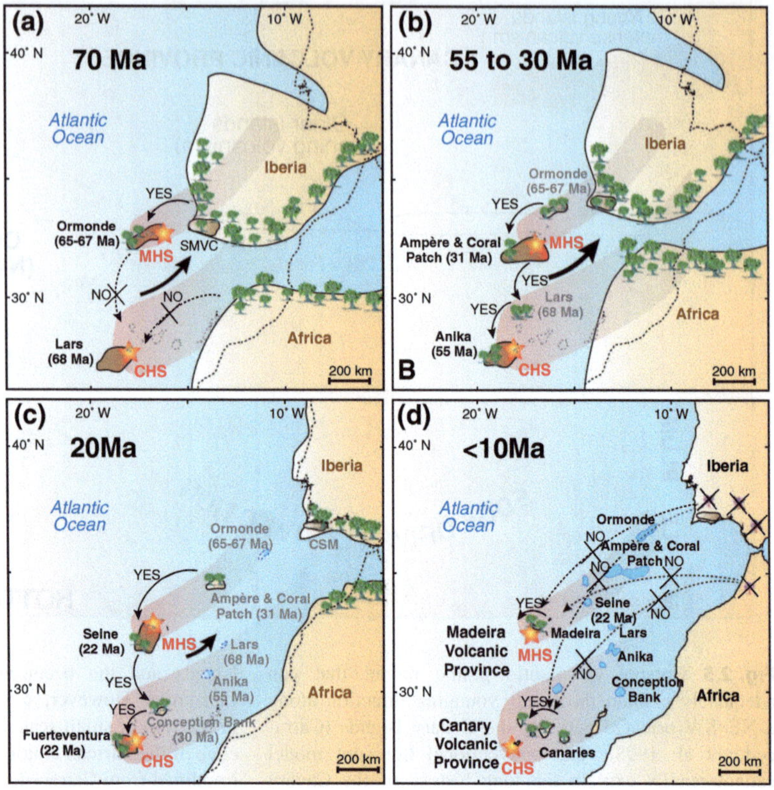

Fig. 2.6 Large islands with high altitudes may have been continuously available in the region for as long as 60 million years (Geldmacher et al. 2005), serving both as stepping stones and as repositories of paleoendemic forms and crucibles of neoendemic radiations of plant and animal groups. This model may represent additional, non-geological evidence in favour of a hot spot origin for the Macaronesian archipelagos. *MHS* Madeira hot spot; *CHS* Canary hot spot; *SMVC* Sierra Monchique volcanic complex (modified from Fernandez-Palacios et al. 2011)

Therefore, this particular feature of the Canarian archipelago, possibly related to the characteristics of the oceanic crust in this area of the NE Atlantic (very old and rigid Jurassic crust), accounts for the existence of Tenerife and Teide Volcano (Fig. 2.8), unfeasible in a scenario of high-rate subsidence as on Hawaii.

2.6 Teide Volcano and the Evolution of the Canaries

The identical source and genetic processes recorded on the islands of the Canarian archipelago in a hot spot context may account for their similarities. However, significant differences between the islands are evident in their volume, elevation, morphology and igneous rock types from W to E, reflecting the increase in age and progression in evolutionary stage.

In contrast with the Hawaiian and most oceanic islands, where subsidence plays a major role, the Canaries show remarkable long-term island stability. Mass wasting and erosion, eventually outpacing volcanic growth, to reduce the size of the islands until they are eroded to sea level, requires periods of time that can exceed 20 My (e.g., Fuerteventura).

The age-dependent ratio of subaerial to submarine volume in the Canary Islands increases from the youngest western to the oldest eastern islands. However, the increase is not constant but shows a maximum in the central island of Tenerife, reflecting that the western islands have not yet attained the mature stage, while the eastern islands are already in an advanced phase of erosive decay (Fig. 2.9).

Therefore, although Gran Canaria, and probably Fuerteventura, also had central differentiated volcanic complexes (e.g., Roque Nublo Volcano), they have been dismantled by erosion

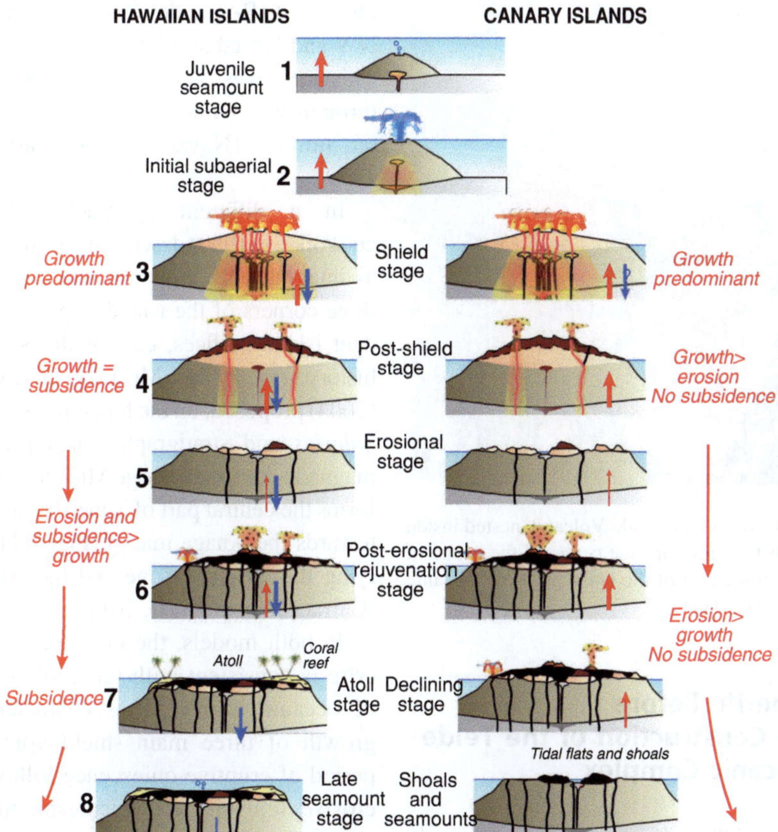

Fig. 2.7 Schematic diagram illustrating significant differences in the evolution of the Hawaiian and the Canary oceanic archipelagos. The former (*left*) typify the life history of oceanic island chains derived from very active and fertile mantle plumes on relatively flexible, fast-moving plates. These islands grow very fast and subside very rapidly into seamounts (the oldest emerged island of the Hawaiian archipelago formed about 6 My ago). In contrast, the Canaries originate from a less active hot spot that penetrates a slow moving old plate, and are composed of long-lived islands with slow growth rates. The main difference is the lack of significant subsidence in the Canaries, with islands remaining emerged until mass-wasted by erosion (modified from Walker 1990; Carracedo et al. 1998)

(Pérez-Torrado et al. 1995; Stillman 1999; Troll et al. 2002). Likewise, the western islands may develop similar central volcanoes in the geological future, but at this stage of evolution of the Canarian archipelago only Tenerife, representing the present evolutionary peak in the development of the Canaries, appears to meet the conditions for an active felsic central complex such as Teide Volcano.

A simplified synthesis of the evolution of the Canary Islands is shown in Fig. 2.10. About 2 My ago a significant change occurred in the sequential development of the islands. The consistent construct of the Canarian archipelago as a single-line chain split after La Gomera into a dual-line configuration. While the onset of each successive island started once the previous one was in decay, La Palma and El Hierro, still in an early stage of shield growth, are being constructed simultaneously. This duality may account for the remarkably slower progress of island construction in the new dual-line configuration compared to the single-line configuration, with an interval of more than 8 My between the onset of La Gomera and that of La Palma and El Hierro.

Fig. 2.8 The 3,718 m high Teide Volcano, nested inside the Las Cañadas Caldera, caps the centre of the island of Tenerife, and forms a part of the latest phase of volcanic construction on the island

2.7 Tenerife Before the Construction of the Teide Volcanic Complex

The Geology of Tenerife has been extensively studied (e.g., Hausen 1955; Fúster et al. 1968; Ridley 1970, 1971; Abdel-Monem et al. 1971; Carracedo 1975, 1979; Schmincke 1982; Wolff 1983, 1987; Ancochea et al. 1990, 1999; Watts and Masson 1995; Bryan et al. 1998, 2002; Thirlwall et al. 2000; Wolff et al. 2000; Edgar et al. 2002; Walter and Schmincke 2002; Guillou et al. 2004; Pittari et al. 2005; Walter et al. 2005; Bryan 2006; Pittari et al. 2006; Carracedo et al. 2007, 2011a; Longpré et al. 2009).

Three main shield volcanoes form the oldest part of the island with compositions ranging from undifferentiated to evolved magmas (basanites to phonolites).

2.7.1 Shield Stage

Fúster et al. (1968) described Tenerife as a large shield volcano mantled by subsequent volcanism, with the core outcropping in the south of the island (Roque del Conde massif), and at the NW and NE edges (Teno and Anaga volcanoes). This idea was supported by later observations through water tunnels excavated for groundwater mining (Navarro 1974; Carracedo 1975, 1979).

In a different approach, Ancochea and co-workers (1990) described the island of Tenerife as integrated by three old massifs located at the three corners of the island, representing independent island edifices, each with its own volcanic history (Fig. 2.11a). Most recently, Guillou et al. (2004) proposed, on the basis of observations from *galerías* and stratigraphic, isotopic, and paleomagnetic data, that a large Miocene shield not only forms the central part of Tenerife, but also extends towards the Anaga massif (Fig. 2.11b, c), underlying the NE Rift Zone and the Anaga volcano (Carracedo et al. 2007, 2011b).

In both models, the eruptive history of Tenerife is consistent with the evolutionary pattern of oceanic islands. It is characterised by the growth of three main shield volcanoes and a period of eruptive quiescence followed by post-erosive rejuvenation volcanism, mainly at the centre of the island.

The first of these old shield volcanoes developed at the central part of Tenerife (the Central Shield, Fig. 2.12a). Erosion and plausibly northbound massive landslides mass wasted the northern, windward flank of the shield, which only outcrops at present in the southwest, leeward flank of the island, and close to the Anaga massif. This geological formation, the oldest outcropping in the island, has been dated by radioisotopic methods (^{40}Ar / ^{39}Ar and K–Ar) between 11.6 and 8.9 million years (Guillou et al. 2004).

About 6 My ago Teno volcano grew attached to the western flank of the Central Shield (Fig. 2.12b), which was probably already in eruptive quiescence at that point. The Teno shield developed in a relatively short period, from ca. 6.11 to about 5.15 My (Guillou et al. 2004; Longpré et al. 2009).

Finally, the shield-building stage of Tenerife was completed with the construction of the

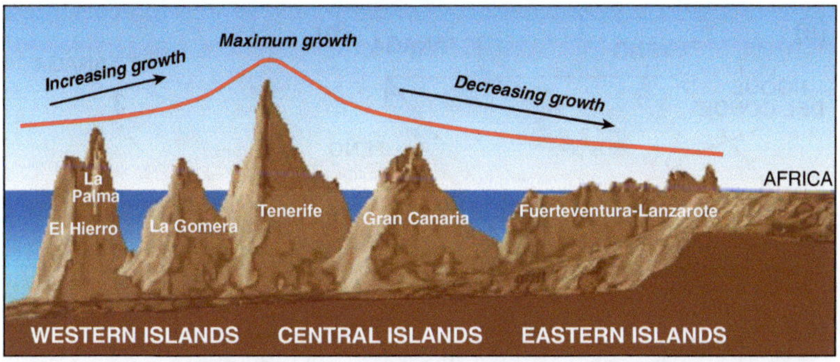

Fig. 2.9 Computer-generated cross section of the Canary Islands, showing age versus height. At present, Tenerife represents the peak of evolutionary development in the Canarian archipelago (Carracedo et al. 1998)

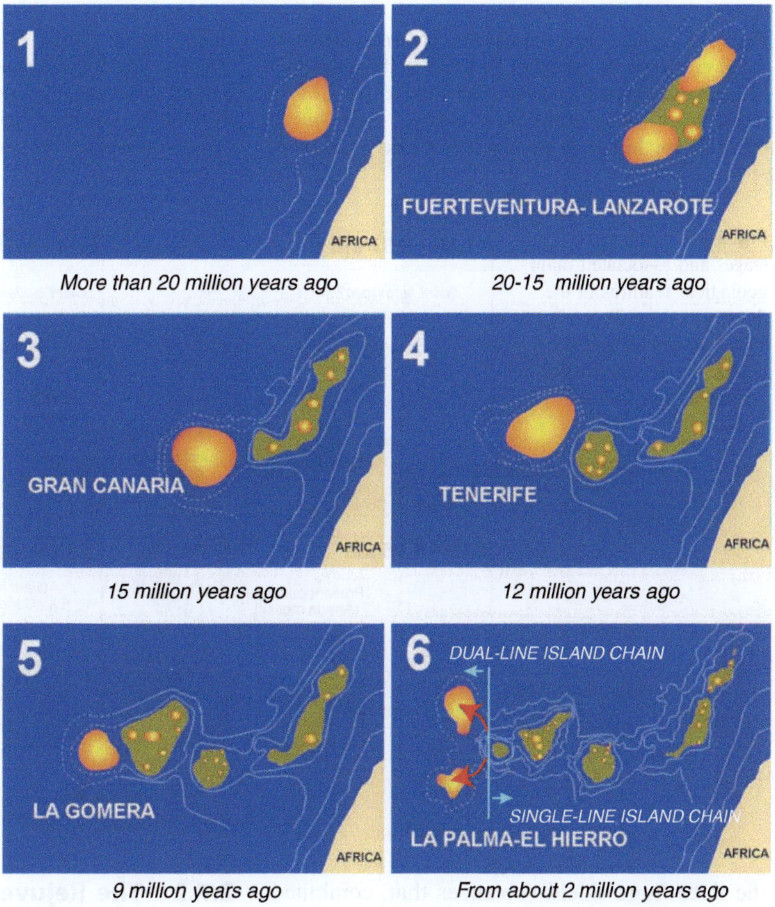

Fig. 2.10 Sequential surfacing of the Canary Islands. An important feature of the Canary Islands is the lack of significant subsidence compared to other hotspot archipelagos, such as the Hawaiian Islands. If the subsidence rate in the Canary Islands were similar to that of the Hawaiian Islands, only La Palma and El Hierro would still exist as islands (modified from Carracedo 1999)

Anaga shield on the opposite side of the island, at the end of the northeast prolongation of the Central Shield (Fig. 2.12c). The Anaga volcano development took place in the interval from about 4.89 to 3.95 My (Guillou et al. 2004; Walter et al. 2005).

The main constructive activity in Tenerife ended about 3.5 My ago with the completion of

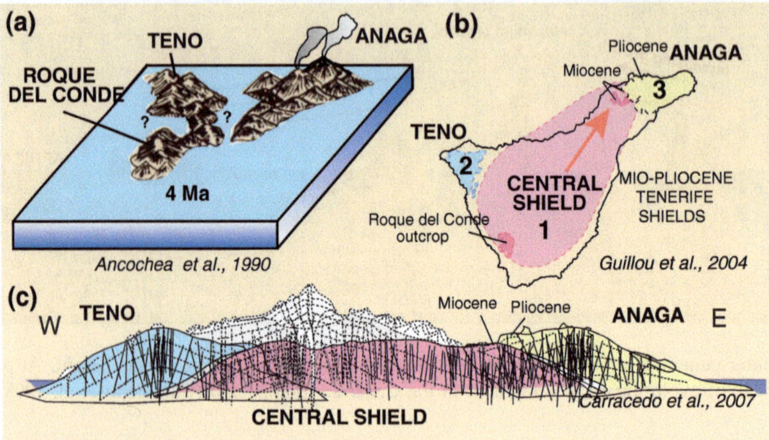

Fig. 2.11 **a** Ancochea and coworkers (Ancochea et al. 1990) described the island of Tenerife as the integration of three old massifs located at the three corners of the island, representing independent edifices, each with its own volcanic history. **b** An alternative idea proposed by Guillou et al. (2004) of the extension of the Central Miocene shield towards the Anaga massif underlying the NE rift zone and the Anaga volcano. **c** Cross-section showing the relative spatial arrangement of Tenerife shield volcanoes (Carracedo et al. 2007)

Fig. 2.12 Successive stages and associated main geological features in the development of Tenerife shield volcanoes and the posterosional rejuvenation central composite Las Cañadas Volcano

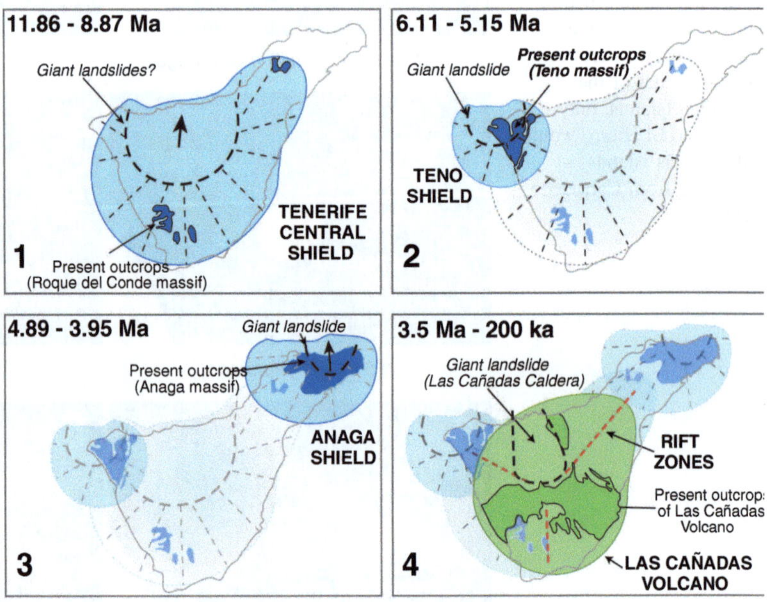

the three large shield volcanoes that, combined, form the bulk (90 %) of the present volume of the island. The main phase of activity of the Central Shield volcano ceased about 9 million years ago, entering a long (5.5 My) interval of volcanic repose and erosion (erosive gap), coinciding with the main phases of construction of the Teno and Anaga shields.

2.7.2 The Rejuvenation Stage of Tenerife: Las Cañadas Volcano

Renewed volcanic activity at the centre of the island formed Las Cañadas Volcano (Fig. 2.12d), from about 3.5 My ago (Ancochea et al. 1990, 1999; Huertas et al. 2002).

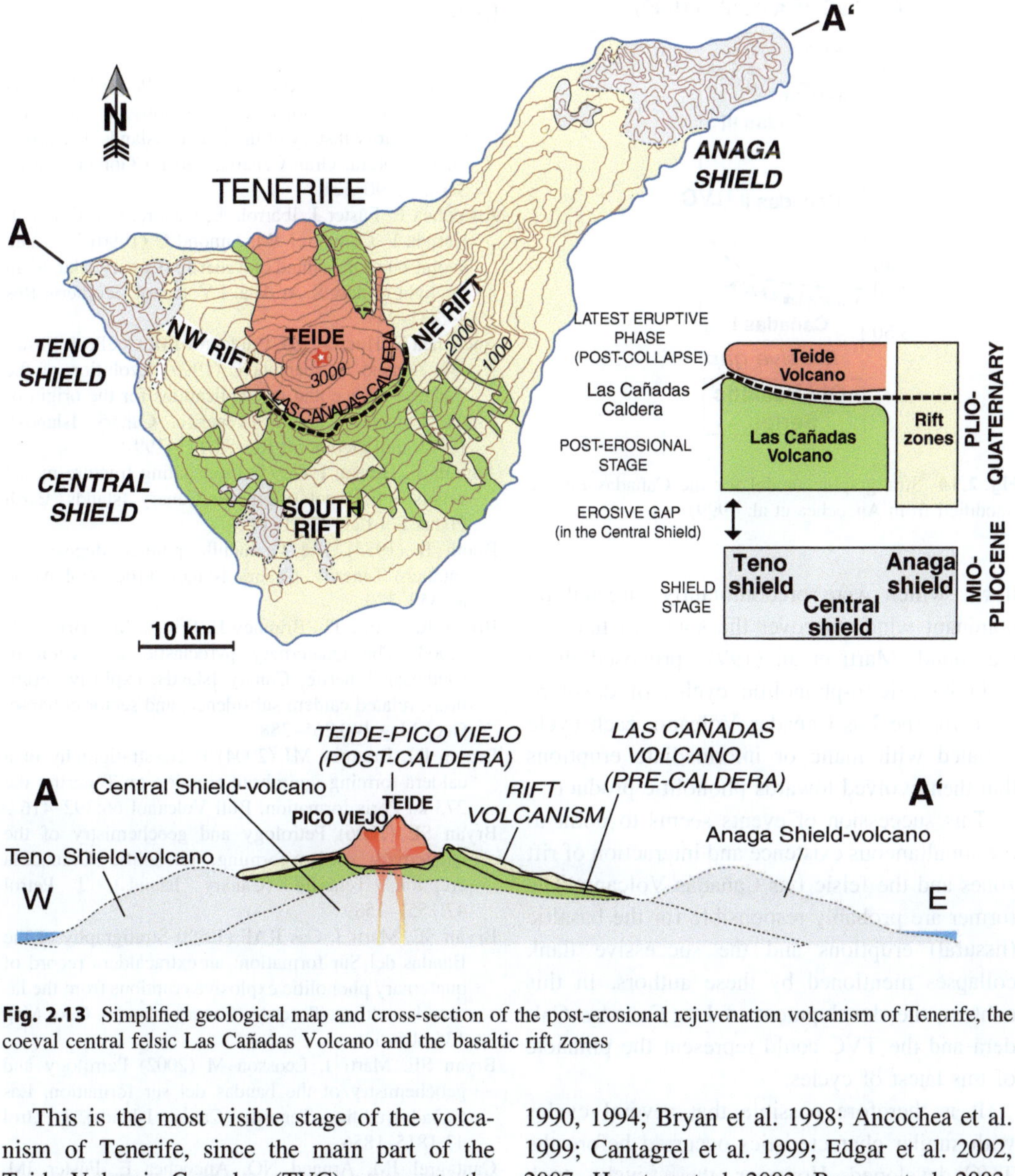

Fig. 2.13 Simplified geological map and cross-section of the post-erosional rejuvenation volcanism of Tenerife, the coeval central felsic Las Cañadas Volcano and the basaltic rift zones

This is the most visible stage of the volcanism of Tenerife, since the main part of the Teide Volcanic Complex (TVC) represents the latest stage of growth of Las Cañadas Volcano (LCV). The coeval activity in the last 3 My of the rift zones (Chaps. 4, 5) and LCV, the latter with abundant central felsic volcanism and the former with predominant fissural basaltic eruptions, cover most of the island's surface, blanketing the outcrops of the shield volcanoes already described (Fig. 2.13).

The LCV has been extensively studied (e.g., Booth 1973; Wolff 1983, 1987; Martí et al. 1990, 1994; Bryan et al. 1998; Ancochea et al. 1999; Cantagrel et al. 1999; Edgar et al. 2002, 2007; Huertas et al. 2002; Brown et al. 2003; Brown and Branney 2004; Pittari et al. 2005, 2006).

According to Ancochea et al. (1999), the LCV developed in three successive phases separated by large scale flank collapses (Fig. 2.14). Phase 1 was predominantly effusive and basaltic, but in phases 2 and 3 eruptions were more differentiated (trachybasalts and phonolites) and more explosive. In these phases, plinian episodes erupted pyroclastic falls and pyroclastic

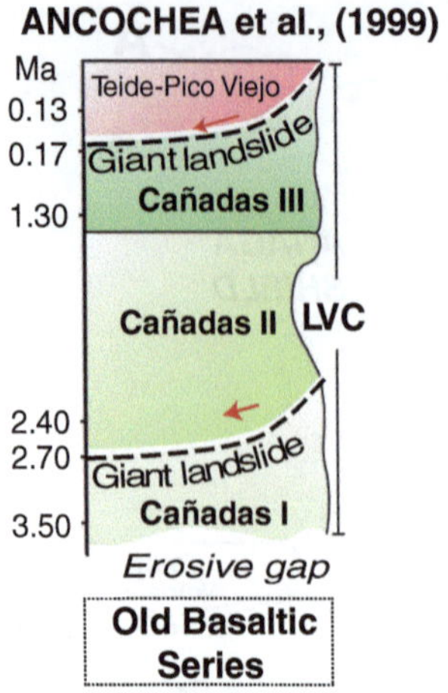

Fig. 2.14 Stratigraphic model for the Cañadas Edifice (modified from Ancochea et al. 1999)

flows, which were predominantly directed by dominant winds to cover the southern flank of the island. Martí et al. (1997) proposed three main basaltic-to-phonolitic cycles of development for the Las Cañadas Volcano, each cycle initiated with mafic or intermediate eruptions that then evolved towards phonolitic products.

This succession of events seems to point to the simultaneous existence and interaction of rift zones and the felsic Las Cañadas Volcano. The former are probably responsible for the basaltic (fissural) eruptions and the successive flank collapses mentioned by these authors. In this context, the development of Las Cañadas Caldera and the TVC could represent the pinnacle of this latest of cycles.

It is therefore possible that several cycles with similar characteristics occurred before the TVC developed. However, these cycles took place in a posterosional island, where rift zones should be expected to have considerably lower energy than the rifts on ocean-island volcanoes in their mainstage of development (e.g., La Palma, El Hierro, Mauna Loa, and Kilauea). Therefore, the most probable future scenario is that their intensity will likely decline, although this does not imply that the TVC will be the last cycle of its kind to take place on the island of Tenerife.

References

Abdel-Monem A, Watkins ND, Gast PW (1971) Potassium-argon ages, volcanic stratigraphy, and geomagnetic polarity history of the Canary Islands; lanzarote, Fuerteventura, Gran Canaria, and La Gomera. Am J Sci 271:490–521

Ancochea E, Fúster J, Ibarrola E, Cendrero A, Coello J, Hernan F, Cantagrel JM, Jamond C (1990) Volcanic evolution of the island of Tenerife (Canary Islands) in the light of new K-Ar data. J Volcanol Geotherm Res 44:231–249

Ancochea E, Huertas MJ, Cantagrel JM, Coello J, Fúster JM, Arnaud N, Ibarrola E (1999) Evolution of the Cañadas edifice and its implications for the origin of the Cañadas Caldera (Tenerife, Canary Islands). J Volcanol Geotherm Res 88:177–199

Anguita F, Hernan F (1975) Propagating fracture model versus a hot spot origin for Canary Islands. Earth Planet Sci Lett 27:11–19

Booth B (1973) The Granadilla pumice deposits of southern Tenerife, Canary Islands. Proc Geol Assoc 84:353–370

Brown RJ, Barry TL, Branney JJ, Pringle MS, Bryan SE (2003) The Quaternary pyroclastic succession of southeast Tenerife, Canary Islands; explosive eruptions, related caldera subsidence, and sector collapse. Geol Mag 140:265–288

Brown RJ, Branney MJ (2004) Event-stratigraphy of a caldera-forming ignimbrite eruption on Tenerife: the 273 ka Poris formation. Bull Volcanol 66:392–416

Bryan SE (2006) Petrology and geochemistry of the Quaternary caldera-forming, phonolitic Granadilla eruption, Tenerife (Canary Islands). J Petrol 47:1557–1589

Bryan SE, Marti J, Cas RAF (1998) Stratigraphy of the Bandas del Sur formation: an extracaldera record of quaternary phonolitic explosive eruptions from the las canadas edifice, Tenerife (Canary Islands). Geol Mag 135:605–636

Bryan SE, Martí J, Leosson M (2002) Petrology and geochemistry of the bandas del sur formation, Las Cañadas edifice, Tenerife (Canary Islands). J Petrol 43:1815–1856

Cantagrel JM, Arnaud NO, Ancochea E, Fúster JM, Huertas MJ (1999) Repeated debris avalanches on Tenerife and genesis of Las Cañadas Caldera wall (Canary Islands). Geology 27:739–742

Carracedo JC (1975) Estudio paleomagnético de la isla de Tenerife. Ph.D. thesis, Universidad Complutense, Madrid

Carracedo JC (1979) Paleomagnetismo e historia volcánica de Tenerife. Aula Cultura Cabildo Insular de Tenerife, Santa Cruz de Tenerife, p 81

Carracedo JC (1999) Growth, structure, instability and collapse of Canarian volcanoes and comparisons with Hawaiian volcanoes. J Volcanol Geotherm Res 94:1–19

Carracedo JC, Day S, Guillou H, Rodríguez Badiola E, Canas JA, Pérez-Torrado FJ (1998) Hotspot volcanism close to a passive continental margin: the Canary Islands. Geol Mag 135:591–604

Carracedo JC, Rodríguez Badiola E, Guillou H, Paterne M, Scaillet S, Pérez Torrado FJ, Paris R, Fra-Palco U, Hansen A (2007) Eruptive and structural history of Teide volcano and rift zones of Tenerife, Canary Islands. Geol Soc Am Bull 119:1027–1051

Carracedo JC, Fernandez-Turiel JL, Gimeno D, Guillou H, Klügel A, Krastel S, Paris R, Perez-Torrado FJ, Rodriguez-Badiola E, Rodriguez-Gonzalez A, Troll VR, Walter TR, Wiesmaier S (2011a) Comment on "The distribution of basaltic volcanism on Tenerife, Canary Islands: implications on the origin and dynamics of the rift systems" by A. Geyer and J. Martí. Tectonophysics 483 (2010) 310–326. Tectonophysics 503:239–241

Carracedo JC, Guillou H, Nomade S, Rodríguez-Badiola E, Pérez-Torrado FJ, Rodríguez-González A, Paris R, Troll VR, Wiesmaier S, Delcamp A, Fernández-Turiel JL (2011b) Evolution of ocean-island rifts: the northeast rift zone of Tenerife, Canary Islands. Geol Soc Am Bull 123:562–584

Edgar CJ, Wolff JA, Nichols HJ, Cas RAF, Martí J (2002) A complex Quaternary ignimbrite-forming phonolitic eruption: the Poris member of the Diego Hernández formation (Tenerife, Canary Islands). J Volcanol Geotherm Res 118:99–130

Edgar CJ, Wolff JA, Olin PH, Nichols HJ, Pittari A, Cas RAF, Reiners PW, Spell TL, Martí J (2007) The late Quaternary Diego Hernandez formation, Tenerife: volcanology of a complex cycle of voluminous explosive phonolitic eruptions. J Volcanol Geotherm Res 160:59–85

Favela J, Anderson D (2000) Extensional tectonics and global volcanism. In: Boschi E, Ekstrom, G, Morelli A (ed) Problems in geophysics for the new millennium. Editrice Compositori, Bologna, pp 463–498

Fernández-Palacios JM, de Nascimento L, Rüdiger O, Delgado JD, García-del-Rey E, Arévalo JR, Whittaker RJ (2011) A reconstruction of palaeo-Macaronesia, with particular reference to the long-term biogeography of the Atlantic island laurel forests. J Biogeogr 38:226–246

Funck T, Dickmann T, Rihm R, Krastel S, Lykke-Andersen H, Schmincke HU (1996) Reflection seismic investigations in the volcaniclastic apron of Gran Canaria and implications for its volcanic evolution. Geophys J Int 125:519–536

Fúster JM, Araña V, Brandle JL, Navarro JM, Alonso V, Aparicio A (1968) Geology and volcanology of the Canary Islands: Tenerife. Instituto Lucas Mallada, CSIC, Madrid

Geldmacher J, Hoernle K, Van den Bogaard P, Zankl G, Garbe-Schönberg D (2001) Earlier history of the ≥70-Ma-old Canary hotspot based on the temporal and geochemical evolution of the Selvagen archipelago and neighboring seamounts in the eastern north Atlantic. J Volcanol Geotherm Res 111:55–87

Geldmacher J, Hoernle K, Van der Bogaard P, Duggen S, Werner R (2005) New Ar-40/Ar-39 age and geochemical data from seamounts in the Canary and Madeira volcanic provinces: support for the mantle plume hypothesis. Earth Planet Sc Lett 237:85–101

Geyer A, Martí J (2010) The distribution of basaltic volcanism on Tenerife, Canary Islands: implications on the origin and dynamics of the rift systems. Tectonophysics 483:310–326

Guillou H, Carracedo JC, Paris R, Pérèz-Torrado FJ (2004) Implications for the early shield-stage evolution of Tenerife from K/Ar ages and magnetic stratigraphy. Earth Planet Sc Lett 222:599–614

Gurenko AA, Hoernle KA, Hauff F, Schmincke HU, Han D, Miura YN, Kaneoka I (2006) Major, trace element and Nd-Sr-Pb-O-He-Ar isotope signatures of shield stage lavas from the central and western Canary Islands: Insights into mantle and crustal processes. Chem Geol 233:75–112

Hausen H (1955) Contributions to the geology of Tenerife (Canary Islands), vol XVIII (1). Societas scientiarum fennica, commentationes physico-mathematicae, geologic results of the Finnish expedition to the Canary Islands 1947–1951. Centraltryckeriet, Helsingfors

Hoernle K, Carracedo JC (2009) Canary Islands, geology. In: Gillespie RG, Clague DA (eds) Encyclopedia of islands (encyclopedias of the natural world). Univ California Press, USA, pp 133–143

Holik JS, Rabinowitz PD, Austin JA (1991) Effects of Canary hotspot volcanism on structure of oceanic-crust off Morocco. J Geophys Res Solid Earth 96:12039–12067

Huertas MJ, Arnaud NO, Ancochea E, Cantagrel JM, Fúster JM (2002) Ar-40/Ar-39 stratigraphy of pyroclastic units from the Cañadas volcanic edifice (Tenerife, Canary Islands) and their bearing on the structural evolution. J Volcanol Geotherm Res 115:351–365

Krastel S, Schmincke HU (2002) Crustal structure of northern Gran Canaria, Canary Islands, deduced from active seismic tomography. J Volcanol Geotherm Res 115:153–177

Krastel S, Schmincke HU, Jacobs GL, Rihm R, Le Bas TP, Alibés B (2001) Submarine landslides around the Canary Islands. J Geophys Res Solid Earth 106:3977–3997

Longpré MA, Troll VR, Walter TR, Hansteen TH (2009) Volcanic and geochemical evolution of the Teno massif, Tenerife, Canary Islands: some repercussions of giant landslides on ocean island magmatism. Geochem Geophys Geosyst 10:Q12017. doi: 10.1029/2009gc002892

Martí J, Mitjavila J, Villa IM (1990) Stratigraphy and K-Ar ages of the Diego Hernández wall and their significance on the Las Cañadas Caldera formation (Tenerife, Canary Islands). Terra Nova 2:148–153

Martí J, Mitjavila J, Araña V (1994) Stratigraphy, structure and geochronology of the Las Cañadas Caldera (Tenerife, Canary Islands). Geol Mag 131: 715–727

Martí J, Hurlimann M, Ablay GJ, Gudmundsson A (1997) Vertical and lateral collapses on Tenerife (Canary Islands) and other volcanic ocean islands. Geology 25:879–882

Martínez W, Buitrago J (2002) Sedimentación y volcanismo al este de las islas de fuerteventura y Lanzarote (surco de Fuster Casas). Geogaceta 32:51–54

McKenzie D, Bickle MJ (1988) The volume and composition of melt generated by extension of the lithosphere. J Petrol 29:625–679

Meco J, Cabrera JM, Carracedo JC, Santos JB, Lozano JFB, Scaillet S, Guillou H, Figueroa ALM, Maire NP, Ramos AJG (2006) Paleoclimatología del neógeno en las Islas Canarias: Geliense. Pleistoceno y Holoceno, Ministerio de Medio Ambiente

Morgan WJ (1971) Convection plumes in the lower mantle. Nat 230:42–43

Navarro JM (1974) La estructura geológica de Tenerife y su influencia en la hidrogeología. In: Simposio Internacional sobre Hidrología de Terrenos Volcánicos, Arrecife, Lanzarote, pp 37–57

Pérez-Torrado FJ, Carracedo JC, Mangas J (1995) Geochronology and stratigraphy of the Roque Nublo cycle, Gran Canaria, Canary Islands. J Geol Soc 152:807–818

Pittari A, Cas RAF, Martí J (2005) The occurrence d origin of prominent massive, pumice-rich ignimbrite lobes within the late Pleistocene Abrigo ignimbrite, Tenerife, Canary Islands. J Volcanol Geotherm Res 139:271–293

Pittari A, Cas RAF, Edgar CJ, Nichols HJ, Wolff JA, Martí J (2006) The influence of palaeotopography on facies architecture and pyroclastic flow processes of a lithic-rich ignimbrite in a high gradient setting: the Abrigo Ignimbrite, Tenerife, Canary Islands. J Volcanol Geotherm Res 152:273–315

Ridley WI (1970) The abundance of rock types on Tenerife, Canary Islands, and its petrogenetic significance. Bull Volcanol 34196–204

Ridley WI (1971) The field relations of the Las Cañadas volcanoes, Tenerife, Canary Islands. Bull Volcanol 35:318–334

Schmincke HU (1982) Volcanic and chemical evolution of the Canary Islands. In: Von Rad U, Hinz K, Sarntheim M, Seibold E (eds) Geology of the northwest African continental margin. Springer, Berlin, pp 273–276

Stillman CJ (1999) Giant Miocene landslides and the evolution of Fuerteventura, Canary Islands. J Volcanol Geotherm Res 94:89–104

Thirlwall MF, Singer BS, Marriner GF (2000) $^{39}Ar-^{40}Ar$ ages and geochemistry of the basaltic shield stage of Tenerife, Canary Islands, Spain. J Volcanol Geotherm Res 103:247–297

Troll VR, Walter TR, Schmincke HU (2002) Cyclic caldera collapse: Piston or piecemeal subsidence? field and experimental evidence. Geology 30:135–138

Urgeles R, Canals M, Baraza J, Alonso B (1998) Seismostratigraphy of the western flanks of El Hierro and La Palma (Canary Islands): a record of Canary Islands volcanism. Mar Geol 146:225–241

Uriarte A (2003) Historia del clima de la tierra. Servicio Central de Publicaciones del Gobierno Vasco, Bilbao

Walker GPL (1990) Geology and volcanology of the Hawaiian Islands. Pacific Sci 44:315–347

Walter TR, Schmincke HU (2002) Rifting, recurrent landsliding and Miocene structural reorganization on NW-Tenerife (Canary Islands). Int J Earth Sci 91:615–628

Walter TR, Troll VR, Cailleau B, Belousov A, Schmincke HU, Amelung F, Bogaard PVD (2005) Rift zone reorganization through flank instability in ocean island volcanoes: an example from Tenerife, Canary Islands. Bull Volcanol 67:281–291

Watts AB (1994) Crustal structure, gravity-anomalies and flexure of the lithosphere in the vicinity of the Canary-islands. Geophys J Int 119:648–666

Watts AB, Masson DG (1995) A giant landslide on the north flank of Tenerife, Canary Islands. J Geophys Res 100:24487–24498

Watts AB, Peirce C, Collier J, Dalwood R, Canales JP, Henstock TJ (1997) A seismic study of lithospheric flexure in the vicinity of Tenerife, Canary Islands. Earth Planet Sc Lett 146:431–447

White RS, McKenzie DP (1989) Volcanism at rifts. Sci Am 261:62–71

Wolff JA (1983) Petrology of Quaternary pyroclastic deposits from Tenerife, Canary Islands. Ph.D. Thesis, University of London, London

Wolff JA (1987) Crystallisation of nepheline syenite in a subvolcanic magma system: Tenerife, Canary Islands. Lithos 20:207–223

Wolff JA, Grandy JS, Larson PB (2000) Interaction of mantle-derived magma with island crust? Trace element and oxygen isotope data from the Diego Hernández formation, Las Cañadas, Tenerife. J Volcanol Geotherm Res 103:343–366

The Teide Volcanic Complex: Physical Environment and Geomorphology

3

Alejandro Rodriguez-Gonzalez, Raphaël Paris, Constantino Criado, and Jose-Luis Fernandez-Turiel

Abstract

Teide volcano, nested inside the Las Cañadas Caldera, offers visitors a view on one of the most dramatic landscapes in the world. This is due to a combination of a long volcanic history that ranges from the Quaternary Las Cañadas volcano to the historical Teide lava flows, as well as to the particular climatic and geomorphological setting in which Tenerife lies. In this chapter we review the morphological imprint of the main volcanic and structural features (massive flank failures, Teide stratovolcano and rift-zone growth) as well as the Late Pleistocene and Holocene non-volcanic landforms (aeolian and periglacial landforms, debris flows and alluvial fans), which provide a useful record of the morphodynamic history of Tenerife and the variable climate influences to which it is subject.

3.1 Introduction

The Canary Islands and especially the present day Teide National Park are, if you will, an open-air museum of volcanic geomorphology, one of the reasons why this area was declared a World Heritage Site by UNESCO in 2007. Despite the moderate frequency of volcanic activity, the volcanic landforms are very well preserved and their diversity is extraordinary. This is due to a combination of varied volcanic processes, a mild climate with rare torrential activity (at least during the last 500 years), and a long and complex volcanic history. This diversity of volcanism has given rise to a wide variety of morphologies: (1) macroscale ($\times 10$–$\times 100$ km^3) volcanic morphologies such as central volcanoes with differentiated lavas (Teide), and basaltic rift zones, (2) macroscale instability features such as giant

A. Rodriguez-Gonzalez (✉)
Departamento de Física (GEOVOL), Universidad de Las Palmas de Gran Canaria, Las Palmas de Gran Canaria, Canary Islands, Spain

R. Paris
Laboratoire Magmas et Volcans UMR 6524 CNRS-UBP-IRD, 5 rue Kessler, Clermont-Ferrand 63038, France

C. Criado
Departmento de Geografía, Universidad de La Laguna, 38071 La Laguna (Tenerife), Canary Islands, Spain

J.-L. Fernandez-Turiel
Institute of Earth Sciences Jaume Almera, ICTJA-CSIC, 08028 Barcelona, Spain

J. C. Carracedo and V. R. Troll (eds.), *Teide Volcano*, Active Volcanoes of the World,
DOI: 10.1007/978-3-642-25893-0_3, © Springer-Verlag Berlin Heidelberg 2013

collapse scars (La Orotava and Güímar, typically $\times 10$ km^3), (3) mesoscale volcanic morphologies represented by lava flows and cones ($\times 10^6$ m^3–1 km^3), the composition (and viscosity) of which influences the morphology of these eruptive landforms, (4) mesoscale landforms resulting from the morphodynamic and climatic history (e.g., alluvial fans, debris flows, nebkhas), and (5) microscale morphologies (e.g., periglacial features: pipkrakes and sorted stripes). Accurate topographical data and detailed geological mapping (Carracedo et al. 2007, 2011) have allowed the characterisation of the chronology and morphology of alluvial fans, debris flows, and almost 50 lava flows emplaced over the last 15 ky. This latter information will be useful for further investigations using numerical modelling to understand lava flow propagation as well as hazard assessment (see Chap. 14).

3.2 Geological Outline

Despite the influence of climate and sea level variations, the landforms and long-term morphological evolution of hotspot islands, such us the Hawaiian, Polynesian, Cape Verde and Canary Islands, are mostly controlled by their volcanic activity and instability. However, as a volcano moves away from above a hotspot, the decline of eruptive rates reduces flank instability and causes lower rates of erosion (Paris 2002). In the Canary Islands, the duration of the hotspot volcanic stage is temporally more extensive than in e.g., Hawaii because of the slow motion of the African plate, thus allowing further volcanic reactivation and landform rejuvenation (Carracedo et al. 1998).

The recent evolution of Tenerife Island has displayed the transition between the hiatus stage (Teno and Anaga shield volcanoes, inactive for 3.9 Ma) and the rejuvenated stage (central Teide and rift zones, still active) of ocean island volcanism. The central part of Tenerife, through which Teide grew, is composed of an eroded shield volcano (Roque del Conde, 11.9–8.9 Ma; Guillou et al. 2004), later covered by several phases of rejuvenation including the Las Cañadas stratovolcano at the island's centre (3–0.13 Ma). Towering above this edifice are Teide (<200 ky) and Pico Viejo (<30 ky) central volcanoes nested inside the Las Cañadas Caldera. Moreover, to the northeast and the northwest are two rift zones, the NERZ and the NWRZ respectively, which converge towards the centre of the island and play a prominent role in its evolution (see Chaps. 4 and 5). Teide National Park, opened in 1954, offers visitors a breathtaking landscape, where landforms reflect the long and unique volcanic history. This however, comes at a price—the view comes hand in hand with the dangers of recurrent flank instability and erosion.

3.3 Massive Flank Failures and Their Morphological Imprint

Flank collapses affecting oceanic shield volcanoes are the largest flank failures on Earth, usually involving tens of km^3 of volcanic material. This is the most important process in the destruction of ocean islands and generally occur at the end of the main shield building stage (Carracedo et al. 2011). The lower submarine flanks of the Canary Islands are covered by 7,500 km^3 of chaotic deposits, mainly produced by debris avalanches, as well as by slumps and submarine debris flows (Masson et al. 2002). This volume represents 6 % of the total volume of the archipelago (Paris et al. 2005), and attests to the large role flank failures play in the morphological evolution of the islands. On the northern flanks of Tenerife in particular, the combined thickness of multiple generations of debris avalanche deposits reaches 700 m with a volume of 500 km^3 (Watts and Masson 1995).

On Tenerife, Teide and Pico Viejo "twin volcanoes" are nested within a large depression open towards the northern shore—the Las Cañadas Caldera. The large-scale morphology of the northeastern rift zone also incorporates similar types of depressions or embayments, i.e., the Güímar and Orotava Valleys. It is now commonly accepted that the origin of these geomorphological imprints is related to flank instability, despite being sometimes associated with other processes as well, e.g., caldera collapse in the Las Cañadas case (e.g., Giachetti et al. 2011).

3.4 Origin of Las Cañadas Caldera

The Las Cañadas Caldera is an asymmetrical, horseshoe-type depression 15 km wide and open to the north, a product of the destruction of Las Cañadas Volcano, a former complex stratovolcano in the centre of Tenerife. The floor of the depression rises to 2,200 m a.s.l. in its eastern part and only 2,000 m in its western part, while the caldera wall rises to an average height of ~600 metres above the caldera floor. Intrusive remnants of the Las Cañadas stratovolcano mark a change in the slope of the caldera floor (Los Roques de García). The highest point of the caldera wall is Montaña de Guajara (2,717 m a.s.l.). However, at the junction with the Orotava Valley the wall is completely dismantled (El Portillo). Indeed, the caldera wall shows topographical, lithological and morphological variations which reflect the magmatic evolution and spatial migration of the active centre of the Las Cañadas Volcano from west to east (Marti and Gudmundsson 2000). The caldera wall also represents the result of destruction through volcano instability and long-term erosion, recorded by several generations of colluvium, alluvial fans, and lacustrine and aeolian deposits.

A key hypothesis regarding the origin of Las Cañadas Caldera is a giant landslide on the northern flanks of Las Cañadas Volcano (Carracedo 1994; Watts and Masson 1995; Cantagrel et al. 1999). This is due to the spatial and chronological link between the Icod "palaeovalley", the large embayment in which Teide volcano and the Las Cañadas Caldera are located, and the offshore debris avalanche deposits found to the north of the island (Masson et al. 2002). Offshore geophysical surveys north of Tenerife underline the existence of several enormous landslide deposits (tens to hundreds of cubic kilometres), the youngest located directly offshore from the Icod palaeovalley (Watts and Masson 1995, 2001). Vertical subsidence affecting the central part of Las Cañadas Volcano has also been a further contributor to the formation of the Las Cañadas Caldera with several medium volume ignimbritc units on the flanks of Las Cañadas Volcano being associated with this process (e.g., Bandas del Sur) (Martí et al. 1994, 1997; Bryan et al. 1998). The vertical caldera collapse history of the Las Cañadas volcano has likely weakened the edifice and thus prepared the structural framework for the Icod lateral collapse (cf. Troll et al. 2002). Thus the Las Cañadas Caldera walls are interpreted as the eroded remnants of a failure headwall, while the lateral scarps of the embayment are partly exposed near Icod and La Guancha. The timing of the Icod failure can be constrained by the age of the oldest lavas which fill the embayment (198 ky in the Salto del Frontón galería) and the uppermost lavas cut by the headwall: 240 ky in Icod (see Chap. 6), 170 ky in La Fortaleza and 180 ky in Diego Hernández (Ancochea et al. 1990, 1999; Martí et al. 1994). Considering the accuracy of the available dates, the Icod failure may hence have occurred around 200 ky. Erosion by numerous debris flows, rockfalls and rockslides on the steep headwall has subsequently enlarged the depression to the south.

3.5 Reconstructing the Icod Landslide and Teide Growth

The investigation and reconstruction of both the depth of the landslide surface and the thickness of volcanic infill in the Las Cañadas Caldera took advantage of the many "*galerías*" (a network of large tunnels excavated to supply the island with water). Thirty-seven of these tunnels penetrate Teide volcano at different depth levels mainly in the northern flank (Márquez et al. 2008). The depth of the base of the breccia deposit related to the Icod landslide and the maximum depth of lava flow infill were used to correlate several geological cross sections, which in turn allowed mapping of the depth of the upper limit of the avalanche deposit. The

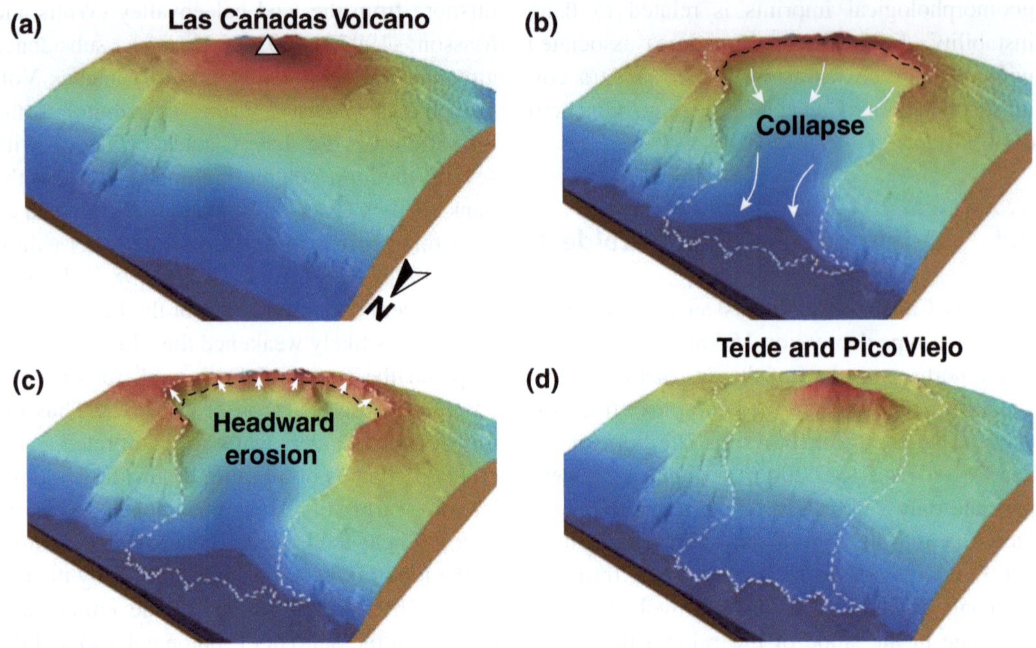

Fig. 3.1 Reconstruction of the formation and evolution of the Las Cañadas Caldera: **a** Las Cañadas Volcano; **b** collapse of Las Cañadas Volcano; **c** erosion of collapse scarp; **d** present-day Teide and Pico Viejo volcanoes

correlation between the off- and onshore breccia distributions shows a U-shaped body covered by several hundred metres of lava flows and extending into the Las Cañadas Caldera and below Teide volcano (Márquez et al. 2008).

New insights into the geomorphological evolution of Teide strata in the central part of Tenerife have recently been achieved by combining detailed field information with GIS-based modelling. Consequently a topographic reconstruction of the Las Cañadas Volcano, the surface after the giant landslide, and the surface of the erosive retrogradation of the collapse scarp is now available.

The pre- and post-collapse maps of the Las Cañadas Volcano show the changes on the terrain surface caused by the different processes, including the collapse *sensu stricto* and those associated with the subsequent (mainly fluvial) erosion. From these topographic surfaces the corresponding Digital Elevation Models (DEMs) can be obtained and from there terrain slopes can be computed (Fig. 3.1). Detailed information about the calibration and validation methodology on the resulting DEMs can be found in Rodriguez-Gonzalez et al. (2010).

A cut and fill analysis process applied to four selected DEMs (the Las Cañadas Volcano, collapsed surface of Las Cañadas Volcano, erosive retrogradation surface of the collapse scarp and the present-day Teide Volcano) allows comparison of two raster surfaces of the same area and identifies locations where elevation values differ. These areas are traced to form polygons in an output vector object. The volume of material added or subtracted can then be calculated for each area.

The total original volume of the landslides from the Las Cañadas Volcano to the coastline can be obtained from the difference between the post- and pre-collapse DEMs (173 km^3). This compares favourably with a volume of about 150 km^3 calculated for the deposit by Masson et al. (2002). Likewise, the volume of erosive retrogradation of the collapse scarp is obtained from the difference between the post- and pre-erosive border scarp (61 km^3), the morphological evolution of the failure scar depending on the temporal and spatial distribution of the subsequent volcanism (Paris 2002). The high erosion rates of a scar, that occur directly after a failure, are associated with debris flows, rockfalls and rockslides from steep slopes. The volumetric comparison between the failure scars and the submarine debris avalanche deposits remains difficult to establish, because of the post-collapse erosion of the scars, hemipelagic sedimentation draping the debris avalanche deposits, and the low accuracy of offshore volume estimates. Using the methodology outlined in Rodriguez-Gonzalez et al. (2010), the volume of eruptive products from Teide Volcano can be determined by calculating the difference between the present-day DEM and the combined DEMs of the post-collapse scarp and erosive retrogradation of the border. This calculates to $\sim 188\ km^3$.

3.6 La Orotava and Güímar Flank Failures

The large-scale morphology of the northeastern rift zone shows that the Güímar and Orotava Valleys, can also be related to giant landslides. The famous "Valle de La Orotava" is a large amphitheatre-shaped coastal embayment (119 km^2), with two lateral scarps almost perpendicular to the north coast of Tenerife (Carracedo et al. 2011 and references therein). This geometry suggests an origin related to a landslide. The entire surface of the depression is covered by post-collapse volcanism and sediments, which reach up to 500 m in thickness in some locations. The Los Organos sub-vertical slopes are considered to be the eroded remnants of the eastern headwall. The Risco Verde northward-facing wall corresponds to a structural discontinuity between Las Pilas lava flows (1.1–0.7 Ma; Ancochea et al. 1999) and the eastward-dipping Diego Hernández pyroclastic deposits (0.4–0.18 Ma; Mitjavila 1990, Mitjavila and Villa 1993, Ancochea et al. 1995), which are heavily eroded, and cannot be related to the La Orotava collapse. Otherwise, a large and arcuate slope break is observed between the southern Tigaiga-Fortaleza scarps and the western Los Organos scarp. The dip of the lava flows erupted from the Pico del Teide and Cordillera Dorsal eruptive vents varies from 0 to 10° in the caldera to 15–30° at the slope break. The ages of these flows range from 11 to 12 ky (Volcán del Portillo) to 37 ky (Montaña del Cerrillar; Carracedo et al. 2003). Thus, the southern headwall of the La Orotava valley is completely overlain by post-landslide lava flows.

The La Orotava flank failure affected the northeastern rift zone and the Las Cañadas edifice, the two volcanic structures being contemporaneously active for approximately 1 Ma. The large difference between the subaerial volume of the scar (57 km^3: Carracedo et al. 2011) and the submarine volume of debris avalanche deposits (80 km^3; Masson et al. 2002) suggests that the failure reached the highest parts of Las Cañadas III volcano (which was 2,500–2,700 m high, Ancochea et al., 1999). La Orotava landslide occurred between 690 ± 10 and 566 ± 13 ka. The maximum age of the failure can be deduced from the age of the topmost materials which are cut by the headwalls and rims of La Orotava Valley. The 270 ky lava flow dated by Ibarrola et al. (1993) is clearly cut by the western rim of the collapse, but Carracedo et al. (2007) found a 566 ky old lava flow overlapping the eastern rim.

The Güímar Valley corresponds to another flank collapse affecting the northeastern rift zone. With an area of 129 km^2 and a missing volume of 44 km^3, it is comparable in size to the La Orotava Valley (Giachetti et al. 2011). The timing of the Güímar collapse is well constrained by the ages of the youngest lavas topping the walls of the depression (866 ky) and oldest lavas filling the failure scar (831 ky; Carracedo et al. 2011). The morphological evolution and erosion rates of both La Orotava and Güímar failure scars are influenced by the temporal and spatial distribution of the subsequent volcanism filling the embayments, as previously demonstrated for other failures on La Palma and La Gomera (Paris and Carracedo 2001; Paris et al. 2005). Areas preserved by post-collapse volcanism are dissected by deep canyons, and retrograde erosion affects the headwalls (e.g., western part of the Güímar Valley).

3.7 Morphology of Teide–Pico Viejo Central Volcano

Teide Volcano (3,718 m a.s.l.) fills the Icod embayment, from the central Las Cañadas Caldera and beyond to the north coast of the island. After ~200 ky of volcanic history, Teide is composed of distinct eruptive vents and morphological features, such as the two stratovolcanoes, many strombolian cones, and also lava domes and associated lava flows. The lava flows are more extensive on the northern flanks of the volcano than on the southern one, due to the geometry of the landslide scar that opens and dips to the north. The length of the flows from the eruptive centres to the north coast is around 12 km, mostly on 5–20° slopes. The transition from the lower flanks to the central volcano itself is marked by a slope break at 1,900 m a.s.l. on the northern flank and at 2,300 m a.s.l. on the southern flank. The central volcano has a cone base 8 km in diameter and displays steep slopes (20–40°). A single shaded relief view can be used as a first order summary of the great variety of volcanic landforms of different ages observed on Tenerife (Fig. 3.2).

The 200 m high summit cone (El Pitón) was formed at the end of the last Teide eruption (1,147 ± 140 BP Lavas Negras; Carracedo et al. 2007) and the material composing the cone is often altered by fumarole activity near the summit crater. The crater is 100 m wide, less than 30 m deep and partially filled by lava and rock falls. The flanks of the volcano are mostly covered by the glassy phonolite of the Lavas Negras flow lobes. The thickest and largest flows are directed towards the southern and the northern flanks, as Pico Viejo has obstructed their path to the west. These materials are well exposed on the southeast and northeast flanks. Erosion has produced gullies between the lava flows, generating debris falls and flows to the base of the volcano, the most relevant debris fans being observed at the bottom of Las Calvas ravine (northwest flank: Corredores Munich) and near the cable car station (south flank). The spatial distribution of the lava flows reveals buried morphologies of "Old Teide", such as (1) a phreatomagmatic vent on the northwest flank at 2,700 m a.s.l., associated with surge deposits found in Las Calvas (see Chap. 12) a 700 m deep depression buried by the terminal cone at 3,500 m a.s.l., open to the east and associated with pyroclastic deposits. Thus, the lavas of "Old Teide" exposed on the east flank could have erupted from this previous depression, which may correspond to a crater or a small graben.

Pico Viejo appears as a secondary volcano on the southwestern flank of Teide, at the junction with the northwest rift zone (NWRZ). Episodically active for 27.5 ky (Carracedo et al. 2007), Pico Viejo produces less differentiated lavas than those of Teide and its flanks are less steep (15–35°). The main crater (800 m wide and 140 m deep) is truncated to the southwest by a deep phreatic vent (300 m wide and 130 m deep). Surge deposits overlying the remnants of a lava lake can be observed on the southern walls of the main crater.

Finally, the majority of the recent volcanic activity of Teide volcano has been associated with peripheral vents on its lower slopes (e.g., Roques Blancos, Pico Cabras, Montaña Abejera, and Montaña Blanca: see Chap. 6 for ages). The activity of these vents involved at least seven voluminous phonolitic flows on the north flank, three of them reaching the coast (see Chap. 14 on hazards). These lava flows have a blocky texture and are channelled by lateral levees. Their thickness is more than 30 m, but can reach 160 m in Roques Blancos (northwest flank of Pico Viejo). Considering their high volume (>0.5 km^3), these eruptions might last for weeks to months. As a side effect, these massive flows seem to locally have improved the flank stability of Teide Volcano.

The pre-historic and historic lava flows located on the west and southwest flanks of Pico Viejo (e.g., Montaña de Chío, 3932 ± 213 BP; Montaña Reventada, 900 ± 150 BP; Las Narices, 1798) represent the junction between the Teide central volcano and the northwest rift zone, as demonstrated by the occurrence of intermediate compositions and magma mixing

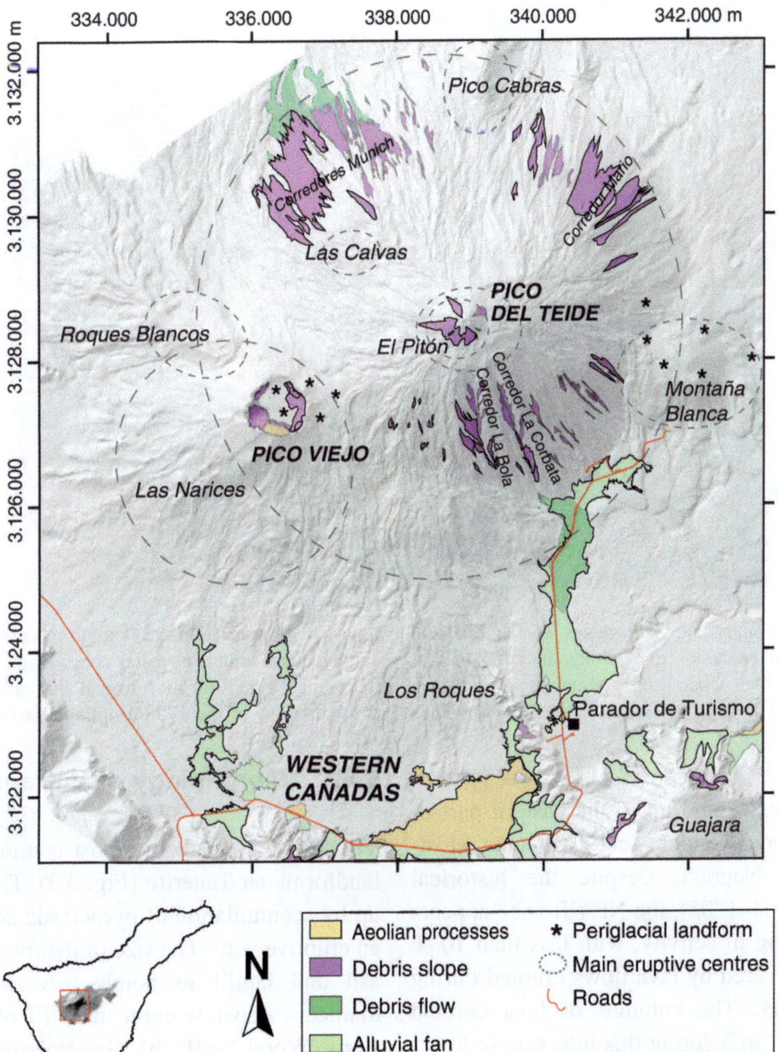

Fig. 3.2 Shaded relief view of Pico del Teide and Pico Viejo volcanoes, showing late Pleistocene and Holocene non-volcanic landforms (UTM coordinates, WGS84 datum)

features between mafic and phonotephrite to phonolite magmas (Wiesmaier et al. 2011; see Chap. 11).

3.8 Young Volcanic Landforms of the Rift Zones

Volcanic rift zones display a characteristic morphology defined by the alignment of eruptive centres along a main axis, a lateral distribution of the lava flows from this axis and the absence of a central volcano (see Chaps. 4 and 5). Volcanic rift zones are 10–20 km long and 1,500–2,000 m high (e.g., El Hierro and La Palma islands). The northeast rift zone of Tenerife (Cordillera Dorsal; see also Chap. 5) is built between the Anaga shield volcano (4.9–3.9 Ma; Guillou et al. (2004) and the Las Cañadas Caldera. The available K–Ar ages suggest that the main volume of the NERZ was accumulated in a relatively short time span

Fig. 3.3 Photographs and 3D images of the different morphologies of volcanic cones along the rift zones of Tenerife. **a** the cone of Montaña Reventada (900 ± 150 BP) is breached towards the west by lava flows; **b** Montaña Samara is a strombolian cone with a single central crater; **c** spatter cones aligned on the south flanks of Teide Volcano, around the cable car station; **d** fissure cone of the 1798 eruption (Narices del Teide)

(1.1–0.83 Ma; Carracedo et al. 2011). The rift zone volcanics extend up to the eastern part of the Las Cañadas Caldera (e.g., Montaña Mostaza, Arenas Negras). Despite the historical eruptions (1704–1705), the NE rift zone appears to be declining in activity, with less than 10 % of its area covered by lava flows erupted during the last 12 ky. The volumes of lava emitted ($\sim 150 \times 10^6$ m^3) during this interval are lower than those of the NW rift zone ($\sim 800 \times 10^6$ m^3) within the same episode, making the NWRZ the most active volcanic structure of the island, with 95 % of its area covered by lava flows emitted during the last 12 ky. In comparison to the Las Cañadas Caldera the landscapes of rift zones vary little and are mainly composed of strombolian lapilli (*picón*) and associated ‘a‘ā lava flows (*malpaís*) (‘a‘ā flows are defined by an extremely irregular surface, usually covered by decimetric fragments of broken crust). The diversity of this landscape is therefore controlled by the relationship between the age of the lavas and the bioclimatic conditions (vegetation, exposure and altitude).

3.8.1 Morphology of Volcanic Cones

Volcanic cones are the most common volcanic landforms on Tenerife (Fig. 3.3). They are built up by accumulation of pyroclastic debris around an eruptive vent. The size of the fragments, from ash and lapilli to bombs several metres in diameter, depends upon the VEI of the explosions (Wood 1980a, b). Gas emissions force the vent open and the ejected pyroclasts follow ballistic trajectories with most falling near the vent to generate a slope with an angle of repose of about 30°. The geometry of cones is mostly controlled by the structure of the vent (fissure or central vent) and the pre-existing topography, but also by the wind direction during the eruption. Cones are rapidly eroded, both by internal processes (hydrothermal activity, instability demonstrated by fissures and slope breaks), further volcanic activity (in the case of polygenic volcanoes) and by external processes (smoothing of the summit ridges, crater filling and accumulation of material downslope). Thus, the morphology of volcanic cones is the result of a combination of volcanic and morphoclimatic

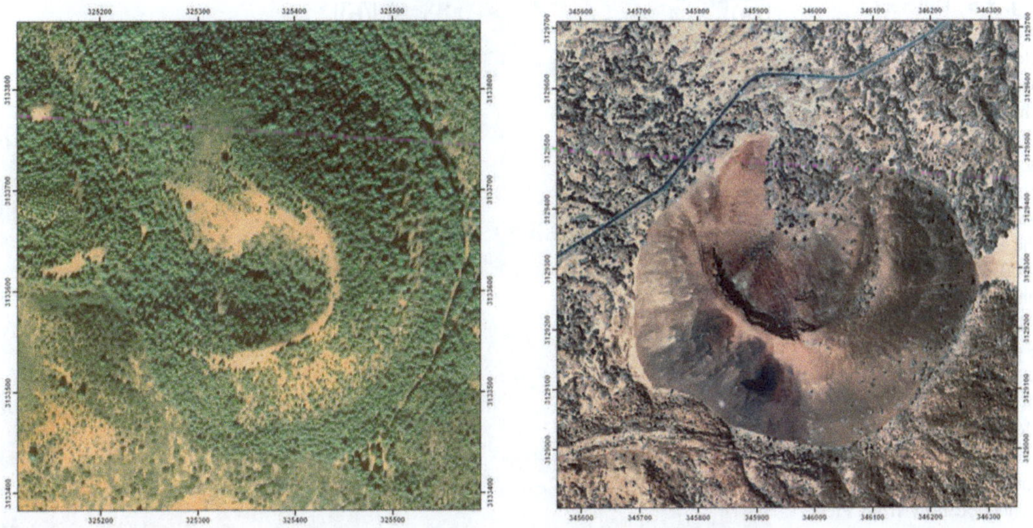

Fig. 3.4 Comparison of vegetation covering two strombolian cones located at different altitudes: *left*, Montaña del Banco (13 ka, 1,280 m a.s.l.); and *right*, Montaña Mostaza (15 ka, 2,240 m a.s.l.)

processes. The degradational evolution of these landforms can be broadly correlated with the time they have been exposed to erosion (Wood 1980a, b; Hooper and Sheridan 1998). However, the possibility of accurately dating cones by morphometric analysis is limited (Rodriguez-Gonzalez et al. 2012) as cone morphometry in Tenerife is controlled by bioclimatic (Fig. 3.4) as well as lithological parameters (hydrothermal alteration, permeability and induration, etc.).

The recent activity of the rift zones of Tenerife has built dozens of volcanic cones, especially along the northeastern and northwestern rift zones. The northwestern rift zone (NWRZ) displays a wide variety of strombolian cone morphologies (Romero 1992): single cones with a central crater or with nested craters, coalescent cones with multiple ridges, the most common case being a simple cone without associated lavas, and hornitos, which are secondary vents from the surface of lava flows. Further, the interaction between the water table and the magma can lead to phreatic explosions (e.g., upper vent of the 1798 fissure) or phreatomagmatic pulses (e.g., late activity of Montaña Reventada, as evidenced by phreatomagmatic deposits on the rim of the main crater). This can also influence the morphology of the volcanic edifice by enlarging the diameter and reducing the height of the cone. Lastly, when the gas content of the magma is low, the pyroclastic fragments are larger, more ductile and tend to weld together on landing, thus building a spatter cone, characterized by steep slopes (e.g., Montaña Majúa spatter cones near the cable car: see Fig. 3.3).

3.8.2 Morphology of Lava Flows

Historical and pre-historical lava flows on Tenerife (Fig. 3.5) are mainly ‘a‘ā flows. Their overall topography masks a chaotic morphology of lava channels, levees and tubes, penitents (tilted fragments of lava crust), lava balls and debris at the front of the flows, islets in the middle of main flows and hornitos (see Chap. 12). Pahoehoe lavas are also exposed in the western part of the caldera (pahoehoe lava flows have a smooth surface, are composed of coalescent tubes and tongues and their propagation is controlled by the equilibrium between the cooling crust and the eruptive rate). They were emitted during the 1798 eruption (Las Narices) and also during the early eruptions of Pico Viejo where lavas overflowed the western rim of the caldera, spread over the northwest and southwest flanks

Fig. 3.5 Different morphologies of lava flows along the rift-zones of Tenerife. **a** 'A'ā flow channelled in a *barranco* (Fasnia, 1705 eruption); **b** 'A'a flow of the 1706 eruption (Garachico-El Tanque); **c** lava tongue of the Reventada eruption (900 ± 150 BP); **d** thick phonolitic flows of Roques Blancos (1712 ± 153 BP); **e** pahoehoe flow of Pico Viejo, covering remnants of Las Cañadas Volcano (Roques de García); **f**: cross-section showing lava tongues and tubes of a pahoehoe flow from Pico Viejo Volcano (27030 ± 430 BP)

and finally reached the coast near Los Gigantes and Icod.

The surface alteration and colour of the lava flows usually reflect their age, the darker flows being the younger (e.g., Lavas Negras—the black lava flows of 1798 in the Las Cañadas Caldera). Nevertheless, in Tenerife, altitude and exposure to rapid temperature changes and humidity are also predominant factors controlling the weathering or preservation of lavas. For instance, the 1706 lava flow in Garachico, on the wet northern flank of the island, carries much denser vegetation than the 1705 lava flow in Arafo, on the southern flank of the island, which is less exposed to humidity. As a result, the wetter areas are carry more vegetation, weathering weathering is more intense and the lavas are very poorly preserved.

The morphological characteristics of lava flows help to constrain relationships between their runout length, area, surface morphology and erupted volume (Walker 1973; Wadge 1978; Pinkerton and Wilson 1994). Mafic and felsic lava flows are clearly distinguished on the basis of their volume vs. aspect ratio. Aspect ratio is often calculated as the ratio of flow length to flow width, with width being sometimes highly variable along the flow. To counteract the errors induced in the aspect ratio calculation by variable flow width, we used the area of a circle equal to the flow area as an alternative value.

Variations in the aspect ratio reflect the petrology and viscosity of lava flows. Basaltic flows give lower aspect ratios compared to the more viscous phonolitic lava flows. Both lava types are able to reach the coast, covering a distance of 10–15 km (Figs. 3.6 and 3.7). The phonolitic flows are thick and more voluminous (1.4 km^3 for Roques Blancos and 0.7 km^3 for the last eruption of Teide), whereas the lava

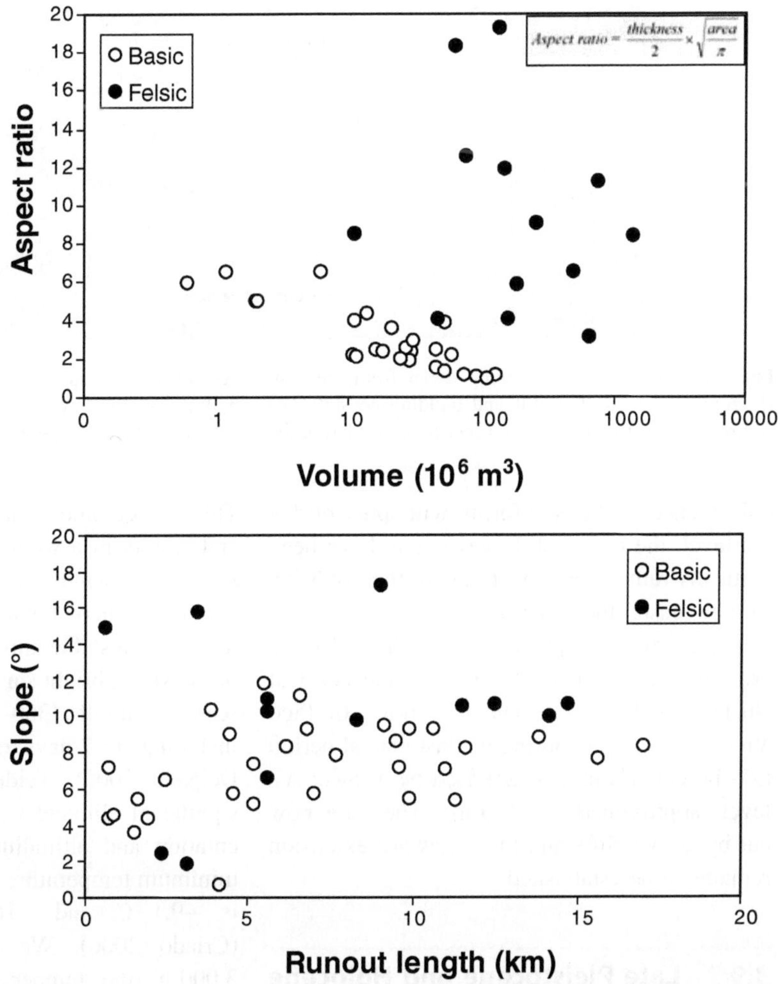

Fig. 3.6 Some morphometric characteristics of Holocene lava flows in Tenerife. Basic and felsic lava flows are clearly distinguished on the basis of their volume vs. aspect ratio. The runout length of flows is not significantly influenced by slope angle, but rather by lava viscosity, eruptive rate and cooling

flows of the rift zones have volumes rarely exceeding 100 million m^3 (Cueva del Ratón, 75 × 10^6 m^3; Montaña Cascajo, 92 × 10^6 m^3; and Montañas de Chío, 110 × 10^6 m^3). The volume and duration of historic eruptions along the NW rift-zone of Tenerife allows determination of their eruption rates. The high eruption rate of the 1706 Garachico eruption (71 m^3/s) is in the range of the 1669, 1980 and 1989 eruptions of Etna, and the 1977–1984 eruptions of Krafla in Iceland (cf. Harris et al. 2000; Crisci et al. 2003). The remaining historic eruptions in Tenerife were less productive (<13 m^3/s), which compares well with examples of the 1983–1987, 1991–1993, 1999 and 2001 eruptions of Etna volcano (Calvari et al. 1994). Considering the range of eruption rates and aspect ratios, the duration of past eruptions of the NW rift zone may have been typically in the order of a week to one month. However, with individual volumes greater than 500 × 10^6 m^3, the phonolitic eruptions of Roques Blancos, Pico Cabras, Abejera Alta and the last eruption of Teide (Lavas Negras) may have lasted several months. The total volume of felsic lava produced in the central area of Tenerife during the Holocene exceeds some 4 km^3.

The northern coast near Buenavista del Norte and Icod de los Vinos is interesting from a morphological point of view (see Fig. 3.7). Holocene lava flows of the NW rift zone fossilised the cliff and built coastal platforms (e.g., Garachico, Liferfe, and Montañas Negras). Considering the ages of these lavas and the

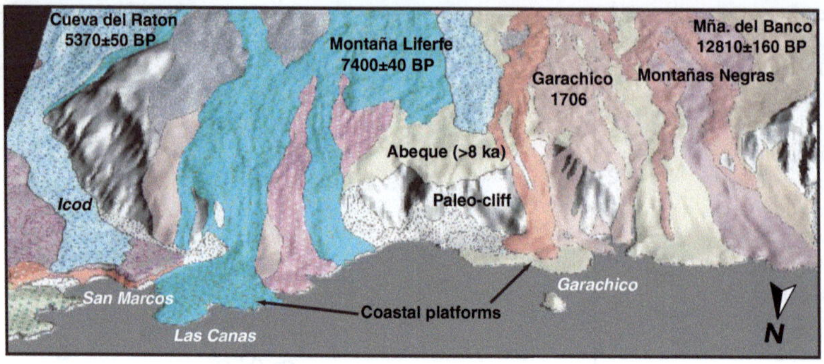

Fig. 3.7 Holocene lavas flowing down from the NW rift-zone, fossilising the cliff and building coastal platforms on the northern coast of Tenerife. Considering the ages of these lavas and the coincidence of the platforms with the present-day sea level, the island of Tenerife may have been vertically stable for at least 8 ka

coincidence of the platforms with present-day sea level, the island of Tenerife may have been vertically rather stable over at least the last 8 ky. In any case, the surface area of the island increased by these processes and the old cliffs are presently located 2.5 km from the coast at Buenavista del Norte. The lava flows of Pico Viejo were emitted during the last glacial period (27–14 ky), a time characterised by a lower sea level (approximately −110 m). They are now cut by active cliffs and their seaward extension remains to be established.

3.9 Late Pleistocene and Holocene Non-Volcanic Landforms and Climatic Influences

Due to recurrent volcanic activity during the Quaternary, most of the landforms on Tenerife are of volcanic origin, but variably reshaped by erosion. However, Holocene landforms produced by periglacial, aeolian and fluvial processes can also be recognized (see Fig. 3.2) and can be used to reconstruct the paleoclimatic history of the high-altitude areas of Tenerife. An aspect of recent interest has been the influence of the growth of Teide volcano and peripheral domes on the microclimate and morphodynamic evolution of Las Cañadas Caldera and its walls.

The Teide edifice is usually above the 'sea of clouds' that accumulates on the northern flanks of Tenerife Island and is exposed to trade winds. This is a common feature of all Canary Islands and results in a wetter climate on the northern side of the islands and a semiarid to arid climate on the southern flanks. The climate is characterized by a stable atmosphere and dry air (90 % of the year), bright sunshine (3448.5 h/year) and scarce rainfall (360–501 mm/year, 43.4 days including 12.7 days of snowfall) (Bustos and Delgado 2000). Teide flanks are exposed to significant climatic variations due to slope orientation and altitudinal changes; the absolute minimum temperature recorded at 2,372 m a.s.l. is −9.1 °C, and −16.8 °C at 3,530 m a.s.l. (Criado 2006). We can assume that above 3,000 m the temperature is below 0 °C for 11 months/year. Thus, the presence of water in rocks and regolith combined with frost activity become important factors in morphogenesis.

Evapotranspiration ranges between 546 and 682 mm/year, causing aridity from July to October. Northwest is the dominant wind direction in the Teide area (50 %), with a mean velocity of 8.1 m/s and a record high of 53.2 m/s. The wind can entrain sand-size and finer particles over the entire area and in some places dust devils are frequent and powerful (Criado et al. 2009; Höllermann 1984) (Fig. 3.8).

3.9.1 Aeolian Landforms

Aeolian landforms are scarce and small in the landscape of Teide National Park. Two types of

Fig. 3.8 Dust devil on Ucanca playa. 7 July 2010 (photo: Nick Brooks)

aeolian landforms can be distinguished. The first derives from the interaction between the aeolian sand fluxes and the shrub vegetation *(Spartocytisus supranubius)*, forming nebkhas, which are natural accumulations of wind-borne sediments within or around the canopies of plants. The second type are 'climbing-dunes' which appear when the wind, carrying sediments, comes up against a steep relief. These aeolian landforms are mostly located in the endorheic areas at the base of the Las Cañadas Caldera walls, reworking the distal facies of alluvial fan systems. Occasionally, the surface consists of a pavement of closely packed, interlocking angular or rounded rock fragments of pebble and cobble size formed by removal of loose, fine-grained particles, a process known as aeolian deflation (Criado et al. 2009).

3.9.2 Periglacial Landforms

Periglacial processes from the highest altitude regions of Tenerife have been described since the 1970s (Höllermann 1978; Morales et al. 1977, 1978). Soils and rocks become sufficiently wet to allow water to freeze after heavy rains and snow melt. Frost layers can reach 8–10 cm in thickness (Höllermann 1978; Martínez de Pisón and Quirantes 1981) and the soil can be frozen for 3–11 months/year near the summit (Quirantes and Martínez de Pisón 1994; Rodríguez et al. 2010). Frost weathering is very active in the Lavas Negras (last eruption products of Teide volcano), producing debris slopes at the front and flanks of these lava flows, especially on the northern slopes (Corredor de la Isla and Corredores Munich). Pipkrakes (needle ices) are very common on flat areas, the active layer being some millimetres to a few centimetres thick. These ice pieces are formed when the liquid water in the soil rises through capillary action to the surface finding air temperatures cold enough to freeze the water. While growing, they push away small soil particles. On sloped surfaces, soil creep may be significantly influenced by needle ice development. Soil movements derived from the frost cycle produce small sorted stripes and polygons. Sorted stripes are frequent in areas occupied by basaltic lapilli (Fig. 3.9), while polygons occur in the Pico Viejo crater, Los Gemelos sector and also in Montaña Rajada crater (Martínez de Pisón and Quirantes 1981; Quirantes and Martínez de Pisón 1994).

Another periglacial process includes gelifluction, a downslope mass movement of soil due to

Fig. 3.9 Pipkrake and small polygons. Playa la Grieta, July 7, 2010

Fig. 3.10 Pumice deposit shaped in lobes by gelifluction, on the northern rim of Pico Viejo crater

the freeze–thaw action upon waterlogged topsoils. The most common result are gelifluction lobes, which always appear in pumice ash-fall deposits, are a few metres long and are similar to terraces (Fig. 3.10). The movement of these lobes may reach ~50 cm/year (Höllermann 1978).

3.10 Fluvial Landforms

3.10.1 Ravines ("Barrancos")

Ravines (locally known as *barrancos*) on the flanks of Teide volcano are almost rectilinear and extend in a radial fashion from the summit. Many more barrancos were in existence around Teide until recently, but were buried by the eruption of the Lavas Negras (1.2 ky BP).

The heads of the *barrancos* are in several cases higher than 1,200 m; their depth normally is less than 100 m with slopes ranging between 30 and 40°. The largest ravines are, from east to west, Corredor Mario (0.34 km^2), Corredor La Corbata (0.74 km^2) and Corredor La Bola (0.50 km^2).

The hydrographical networks of the *barrancos* are very simple and the head of these *barrancos* do not display widening. They have developed since at least the Late Pleistocene, as the ravines cut rocky outcrops of Teide older than 32 ky (Carracedo et al. 2007). Fluvial processes are sporadic within these *barrancos* but there is evidence of efficient geomorphological activity during the last millennium. The Barranco de Corredor La Bola is partially filled by the Lavas Negras eruption (Fig. 3.11). The entrance of a branch of the lava flow inside the *barranco* has modified the geomorphic system, and a new channel, 10 m wide and 5 m deep, was scoured.

3.10.2 Alluvial Fans and Debris Flows

Alluvial fans are depositional landforms occurring where confined streams emerge from mountain catchments into zones of reduced stream power (Harvey 1997). There are three groups of alluvial fans around Teide and Pico Viejo volcanoes. The first group is located on the lower most southern flank of Teide volcano (Bravo and Bravo-Bethencourt 1989; Martínez de Pisón and Quirantes 1981), the second one on the southern flank of Pico Viejo volcano, and the third on the northern flank of Teide volcano (Corredores Munich). Alluvial fans located on the lowermost slopes of the Las Cañadas Caldera wall are not included in this chapter.

The first group of alluvial fans is produced by the erosive dismantling of the southern slopes of Teide volcano. The total area of the fan is poorly constrained because it is partly buried by lava flows and domes. Its present-day area is 1.83 km^2, with a maximum length of 3.5 km. In the distal sector (close to the Parador building), the alluvial fan rests upon tephriphonolite flows from Pico Viejo (20.7 ky BP, Carracedo et al. 2007). Evidence of torrential rain activity after the emplacement of Late Pleistocene lavas comes

Fig. 3.11 The Corredor La Bola partially filled by the Lavas Negras eruption

Fig. 3.12 Alluvial fan at the foot of teide resting on a trachyte lava flow from phase VIII Montaña Blanca (~2 ka BP, Ablay et al. 1995); the associated channel was filled by the Lavas Negras flow (1.2 ka BP)

Fig. 3.13 Debris flow facies on Cañada Blanca alluvial fan. The cross-section is located on the TF-21 road. Blocks with alunite (*white arrow*) come from fumarolic areas of the Teide summit

in the form of pockets of rounded pumice gravels belonging to Montaña Majúa (relatively dated 5–4 ky BP, Carracedo 2006). The eastern sector of the alluvial fan covers phonolite flows of the VIII phase of the Montaña Blanca eruption (2 ky BP, Ablay et al. 1995) whereas the former *barranco* was filled by the Lavas Negras eruption (1.2 ky BP, Carracedo et al. 2007) (Fig. 3.12).

At the apex of the Cañada Blanca alluvial fan there are sectors occupied by debris flow deposits, including boulders up to 18 metric tons. Associated with these boulders are Guanche dwellings and abundant pottery remains, providing evidence of debris flows before the sixteenth century AD. In addition, in a trench next to road TF-21, the debris flow deposits are overlying the Lavas Negras eruption (Fig. 3.13). Field surveys of the levées of more recent debris flows, resting on debris older than the Lavas Negras eruption, did not reveal any pottery or tools from the Guanche culture. This suggests that these debris flow deposits were generated during the last 500 years, by extreme events of heavy rains (Fig. 3.14). The most important storm in historical times in the Canary Islands was the 1826 hurricane (Bethencourt-Gonzalez and Dorta-Antequera 2010), which may have been the very occasion on which this fan formed.

Today torrential activity occurs in channels with a restricted annual period of activity, due to limited supplies of melt water and heavy rains. Stump trees (*Pinus radiata*) partially covered by recent sediments are solid evidence for sporadic torrential activity. Nevertheless, on September

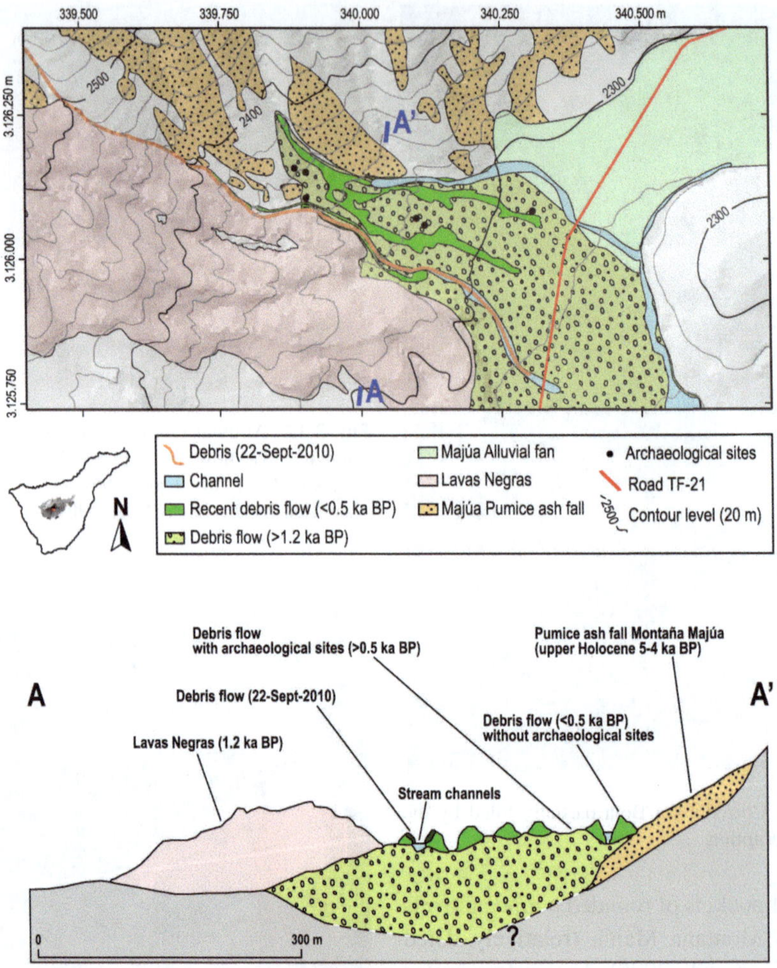

Fig. 3.14 Geomorphological map of debris flow emplaced at the junction of Corredor La Bola and La Corbata ravines and detailed cross section along A-A' (UTM coordinates, WGS84 datum)

22, 2010 an unusual summer storm that produced heavy rains (>90 mm/day) and powerful erosion of fine hydrothermally-altered material (by fumaroles) occurred at the head of the Corredor La Bola. The resulting debris flow transported sediment 2.5 km downstream and stopped only a few metres from the TF-21 road (Table 3.1).

The second group of alluvial fans is located at the mouth of *barrancos* on the southern flanks of Pico Viejo volcano. They are less important in terms of volume. Present-day activity is limited to small amounts of sediment reworking inside these channels. In this area, archaeological remains from the Guanche culture are frequent but never buried by recent deposits.

The alluvial fans associated with the Corredores Munich are very young (<last millennium). In the lower part of the debris cones there is a system of debris flows with sharp levées. The lowest ones have been colonized by a natural pine forest over the last century. The highest debris flow shows evidence of recent sediment transport, especially blocks and gravels produced by frost weathering on Late Pleistocene phreatomagmatic breccia (Calvas del Teide) and Lavas Negras flow. The influence of the Little Ice Age on the activity of these channels and related debris flows is questionable (Martin Moreno 2011), and further investigations are needed to corroborate and affirm any correlation (Table 3.2).

Table 3.1 Chronology of different alluvial fans and debris flows from the southern side of Teide Volcano based on their relationship with volcanic landforms (Ablay et al. 1995; Carracedo et al. 2007)

Volcanic event	Alluvial event	Location	Lithology/facies	Geomorphology	Age ^{14}C	Archaeological remains
–	Present-day activity	Corredor La Bola ravine	Debris flow	Very narrow debris flow 2,500 m long	22 September 2010	–
–	Present-day activity	On the channels	Alluvial sediment (gravel and sand)	Very narrow and shallow alluvial channel scarred tree stumps	Present	No evidence
–	Alluvial fan of 1826?	Intersection of Corredor La Bola and La Corbata ravines	Debris flow	Debris flow with very well preserved levées	Nineteenth century AD?	No evidence
Lavas Negras	–	Lavas from the top of Teide volcano	Phonolite	Very long a'a' lavas	663–943 AD*	–
–	Roquillo alluvial fan	Inactive alluvial fan (channel filled by Lavas Negras)	Alluvial fan	Dense stone pavement, sparse boulders, aeolian processes	65–95 AD to 663–943 AD	Pottery shards and obsidian artifacts from Guanche culture
–	Cable railway station alluvial fan	Mainly inactive, weak activity in the lowest part of small ravines beside Majúa alluvial fan	Alluvial fan	Dense stone pavement, sparse boulders	Older than sixteenth century AD	Very abundant pottery shards and obsidian artifacts from Guanche culture
–	Cañada Blanca alluvial fan	In the former confluence of Corredor La Bola and La Corbata ravines	Debris flow, including alunite blocks issued from summit crater	Dense stone pavement, abundant large boulders	Older than sixteenth century AD and older 663–943 AD	Very abundant pottery shards and obsidian artifacts from Guanche cave dwellings
Montaña Blanca (phase VIII)	–	Dome volcanic event in the eastern flank of Teide volcano	Phonolite	Dome lavas, short spreading and broad surfaces covered by pumice ash-fall	65–95 AD**	–

* After Carracedo et al. (2007); ** After Ablay et al. (1995)
Archaeological remains from the Guanche culture and their resultant effects on historical forestry

Table 3.2 Chronology of debris flows from the northern side of Teide Volcano based on their relationship with volcanic landforms (Carracedo et al. 2007), archaeological remains from the Guanche culture and their resultant effects on historical forestry

Volcanic event	Alluvial event	Location	Lithology/ facies	Geomorphology	Age ^{14}C	Archaeological remains
–	Present-day activity	Corredores Munich	Debris flow	Very well-developed debris flow system	Present	No evidence
–	Little Ice Age	Corredores Munich	Debris flow	Very well-developed debris flow system (today under forest cover)	1.2–0 BP	No evidence
Lavas Negras	–	Lavas from the top of Teide Volcano	Phonolite	Very long a'a' lavas	663–943 AD	–

The volume of the alluvial fans on La Corbata debris flow and the large sizes of clasts (up to boulders), which were produced before 1.2 ky BP, can be explained as the result of extreme events of heavy rains, such as the hurricane of 1826 (Table 3.2).

Acknowledgments Authors acknowledge the Teide National Park for access to the areas investigated.

References

Ablay GJ, Ernst GGJ, Marti J, Sparks RSJ (1995) The ~2 ka subplinian eruption of Montaña Blanca, Tenerife. Bull Volcano 57(5):337–355. doi: 10.1007/bf00301292

Ancochea E, Fuster J, Ibarrola E, Cendrero A, Coello J, Hernan F, Cantagrel JM, Jamond C (1990) Volcanic evolution of the island of Tenerife (Canary Islands) in the light of new K-Ar data. J Volcanol Geotherm Res 44(3–4):231–249

Ancochea E, Huertas MJ, Fúster JM, Cantagrel JM, Coello J, Ibarrola E (1995) Geocronología de la pared de la Caldera de Las Cañadas Tenerife, Islas Canarias. Boletín de la Real Sociedad Española de Historia Natural, Sección Geología 90:107–124

Ancochea E, Huertas MJ, Cantagrel JM, Coello J, Fúster JM, Arnaud N, Ibarrola E (1999) Evolution of the Cañadas edifice and its implications for the origin of the Cañadas Caldera (Tenerife, Canary Islands). J Volcanol Geotherm Res 88(3):177–199

Bethencourt-Gonzalez J, Dorta-Antequera P (2010) The storm of November 1826 in the Canary Islands: possibly a tropical cyclone? Geografiska Annaler Series a-Physical Geography 92A(3):329–337

Bravo T, Bravo-Bethencourt J (1989) Mapa volcanológico de Las Cañadas y Teide-Pico Viejo. In: Araña V, Coello J (eds) Los volcanes y la caldera del Parque Nacional del teide. ICONA, Madrid

Bryan SE, Marti J, Cas RAF (1998) Stratigraphy of the Bandas del Sur formation: an extracaldera record of Quaternary phonolitic explosive eruptions from the Las Cañadas edifice, Tenerife (Canary Islands). Geol Mag 135:605–636

Bustos JJ, Delgado F (2000) Climatología del Parque Nacional de Las Cañadas del Teide. http://www.iac.es/proyect/sitesting/TOPSITES/principal.html

Calvari S, Coltelli M, Neri M, Pompilio M, Scribano V (1994) The 1991–1993 Etna eruption: chronology and lava flow-field evolution. Acta Vulcanol 4:1–14

Cantagrel JM, Arnaud NO, Ancochea E, Fúster JM, Huertas MJ (1999) Repeated debris avalanches on Tenerife and genesis of Las Cañadas Caldera wall (Canary Islands). Geology 27(8):739–742. doi:

10.1130/0091-7613(1999)027<0739:rdaota>2.3.co;2

Carracedo JC (1994) The Canary-Islands—an example of structural control on the growth of large oceanic-island volcanoes. J Volcanol Geotherm Res 60(3–4):225–241. doi:10.1016/0377-0273(94)90053-1

Carracedo JC (2006) El Volcán Teide. CajaCanarias, Santa Cruz de Tenerife

Carracedo JC, Day S, Guillou H, Badiola ER, Canas JA, Pérez-Torrado FJ (1998) Hotspot volcanism close to a passive continental margin: the Canary Islands. Geol Mag 135(5):591–604. doi:10.1017/s0016756898001447

Carracedo JC, Paterne M, Guillou H, Pérez Torrado FJ, Paris R, Rodríguez Badiola E, Hansen A (2003) Dataciones radiométricas (C^{14} y K-Ar) del Teide y el Rift NO, Tenerife, Islas Canarias. Estudios Geológicos 59(1–4):15–29

Carracedo JC, Rodríguez Badiola E, Guillou H, Paterne M, Scaillet S, Pérez Torrado FJ, Paris R, Fra-Paleo U, Hansen A (2007) Eruptive and structural history of Teide Volcano and rift zones of Tenerife, Canary Islands. Geol Soc Am Bull 119(9):1027–1051

Carracedo JC, Guillou H, Nomade S, Rodriguez-Badiola E, Pérez-Torrado FJ, Rodriguez-Gonzalez A, Paris R, Troll VR, Wiesmaier S, Delcamp A, Fernandez-Turiel JL (2011) Evolution of ocean-island rifts: the northeast rift zone of Tenerife, Canary Islands. Geol Soc Am Bull 123(3–4):562–584

Criado C (2006) Geomorfología del Teide y de las dorsales activas. Parte II: geomorfología climática. In: Carracedo JC (ed) Los volcanes del Parque Nacional del Teide: el Teide, Pico Viejo y las dorsales activas de Tenerife. OAPN-Ministerio de Medio Ambiente, Madrid

Criado C, Machado C, Afonso J (2009) Geomorfología eólica del Parque Nacional del Teide (Tenerife). In: Beltrán Tejera E, Afonso-Carrillo J, Gg A, Rd O (eds) Homenaje al Prof. Dr. Wolfredo Wildpret. Instituto de Estudios Canarios, La Laguna, pp 685–704

Crisci GM, Gregorio S, Rongo R, Scarpelli M, Spataro W, Calvari S (2003) Revisiting the 1669 Etnean eruptive crisis using a cellular automata model and implications for volcanic hazard in the Catania area. J Volcanol Geotherm Res 123(1–2):211–230. doi:10.1016/s0377-0273(03)00037-4

Giachetti T, Paris R, Kelfoun K, Pérez-Torrado F (2011) Numerical modelling of the tsunami triggered by the Güímar debris avalanche, Tenerife (Canary Islands): comparison with field-based data. Mar Geol 284(1–4):189–202. doi:10.1016/j.margeo.2011.03.018

Guillou H, Carracedo JC, Paris R, Pérez Torrado FJ (2004) Implications for the early shield-stage evolution of tenerife from K/Ar ages and magnetic stratigraphy. Earth Planet Sci Lett 222(2):599–614

Harris AJL, Murray JB, Aries SE, Davies MA, Flynn LP, Wooster MJ, Wright R, Rothery DA (2000) Effusion rate trends at Etna and Krafla and their implications for eruptive mechanisms. J Volcanol Geotherm Res 102(3–4):237–270. doi:10.1016/s0377-0273(00)00190-6

Harvey A (1997) The role of alluvial fans in arid zone fluvial systems. In: Thomas DSG (ed) Arid zone geomorphology: process form and change in drylands. Wiley, UK

Höllermann P (1978) Soil movements in the subtropical mountain environment of high Tenerife (Canary Islands). In: Colloque sur le périglaciaire d'altitude actuel et herité dans le domain mediterranéen et ses abords, pp 91–112

Höllermann P (1984) Studien zur aktuellen morphodynamik un geökologie der Kanareninseln Teneriffa und Fuerteventura. Vancenlocs and Roprecht Göttingen

Hooper DM, Sheridan MF (1998) Computer-simulation models of scoria cone degradation. J Volcanol Geoth Res 83(3–4):241–267. doi:10.1016/s0377-0273(98)00031-6

Ibarrola E, Ancochea E, Fuster JM, Cantagrel JM, Coello J, Snelling NJ, Huertas MJ (1993) Cronoestratigrafía del Macizo de Tigaiga: evolución de un sector del edificio Cañadas (Tenerife, Islas Canarias). Boletín de la Real Sociedad Española de Historia Natural (Sec Geol) 88(1–4):57–72

Márquez A, López I, Herrera R, Martín-González F, Izquierdo T, Carreño F (2008) Spreading and potential instability of Teide Volcano, Tenerife, Canary Islands. Geophy Res Lett 35:L05305, 5. doi:10.1029/2007GL032625

Marti J, Gudmundsson A (2000) The Las Canadas Caldera (Tenerife, Canary Islands): an overlapping collapse caldera generated by magma-chamber migration. J Volcanol Geotherm Res 103(1–4):161–173. doi:10.1016/s0377-0273(00)00221-3

Martí J, Mitjavila J, Araña V (1994) Stratigraphy, structure and geochronology of the Las Cañadas Caldera (Tenerife, Canary Islands). Geol Mag 131(6):715–727. doi:10.1017/S0016756800012838

Martí J, Hurlimann M, Ablay GJ, Gudmundsson A (1997) Vertical and lateral collapses on Tenerife (Canary Islands) and other volcanic ocean islands. Geology 25(10):879–882. doi:10.1130/0091-7613(1997)025<0879:valcot>2.3.co;2

Martin Moreno R (2011) La Pequeña Edad de Hielo en el alto Teide (Tenerife, Islas Canarias). Menciones históricas y morfogénesis periglaciar. Ería 83:331–342

Martínez de Pisón E, Quirantes F (1981) El Teide. Estudio Geográfico. Ed, Interinsular Canaria, Santa Cruz de Tenerife

Masson DG, Watts AB, Gee MJR, Urgeles R, Mitchell NC, Le Bas TP, Canals M (2002) Slope failures on the flanks of the western Canary Islands. Earth Sci Rev 57(1–2):1–35. doi:10.1016/s0012-8252(01)00069-1

Mitjavila J (1990) Aplicación de técnicas de geoquímica isotópica y de geocronología al estudio vulcanológico del edificio de Diego Hernández y su relación con la Caldera de Las Cañadas Tenerife, PhD Thesis, Universidad de Barcelona

Mitjavila J, Villa I (1993) Temporal evolution of Diego Hernández formation Las Cañadas, Tenerife and confirmation of the age of the Caldera using the ^{40}Ar-^{39}Ar method. Revista de la Sociedad Geológica de España 6:61–65

Morales A, Martín F, Quirantes F (1977) Formas periglaciares en Las Cañadas del Teide. ACT, Santa Cruz de Tenerife

Morales A, Martín F, Quirantes F (1978) Formas periglaciares en Las Cañadas del Teide. In: Colloque sur le périglaciaire d'altitude actuel et herité dans le domain mediterranéen et ses abords, pp 79–90

Paris R (2002) Rythmes de construction et de destruction des édifices volcaniques de point chaud: l'exemple des Iles Canaries (Espagne). Ph. D., University Paris 1 Panthéon-Sorbonne

Paris R, Carracedo JC (2001) Formation d'une caldera d'érosion et instabilité récurrente d'une île de point chaud: la Caldera de Taburiente, La Palma, Iles Canaries. Géomorphologie 2:93–106

Paris R, Guillou H, Carracedo JC, Perez-Torrado FJ (2005) Volcanic and morphological evolution of La Gomera (Canary Islands), based on new K-Ar ages and magnetic stratigraphy: implications for oceanic island evolution. J Geol Soc (London, United Kingdom) 162(3):501–512

Pinkerton H, Wilson L (1994) Factors controlling the lengths of channel-fed lava flows. Bull Volcanol 56(2):108–120. doi:10.1007/bf00304106

Quirantes F, Martínez de Pisón E (1994) El modelado periglaciar de Canarias. In Gómez-Ortiz, Simón, Y Salvador Periglaciarismo en la Península ibérica, anarias y Baleares. SEG, Granada

Rodriguez-Gonzalez A, Fernandez-Turiel JL, Perez-Torrado FJ, Gimeno D, Aulinas M (2010) Geomorphological reconstruction and morphometric modelling applied to past volcanism. Int J Earth Sci 99(3):645–660. doi:10.1007/s00531-008-0413-1

Rodriguez-Gonzalez A, Fernandez-Turiel JL, Perez-Torrado FJ, Paris R, Gimeno D, Carracedo JC, Aulinas M (2012) Factors controlling the morphology of monogenetic basaltic volcanoes: the Holocene volcanism of Gran Canaria (Canary Islands, Spain). Geomorphology 136(1):31–44. doi:10.1016/j.geomorph.2011.08.023

Rodríguez M, Jiménez C, Tejedor M (2010) Estudio de la temperatura de los suelos de la isla de Tenerife. Fundación Canaria Salud y Sanidad de Tenerife, Santa Cruz de Tenerife

Romero C (1992) Estudio geomorfológico de los volcanes históricos de Tenerife. ACT, Santa Cruz de Tenerife

Troll VR, Walter TR, Schmincke H-U (2002) Cyclic caldera collapse: piston or piecemeal subsidence? field and experimental evidence. Geology 20(2):135–138

Wadge G (1978) Effusion rate and the shape of 'A'ā lava flow-fields on Mount Etna. Geology 6:503–506

Walker GPL (1973) Lengths of lava flows. Philos Trans R Soc Lond Ser a-Math Phys Eng Sci 274(1238):107–118. doi:10.1098/rsta.1973.0030

Warren SD, Hohmann MG, Auerswald K, Mitasova H (2004) An evaluation of methods to determine slope using digital elevation data. Catena 58(3):215–233. doi:10.1016/j.catena.2004.05.001

Watts AB, Masson DG (1995) A giant landslide on the north flank of Tenerife, Canary Islands. J Geophys Res 100(B12):24487–24498

Watts AB, Masson DG (2001) New sonar evidence for recent catastrophic collapses of the north flank of Tenerife, Canary Islands. Bull Volcanol 63(1):8–19

Wiesmaier S, Deegan F, Troll V, Carracedo JC, Chadwick J, Chew D (2011) Magma mixing in the 1100 AD Montaña Reventada composite lava flow, Tenerife, Canary Islands: interaction between rift zone and central volcano plumbing systems. Contrib Miner Petrol 162(3):651–669. doi:10.1007/s00410-010-0596-x

Wood CA (1980a) Morphometric analysis of cinder-cone degradation. J Volcanol Geotherm Res 8(2–4):137–160. doi:10.1016/0377-0273(80)90101-8

Wood CA (1980b) Morphometric evolution of cinder cones. J Volcanol Geotherm Res 7(3–4):387–413. doi:10.1016/0377-0273(80)90040-2

Structural and Geological Elements of Teide Volcanic Complex: Rift Zones and Gravitational Collapses

4

Juan Carlos Carracedo and Valentin R. Troll

Abstract

Initially recognised in the Hawaiian Islands, volcanic rift zones and associated giant landslides have been extensively studied in the Canaries, where several of their more significant structural and genetic elements have been established. Almost 3,000 km of water tunnels (*galerías*) that exist in the western Canaries provide a unique possibility to access the deep structure of the island edifices. Recent work shows that rift zones to control the construction of the islands, possibly from the initial stages of island development, form the main relief features (shape and topography), and concentrate eruptive activity, making them crucial elements in defining the distribution of volcanic hazards on ocean islands.

4.1 Introduction

Rift zones constitute the most pronounced and persistent structures in the development of oceanic volcanic islands because they: (1) control the construction of the insular edifices, possibly from the initial stages; (2) form the main relief features (shape and topography); (3) concentrate eruptive activity; (4) frequently play a key role in the generation of flank collapses and the catastrophic disruption of well-established volcano plumbing systems; (5) are crucial structures in the distribution of volcanic hazards; and (6) condition the storage of natural resources, such as groundwater (Navarro and Farrujia 1989).

Although rifts were initially recognized on the Hawaiian Islands (Fiske and Jackson 1972; Swanson et al. 1976; Walker 1986, 1987, 1992; Dieterich 1988), a good part of the progress made in understanding their genesis and structure has been achieved through their study in the Canary Islands (Carracedo 1975, 1979, 1994, 1996, 1999; Carracedo et al. 1992, 1998, 2001, 2007, 2011; Guillou et al. 1996; Walter and Schmincke 2002; Delcamp et al. 2010).

Compared with those of the Hawaiian Islands, the rifts of the Canaries are considerably longer lasting, exert greater overall control on the construction of the islands, and present more pronounced elements of relief. The lower magmatic activity of the mantle plume or hotspot

J. C. Carracedo (✉)
Departamento de Física (GEOVOL), Universidad de Las Palmas de Gran Canaria, Las Palmas de Gran Canaria, Canary Islands, Spain
e-mail: jcarracedo@proyinves.ulpgc.es

V. R. Troll
Department of Earth Sciences, CEMPEG, Uppsala University, Uppsala 75236, Sweden
e-mail: Valentin.Troll@geo.uu.se

J. C. Carracedo and V. R. Troll (eds.), *Teide Volcano*, Active Volcanoes of the World,
DOI: 10.1007/978-3-642-25893-0_4, © Springer-Verlag Berlin Heidelberg 2013

Fig. 4.1 Panoramic view from the top of Pico Viejo Volcano onto the North West Rift Zone of Tenerife, an excellent example of the evolution of a recent volcanic rift. The Teno Miocene Shield outcrops in the far distance (about 20 km)

that has generated the Canaries produces much lower eruptive rates (Geldmacher et al. 2001). This favours higher-aspect-ratio rift zones by accumulation of relatively short flows, promoting the growth of prominent ridges in the relief of these islands (Fig. 4.1). The very low drift velocity of the African plate and the apparent lack of significant subsidence of the Canaries allow for long periods of subaerial activity of the islands (at least 22 My), with corresponding long-lasting rifts that frequently display recurrent activity (Carracedo et al. 1998, 2011).

4.2 Oceanic Rift Zones. What are They and What Do They Represent?

Elongate zones where eruptive vents concentrate to form ridges are common and very pronounced features of oceanic volcanoes. Where erosion has incised sufficiently deeply into these features, their internal structure appears as a dense swarm of dykes broadly parallel to the axis of the ridge, forming "coherent intrusion complexes" (Walker 1992) or "rift zones" (Fiske and Jackson 1972; Carracedo 1975, 1994; Swanson et al. 1976; Wyss 1980; Stillman 1987). This swarm of dykes generally shows a gaussian distribution, with the intrusion density falling rapidly to near zero at the margins of the complexes. A similar pattern is apparent in the distribution of eruptive vents in the ridges (Fig. 4.2).

A high concentration of dykes in the rift zones was first deduced by MacFarlane and Ridley (1968) from conspicuous gravity ridges in the Bouguer anomaly map of Tenerife (Fig. 4.3). According to these authors, the growth of the island was largely controlled (both subaerial and submarine parts) by dyke injection along three major rift zones, with angles of about 120° between them. This idea was also applied by Macdonald (1972) to explain the ground plan, shape and internal structure of the Hawaiian shields.

Detailed studies of these features have been carried out on the Hawaiian volcanoes since the 1960s (Macdonald 1965; Fiske and Jackson 1972; Macdonald 1972; Swanson et al. 1976; Walker 1986, 1987, 1992; Dieterich 1988). Eventually, Walker (1992) defined rift zones as the surficial expression of vents and eruptive sites fed by dyke complexes at depth, pointing out that these structures may be an invariable characteristic of ocean volcanoes.

A significant advancement in the understanding of oceanic rifts has been attained in the Canary Islands, particularly on El Hierro, La Palma and Tenerife from the 1990s onward (Carracedo 1994, 1996, 1999; Guillou et al. 1996; Carracedo et al. 1999, 2007, 2011; Gee et al. 2001; Walter and Schmincke 2002; Walter and Troll 2003; Walter et al. 2005; Delcamp et al. 2010). This work took advantage of the numerous water tunnels in Tenerife and La Palma used for groundwater mining (locally called "*galerías*", 2 × 2 m and several kilometres long, with a combined length for both islands exceeding 3,000 km). These *galerías* facilitate access to the deep structure of the rift zones, providing a unique opportunity for direct observations and sampling (see Fig. 4.3 in Carracedo 1994).

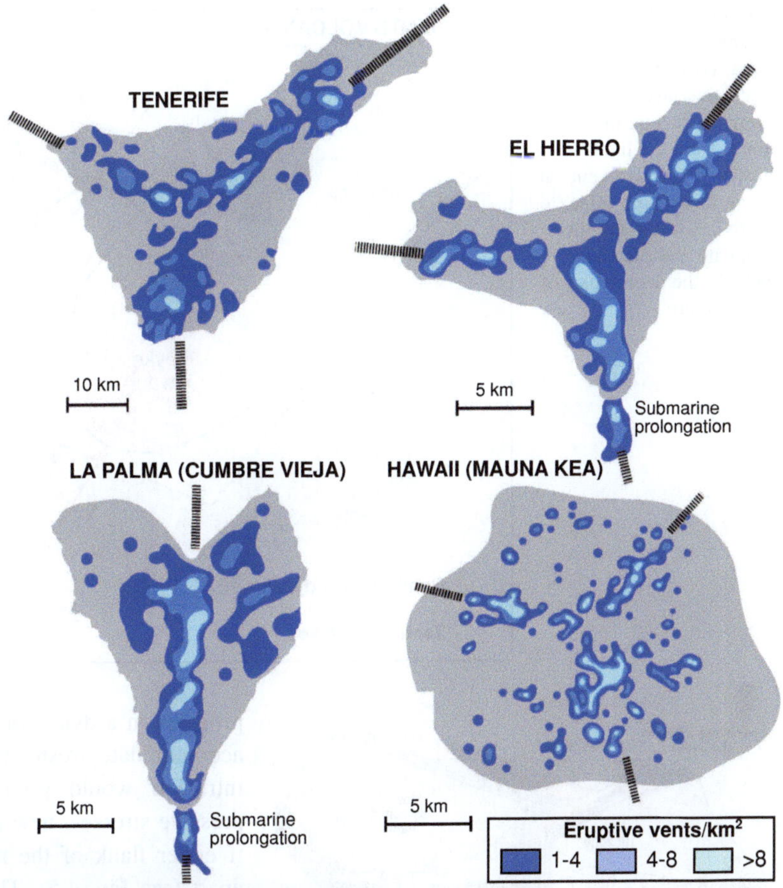

Fig. 4.2 Concentration of Quaternary eruptive centres in Tenerife, El Hierro, La Palma (Cumbre Vieja), and Hawaii (Mauna Kea). Mauna Kea data from Porter (1972)

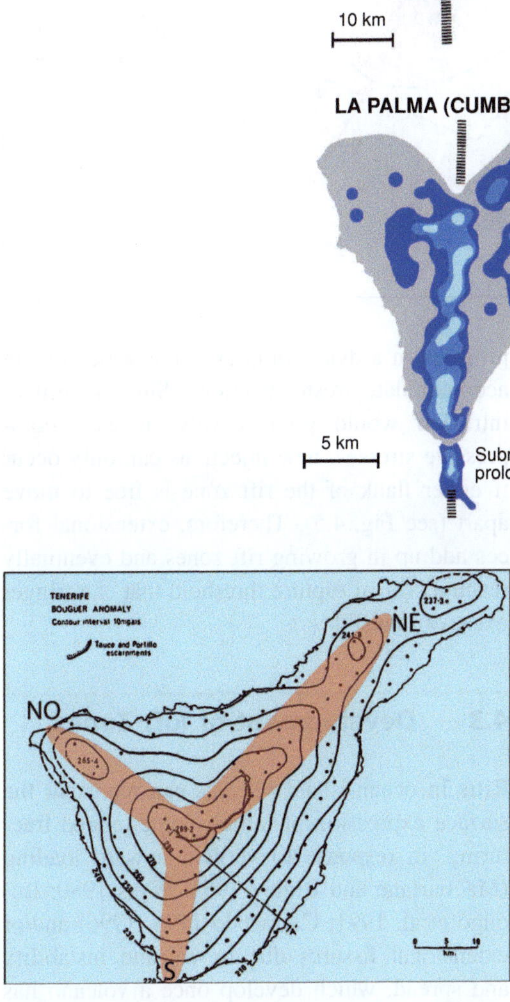

Fig. 4.3 Bouguer anomaly map of Tenerife showing a three-pointed star shape (from MacFarlane and Ridley 1968)

The Taburiente shield and Cumbre Vieja Volcano, both on the island of La Palma, are end-members in the evolutionary stages of rift zones. There, an old and extinct (Plio-Pleistocene) deeply eroded dyke complex (Taburiente), and a recent (<125 ky), active rift zone (Cumbre Vieja) make up the key architectural elements of the island. The latter allows observation of the surface distribution of eruptive vents in these situations, and their main eruptive facies (1 in Fig. 4.4). This comprises a volcanoclastic facies (Fig. 4.4) at the central axis of the rift, and a lava facies (lf) at the flanks of the structure. Deeper in the rift zone, there appears to be a dense group of dykes, oriented approximately along the rift axis (2 in Fig. 4.4). These dykes are the conduits feeding the eruptive vents of the rift, although part of them probably never reaches the surface (Gudmundsson et al. 1999). The internal organisation of the dyke complex can be observed at the floor of the Caldera de Taburiente, where a lateral collapse exposed the core of the shield (3 in Fig. 4.4). The root of the dyke complex is

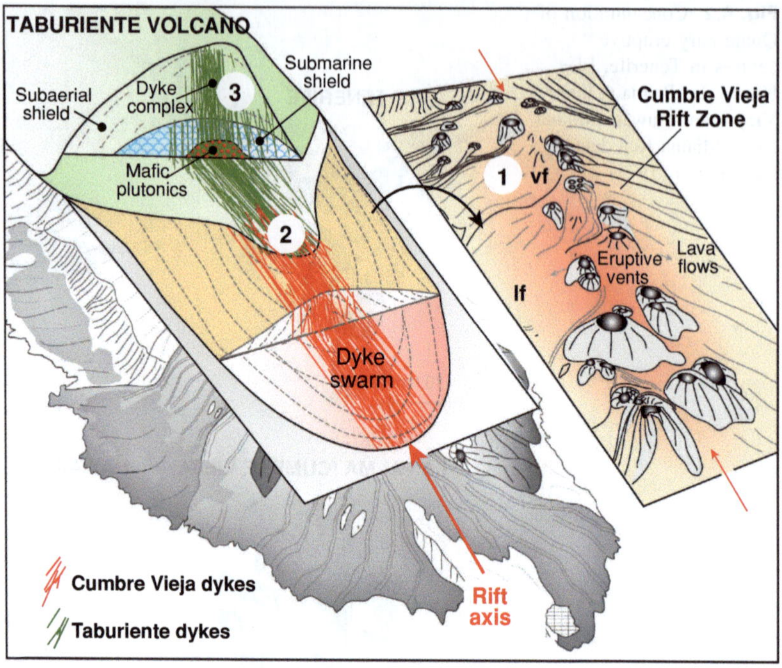

Fig. 4.4 Anatomy of oceanic rift zones: Cumbre Vieja, La Palma. The successive layers show the internal structure of rift zones, from the tight cluster of eruptive vents at the surface of the ridge, to the dyke swarm and the cumulate and plutonic rocks in the deeper part of the structure

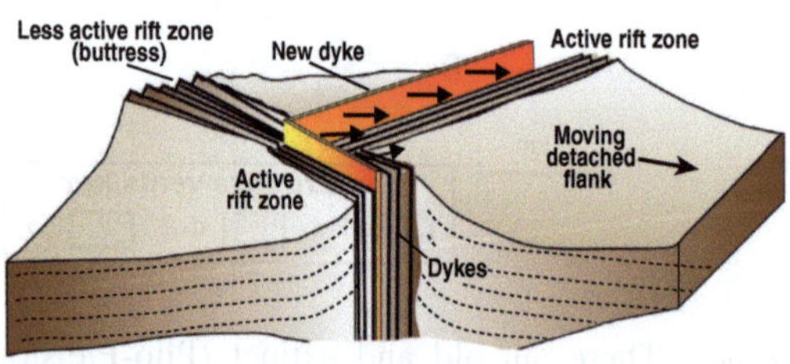

Fig. 4.5 In triaxial rift zones, two of the three arms are usually more active, the third acting as a buttress. Repetitive injections into the active rifts force the enclosed block between these rift arms outwards opposing the buttress and, eventually cause collapse

formed by a plexus of mafic plutonics and cumulates related to the magma chambers and pockets that supply the overlying rift eruptions.

Repetitive injection of blade-like dykes progressively increases the anisotropy of the complex, forcing new dykes to wedge their path parallel to the intrusions (like a knife between the pages of a book, Fig. 4.5). If this process is sustained and if injections are sufficiently frequent, parts of the rift zones may remain hot (thermal memory) to preferentially guide the path of successive intrusions (e.g., Vogt and Smoot 1984). However, intrusion can only progress in a dyke complex if the structure can accommodate fresh injections. Since repetitive intrusion would progressively increase compressive stresses, new injections can only occur if either flank of the rift zone is free to move apart (see Fig. 4.5). Therefore, extensional forces add up in growing rift zones and eventually reach a critical rupture threshold that can trigger massive landslides.

4.3 Development of Rift Zones

Rifts in ocean-island settings can represent the surface expression of initial plume-related fracturing, in response to vertical upward loading (MacFarlane and Ridley 1968; Wyss 1980; Luongo et al. 1991; Carracedo 1994, 1996) and/or extensional fissures due to volcano instability and spread, which develop once a volcano has grown to a certain height and instability (Walter and Troll 2003; Walter et al. 2005; Delcamp et al. 2010, 2012).

Despite advances in the understanding of volcano deformation, it remains unclear how particular rift zones develop. Fractures and rift zones in Tenerife have repeatedly developed in triaxial patterns. These triple-armed rifts are thought to result from magmatic doming, and thus slight upward bending of the crust (Carracedo 1994), or gravitational spreading effects (Walter 2003; Walter and Troll 2003; Walter et al. 2005). Several such "triaxial rift zones" exist on the island (as also on El Hierro), some of which were active simultaneously.

Endogenously driven mechanisms are thought to play a major role in establishing axial volcano architectures. Plumes typically cause uplift that ruptures the rigid oceanic plate along three rifts meeting at triple junctions. Commonly, two of these rifts become a plate boundary (either a ridge or a ridge/transform) while the third does not spread and becomes a failed arm. A similar mechanism was postulated by D'Albore and Luongo (2009) and Luongo et al. (1991) for the tectonic structures of the Neapolitan area, with the Phlegraean Fields occupying the centre of a triple junction generated by a rising crustal tumescence (a plume). The regular triple-armed junctions and triaxial rift zones on volcanoes would then result from the least-effort fracturing of the brittle crust at 120° angles (Luongo et al. 1991; Carracedo 1994, 1996). This least-effort model (Fig. 4.6) is considered to explain (a) the aligned concentration of eruptive sites on the Canaries (Tenerife, El Hierro and La Palma), (b) the longevity and direction of rift zones, and (c) the genesis of volcano sector collapses located in-between 2–120° rift arms (Carracedo 1994, 1996). In this model, the rift zones are thought to have initiated early in the history of the islands and form their deep inner structure.

However, important objections to this model have been raised. If triaxial rift zones developed simultaneously on particular islands (e.g., Tenerife, Hawaii) the location of the centres of those rift systems should be sufficiently distant from each other considering the highly viscous relaxation behaviour of the upper mantle and flexure wavelengths of the crust (Watts and Masson 2001). If Tenerife shield volcanoes (Teno, Anaga and Central shields) are thought to be triaxial structures, they are probably located too close to one another to meet those conditions (Walter and Troll 2003).

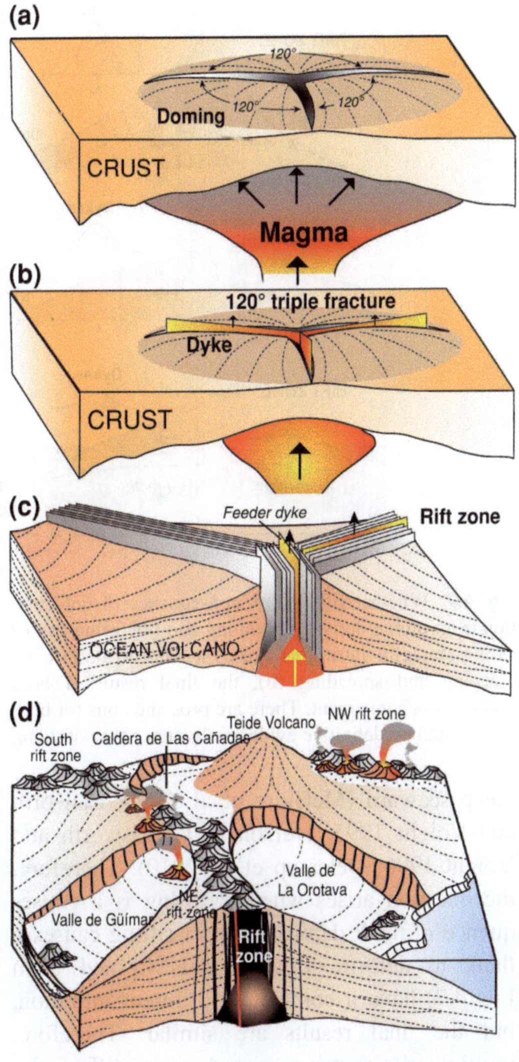

Fig. 4.6 Model proposed by Carracedo (1994, 1996) linking volcanic rift zones and landsliding in the Canary Islands. Three-armed rifts, spaced at ~120°, seem to be the naturally preferred configuration, as in the case of El Hierro and Tenerife. This architecture is thought to be a response to least-effort fracturing. The resulting three-sided base pyramidal edifice geometry may be further enhanced by landsliding between the rift arms, propagating perpendicular to the rift direction

An alternative process is that flank deformation is caused by rifting, once a volcano becomes sufficiently unstable for dyke intrusions to force the flanks of the volcano to spread and

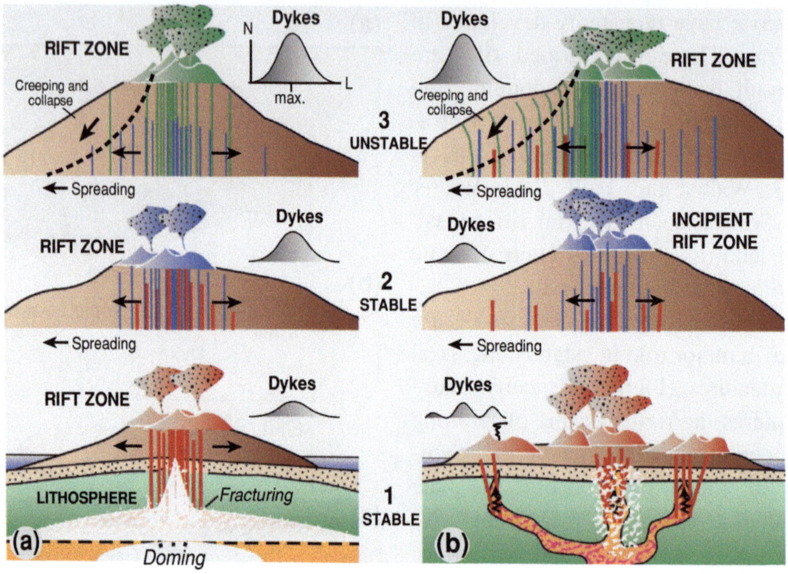

Fig. 4.7 Whether rifting is a consequence of deformation from plume-derived updoming and fracturing (**a**), or rifting (forceful intrusion) causes a flank to deform by creeping and spreading (**b**), the final result of both processes is convergent. There are pros and cons for both models and no definitive evidence favours either of them. In fact, both types of rift zones may be present in the Canaries, with type A predominant in the early stages of construction of the island volcanoes and type B becoming more prevalent in the latter stages of rift development. *N* number of dykes; *L* distance across the rift

creep seaward (McGuire et al. 1990; Elsworth and Voight 1995; Iverson 1995; Elsworth and Voight 1996; Delcamp et al. 2010). Therefore, the question arises whether rifting is a consequence of flank deformation, or rifting causes a flank to deform. Both models (a and b in Fig. 4.7) have a completely different initiation, but the final results are similar. Therefore, multiple rift systems may develop differently. Triple-armed rift zones can result from the least-effort fracturing of the brittle crust (see a in Fig. 4.7), at the initial stages of development of a particular island (e.g., the Central Shield of Tenerife) where plume-related or oceanic fractures may provide important magma pathways (e.g., Carracedo 1994; Geyer and Martí 2010; Carracedo et al. 2011). Alternatively, ridge-like volcanoes have been shown to develop a third arm once the edifice has matured and developed instabilities. Then, a more passive rift arm may open opposite the collapse scar due to extensional stresses (e.g., Walter and Troll 2003; Walter et al. 2005).

Observations on Tenerife and El Hierro shields as well as in analogue gelatine experiments have shown that slight eccentricity of the creeping sector focuses dyke intrusion along two curved axes tangential to the stable/unstable interface. In contrast, strong eccentricity results in only one main tangential rift, while other rifts remain poorly developed (Walter and Troll 2003; Walter et al. 2005). With initiation of a creeping sector, an initially radial or ridge-like geometry is likely to reconfigure and produce rift-zones that will lead to additional rift arms. The most common arrangement resulting from such geometry would be another (third) arm to form the frequent triple-armed systems. Intrusion into the margin between stable and unstable sectors may thus favour the triple-armed configuration.

This architectural evolution may be illustrated in the development of the Taburiente shield in the early subaerial construction of La Palma, where rift zones seem to have progressed from an initial disperse radial distribution of eruptive vents (Fig. 4.8). Southward migration

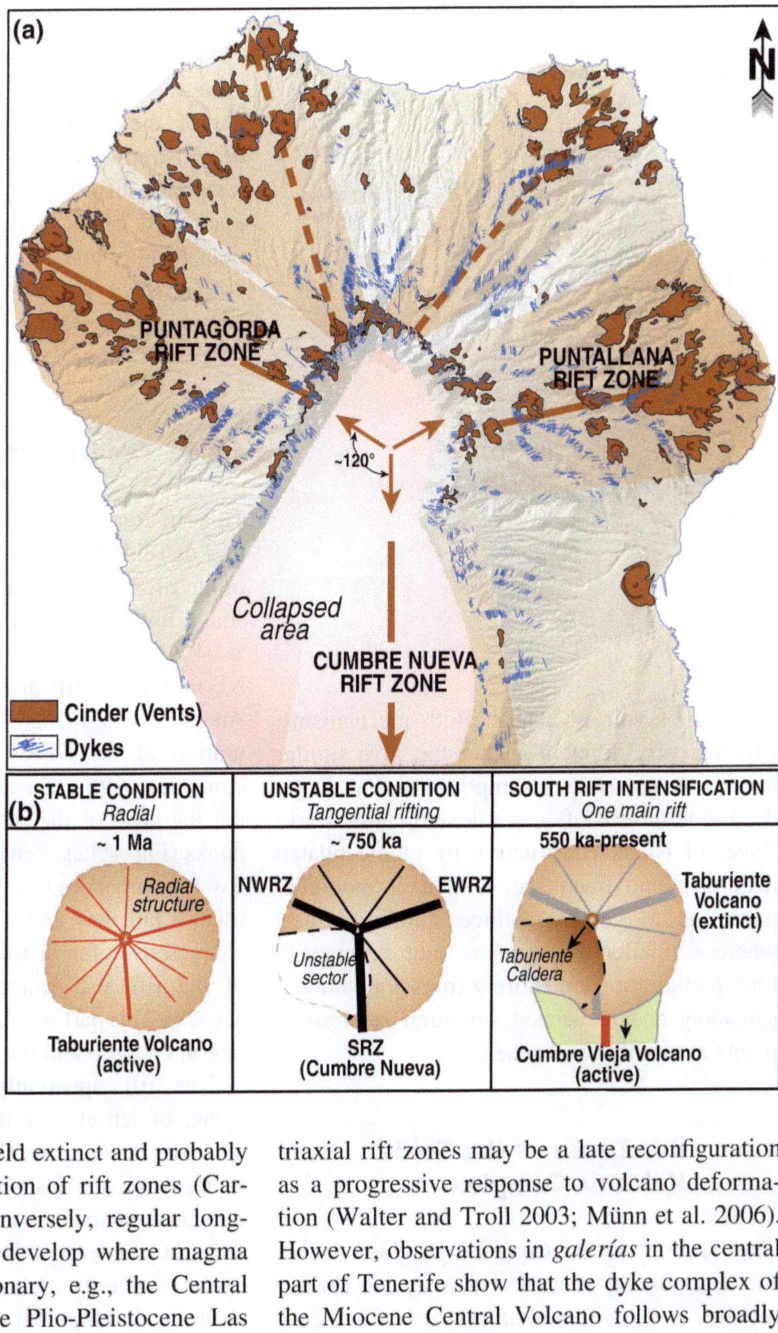

Fig. 4.8 **a** Eruptive vents and dyke outcrops of the Taburiente shield volcano (~0.77–0.4 Ma), La Palma, with rift zones forming a radial structure. The incipient three-armed rift organisation (*solid lines*) was apparently left incomplete by the extinction of Taburiente Volcano at an early stage of development (from Carracedo et al. 2001). **b** Stages of structural evolution of La Palma from an initial radial structure. The position and direction of the creeping flank favoured extension in an east–west direction on the southern flank, and thus the formation of a north–south rift zone. Once formed, the main south rift stabilized by the alternation of constructive and destructive processes such as volcanism, landsliding and erosion (modified from Walter and Troll 2003)

of volcanism left the shield extinct and probably interrupted the organisation of rift zones (Carracedo et al. 2001). Conversely, regular long-lived triaxial rift zones develop where magma plumbing remains stationary, e.g., the Central Miocene Shield and the Plio-Pleistocene Las Cañadas Volcano, in Tenerife (Fig. 4.9).

Analogue gelatine and sand-box experiments confirm the generation of a triangular system of conjugate graben axes in settings reproducing the steady conditions of El Hierro (Fig. 4.10), where magma plumbing apparently has remained stationary, suggesting that these triaxial rift zones may be a late reconfiguration as a progressive response to volcano deformation (Walter and Troll 2003; Münn et al. 2006). However, observations in *galerías* in the central part of Tenerife show that the dyke complex of the Miocene Central Volcano follows broadly the very same orientation as the rift zones that developed during the formation of Las Cañadas Volcano and those of the present day rift zones (Carracedo 1975, 1979).

At present there is no definitive evidence in favour of either of these models—endogenously driven mechanisms or rifting by spreading and

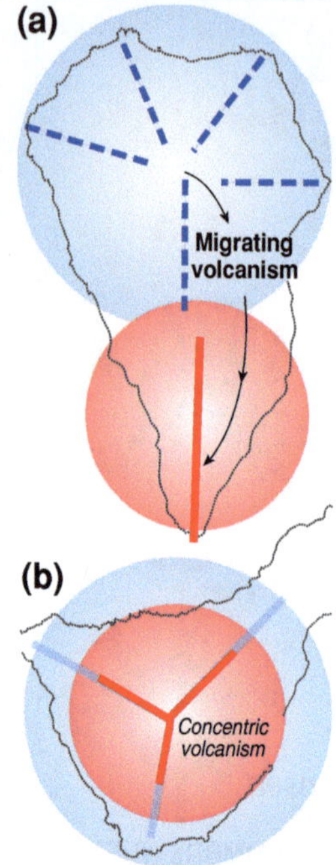

Fig. 4.9 Classical triple-armed rift zones are usually not well developed when moving magmatic sources are involved (e.g., **a** La Palma). A stationary magma supply, however, gives rise to concentrically overlapping volcanoes and well-developed triple-armed rift zones (e.g., **b** Central shield in Tenerife) (modified from Carracedo et al. 2001)

creeping of volcano flanks. Both mechanisms, although very different at the start give similar results. A plausible assumption is that large, deep triple-armed rift zones develop at the early stages of island construction by plume related updoming and fracturing, with later modifications due to volcano edifice stability issues, whereas smaller rift systems (not necessarily multiple) might form entirely from gravitational spreading and associated structural re-arrangements at unstable volcanoes.

4.4 Rift Zones of the Teide Volcanic Complex

The Teide Volcanic Complex provides one of the best possible scenarios to study the characteristics and evolution of rift zones in ocean volcanoes. The North East Rift Zone (NERZ) presents a superb opportunity to study the entire cycle of activity of an oceanic rift zone. This rift, inactive for hundreds of thousands of years along most of its length, has been deeply mass-wasted by erosion and massive landsliding, allowing an in-depth study of its internal structure, including the complex network of dykes exposed (Delcamp et al. 2010; Carracedo et al. 2011). On the other hand, the North West Rift Zone (NWRZ) represents an outstanding example of the latest stages of rift development, demonstrating interesting patterns of spatial and temporal distribution of eruptive vents and associated geochemical and petrological variations (Ablay and Martí 2000; Carracedo et al. 2007), including rare examples of complex magma mixing (Wiesmaier et al. 2011).

4.4.1 The NE Rift Zone

This rift zone extends for about 35 km, from the foot of Teide to the Anaga massif. The deep core of the rift is an extension of the Central Miocene shield towards the Anaga massif (Guillou et al. 2004; Carracedo et al. 2011), outcropping at the NE end of the rift and underlying the Pliocene Anaga Volcano (Fig. 4.11). The present configuration of the NERZ is characteristic of rift structures, with a cluster of eruptive vents forming the crest of the ridge and lava flows at the flanks (Fig. 4.12). Vents are tightly packed at the SW (proximal) end of the rift, whereas at the SE (distal) tail they end and appear dispersed in a characteristic fan distribution. The proximal end of the rift also concentrates the most recent activity. This part actively contributed to the Icod lateral collapse and the evolution of the TVC.

The rift apparently had three successive cycles of activity—in the Miocene, the Pliocene and the Pleistocene (Fig. 4.13). The last one (comprising the last million years) is the best documented and is the only one that is related to the TVC, at least in its final stages. This latest cycle of activity of the NERZ has been coeval with the development of Las Cañadas Volcano, but both volcanoes were clearly interacting, as suggested by sequences of basaltic lapilli from the NERZ alternating with beds of phonolitic pumice from Las Cañadas Volcano. It appears that most recent age dates, in fact, imply that the Anaga shield is younger than the central edifice, making an arrangement of shields to form rift-zones as shown in Fig. 4.10 somewhat unlikely.

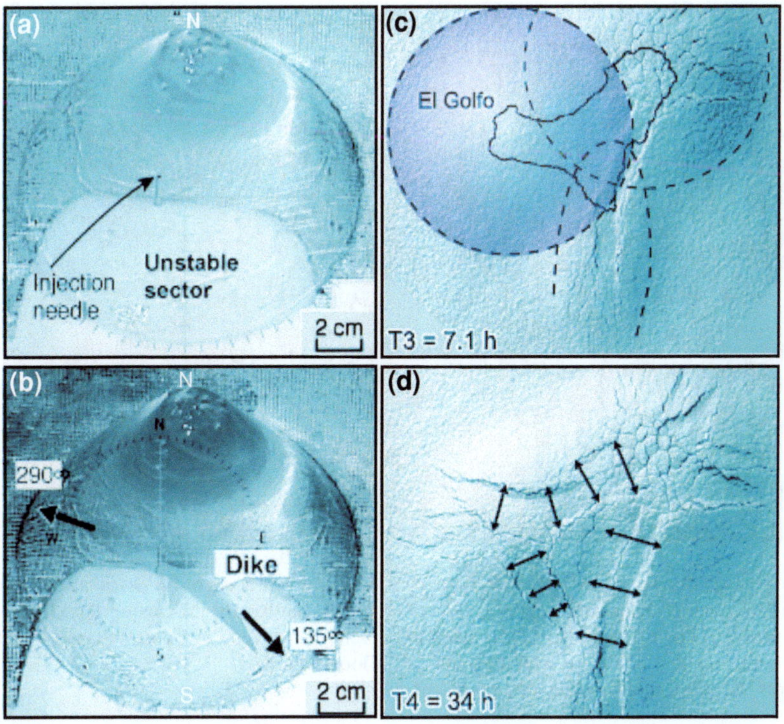

Fig. 4.10 **a**, **b** Scaled analogue experiment with gelatine models. **a** Gelatine cone before injection of a liquid (the magma) into the interface creeping/non-creeping sector and a slight southwestward eccentricity of the lubricated base. **b** After injection, 80 % of the experiments produced a triple-arm intrusion arrangement (Walter and Troll, 2003). **c**, **d** analogue experiment with sand cones simulating the overlapping "Tiñor cone" and the "Southern Ridge" (El Hierro) emplaced simultaneously. After 7.1 h, the "El Golfo cone" was added overlapping the 'Tiñor cone' and the ridge. In **d**, the two cones and the ridge have spread for 34 h showing a triangular system of conjugate graben axes (Munn et al. 2006)

Three reasonably well-dated and documented successive giant landslides in the latest active cycle of the NERZ provide relevant information to understand the genesis and characteristics of mass-wasting processes in oceanic volcanoes, and help to clarify the succession of events giving rise to the formation of the Las Cañadas-Icod-La Guancha collapse depression and the subsequent nested Teide Volcano.

4.4.2 Evolution of the NE Rift Zone

The initial, pre-collapse stages of the latest cycle of activity of the NERZ developed a volcanic ridge that may have reached an altitude of about 2,000 m a.s.l. (Fig. 4.14a). The critical phase of construction was between ca. 1,100 and 860 ky, when the growth rate may have reached 3.5 m/ky, indicating an intense episode of intrusive and eruptive activity leading to the progressive instability of the volcano. This, in turn, led to dykes changing direction in response to the increasing instability at this stage (see e.g., Walter and Troll 2003; Delcamp et al. 2010) from 20° to 40°, the main orientation of intrusions in the NERZ, to 0°–10° at the final stages.

The main constraint for the time of occurrence of the first lateral collapse (Micheque), with an estimated volume assessed from digital elevation model analysis of $\sim$60 km^3, is primarily based on the ages obtained in the Los Dornajos galería (see upper section in Fig. 4.13), which suggests that this collapse must have occurred ca. 830 ky, the age of the first nested lavas above the avalanche breccia. The landslide generated a basin in the north flank of the rift,

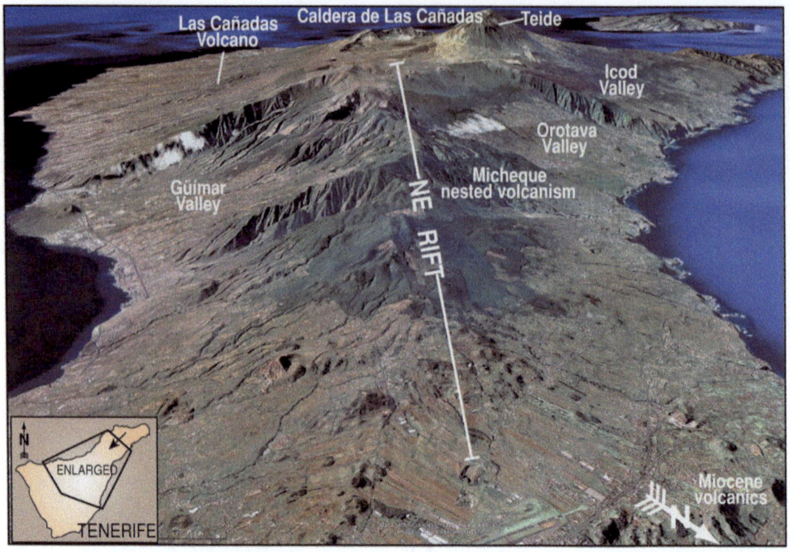

Fig. 4.11 Google Earth image of the NE Rift Zone of Tenerife viewed from the Anaga massif (oblique view of Tenerife from the NE). The rift had already extended in the Miocene from the central edifice of what is now Las Cañadas towards the Miocene-Pliocene Anaga massif. The landslide scars of La Orotava and Güímar are clearly visible, unlike the Micheque landslide, which is completely covered by post-collapse volcanism (image Google Earth)

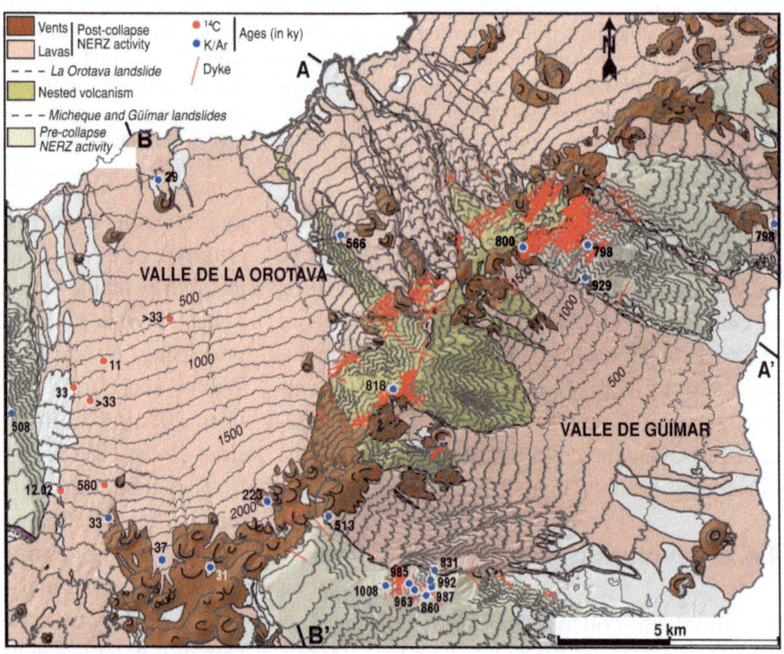

Fig. 4.12 Simplified geological map of the NE Rift Zone of Tenerife showing the distribution of eruptive vents and lava flows. Ages (in ky) from Carracedo et al. (2011)

probably extending towards the present-day valley of La Orotava (Fig. 4.14b). Subsequent volcanism filled large parts of the collapse basin, extending beyond the coastline, concealing the scar and the avalanche breccia to be only found in *galerías* in the northern flank of the rift zone.

A second landslide (the Güímar lateral collapse, estimated volume: 47 km^3), at the east

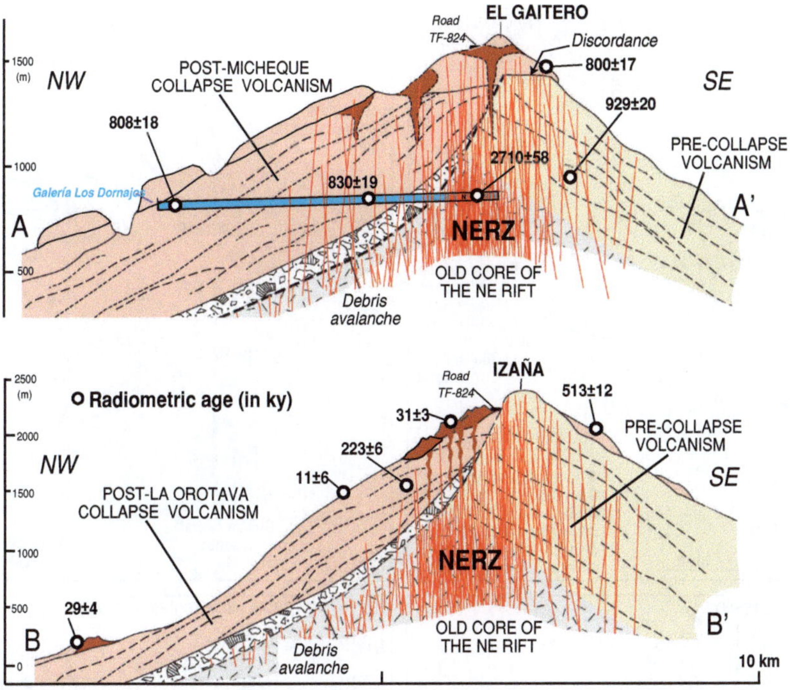

Fig. 4.13 Geological cross-sections of Tenerife (NERZ) perpendicular to the rift axis (compare with Fig. 4.12 for section lines). Two of the lateral collapses (Micheque and La Orotava) are crossed by the sections, showing that the rift zone has been operating for at least for 2.7 Ma. Ages in ky (from Carracedo et al. 2011)

flank of the NERZ, formed a pronounced (10 × 10 km) depression (Fig. 4.14b). The timing of this collapse is constrained by the age of 860 ± 18 ky obtained from lava flows topping the southern collapse scar (Pared de Güímar), and that of the first volcanism nested inside the landslide embayment, dated at 831 ± 18 ky.

The eruptive rate and volume of the Güímar in-fill formations seem much lower than those of the Micheque event. This suggests that, although roughly contemporaneous, the Micheque collapse may have been the first of the two to occur, coinciding with a phase of intense volcanic and intrusive activity. This may point to a fundamental difference in the mechanism that caused the two flank failures: distensive stresses associated with intense eruptive and intrusive activity in the Micheque collapse, and gravitational instability increased by the response to the earlier collapse in the case of the Güímar landslide. This would explain the observation that, by far, the greater part of volcanism continued to be concentrated in the interior of the first, the Micheque collapse, even after the Güímar landslide took place. This caused the total infilling of the Micheque depression and the evolution of significant volumes of magma (0.5–1.0 km^3) towards highly differentiated compositions in this sector (Fig. 4.14c, d).

A third collapse at the northern flank of the NERZ (Orotava lateral collapse, estimated volume: 57 km^3) formed the Orotava Valley (Fig. 4.14d). The relatively accurate dating of the previous collapses has not been achieved in this last case. Its age is constrained by a minimum age of 566 ± 13 ky from lavas of felsic compositions of the Micheque nested volcanism cascading over the eastern scar of the Orotava Valley (Carracedo et al. 2011), and the age of 690 ± 10 ky, obtained by Abdel-Monem et al. (1971) from the lower part of the collapsed sequence at the southern (Tigaiga) scar (Fig. 4.14d). It seems therefore that the Orotava collapse occurred between 690 ± 10 and

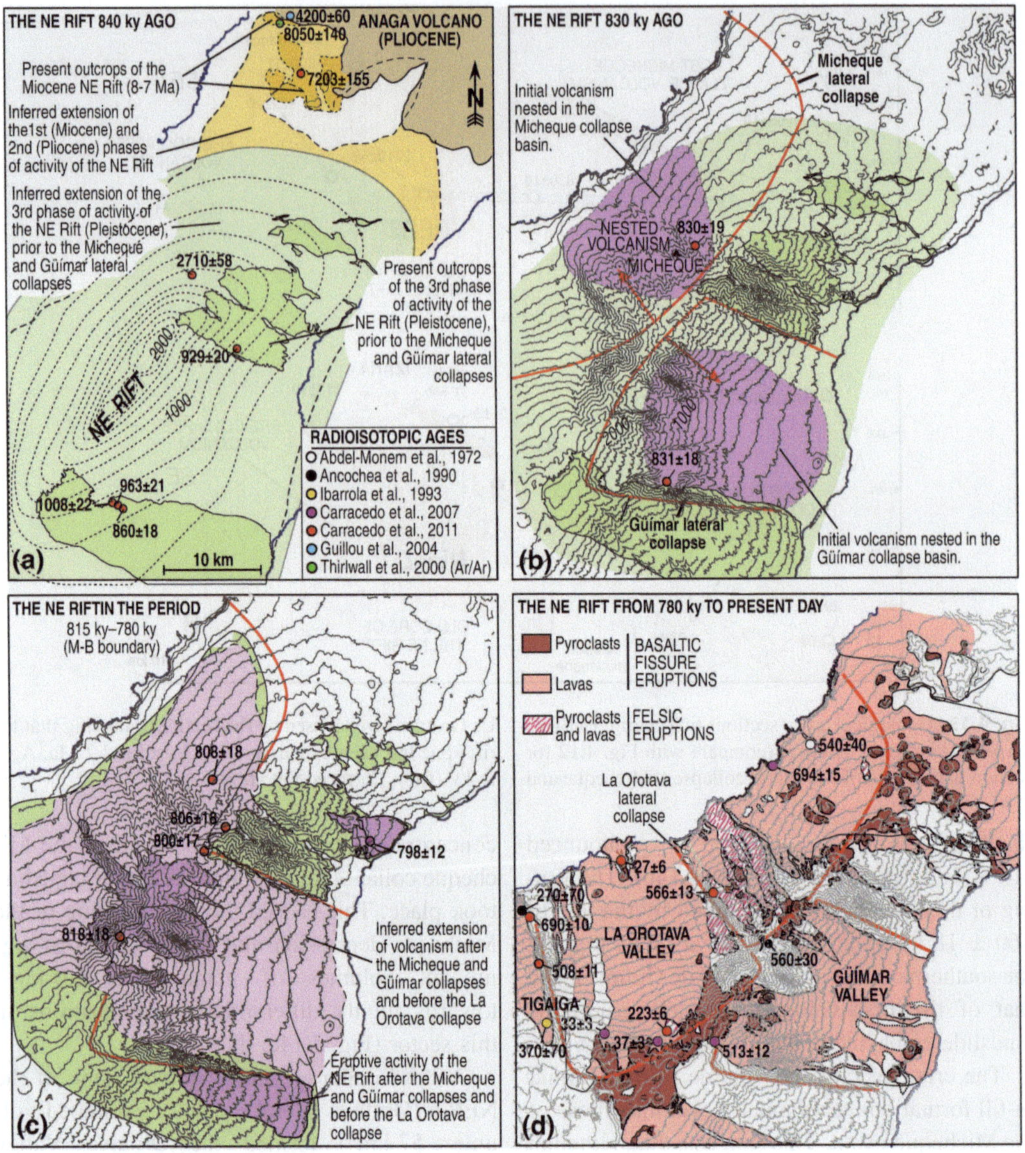

Fig. 4.14 Successive stages of development of the NE Rift Zone of Tenerife (modified from Carracedo et al. 2011; Abdel-Monem et al. 1972; Ibarrola et al. 1993; Thirlwall et al. 2000)

566 ± 13 ky, which places it significantly after the Micheque and Güímar landslides.

4.4.3 Decline and Dispersed Activity of the NERZ

Following the three collapses, the rift entered into a stage of stabilisation and progressively decreasing eruptive activity. Simultaneously, the dispersion of the eruptive centres, previously grouped preferentially at the crest of the rift, increased, particularly at the distal NE end (see Fig. 4.14d). These eruptions, all of normal polarity, have given ages of 513 ± 12 ky (Carracedo et al. 2007), 540 ± 40 ky (Abdel-Monem et al. 1971) and 560 ± 30 ky (Ancochea et al. 1990). NERZ eruptive activity, although

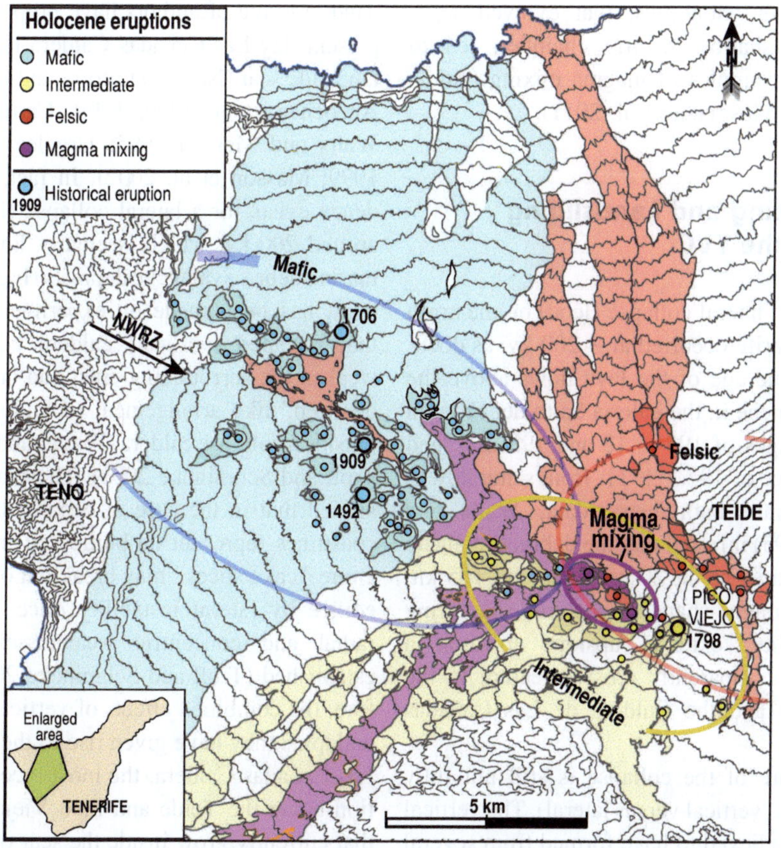

Fig. 4.15 Holocene volcanism in the NW rift zone demonstrating the characteristic groupings of eruptive vents along the crest of the ridge. Assuming a common mafic parent from the uppermost mantle, eruptions are spatially arranged according to composition, with basanites at the western (distal) end, and phonolites at the eastern (proximal) end, close to the shallow and differentiated magma reservoirs of Teide Volcano

attenuated, has continued until recent times, particularly in the proximal (SW) area of the rift, as underlined by ages of 37 ± 3, 33 ± 3, and 33 ± 1 ky (Carracedo et al. 2007), and even to historic times (e.g., the Fasnia and Arafo eruptions in 1705 A.D.).

4.4.4 The NW Rift Zone

Just as the deeply eroded NERZ provides relevant information for the understanding of the entire cycle of growth and mass destruction of rift zones, the NW rift, very active in the Holocene, gives significant details of the temporal and spatial distribution of surface volcanism and thus provides indirect information about the evolution and internal structure of the TVC magma system during its most recent volcanic cycle (Ablay et al. 1998; Carracedo et al. 2007; Wiesmaier et al. 2011).

The eruptive vents cluster in the characteristic pattern of rift zones at the crest of the rift, while lava flows extend down both flanks (Fig. 4.15). One of the most interesting features is the compositional distribution of eruptions, showing a distinct bimodal series, with basanite and phonolite, respectively as the distal and proximal end-members, and intermediate eruptions in the central part of the rift zone (Fig. 4.15). The petrologically distinct magmas evolved from a common primitive basanite parent by crystal fractionation (Ablay et al. 1998). The interaction of these two magmas, i.e., basanites with phonolites, that evolved separately in a shallow central chamber, led to spectacular examples of magma mixing (Araña et al. 1994; Wiesmaier et al. 2011).

4.5 Rifting and Landsliding in the TVC

The youngest lateral collapse (Icod) on the north flank of Tenerife occurred at ~200 ky, as documented by the age of the lava flows above the debris avalanche in the galería Salto del Frontón (two K/Ar ages of 195 ± 12 and 198 ± 5, and one Ar/Ar age of 192.3 ± 11, from same flow).

From swath bathymetry data, Watts and Masson (1995) inferred that the debris avalanche covered an area about 20 km wide and 105 km long. The tongue-shaped structure, suggesting high mobility, extends upslope towards a "chute-like", apparently erosive feature in the Icod Valley and the Caldera de Las Cañadas (Fig. 4.16).

The nature of the collapse is still not fully resolved (i.e., vertical versus lateral). The vertical collapse is believed to have formed from several classical (vertical) caldera collapses between 1.2 and 0.17 My (e.g., Ridley 1971; Booth 1973; Martí et al. 1994; Bryan et al. 1998). On the other hand, a range of authors have proposed that the present day Las Cañadas Caldera is primarily a landslide scar (Navarro Latorre and Coello 1989; Ancochea et al. 1990, 1999; Carracedo 1994; Watts and Masson 1995; Urgeles et al. 1997, 1999; Masson et al. 2002). In fact, strong evidence exists for a lateral collapse (landslide) at around 200 ky, which is clearly linked to submarine debris avalanche deposits (Fig. 4.16). The point has been made on experimental grounds that repeated (vertical) caldera collapses can weaken the surrounding crust and create a "spider-web"-like arrangement of faults inside and outside a collapse caldera (Walter and Troll 2001; Troll and Schmincke 2002). These authors have argued that in the case of ocean islands, where coastlines represent un-buttressed free surfaces, entire "cake slices" may break out of an island's edifice by lateral instability once a system of radial and concentric weaknesses has been established (Troll and Schmincke 2002). Therefore, the combined effects of vertical and lateral collapses may have given rise to the present-day Las Cañadas Caldera, the most recent modification being the Teide and Pico Viejo complexes that currently grow inside the scar of the 200 ky (lateral) Icod collapse, which in turn, likely exploited older instabilities in the Las Cañadas edifice.

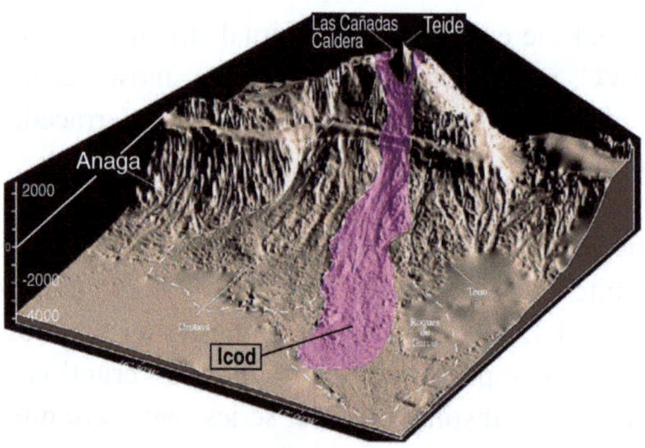

Fig. 4.16 3D representation of the north flank of Tenerife, viewed from the northwest, showing the successive lateral collapses. The youngest (Icod) is indicated in a different colour. Note the extension upslope of the debris avalanche towards the southern wall of the Cañadas Caldera, interpreted as the headwall of this giant landslide. Teide Volcano is nested in the Icod collapse depression (from Watts and Masson 1995)

A continuous layer of debris avalanche deposits extends inside the Las Cañadas Caldera below the present Teide stratocone (Márquez et al. 2008), providing strong support for a landslide origin of the currently visible depression.

The relevant aspect of this collapse event is that it formed a general spatial and temporal basis for the TVC and had a direct role in its construction and in promoting the magmatic variability present in the current volcanic complex.

4.6 Rifting, Landsliding and Magmatic Variation

A comparative analysis of the evolution of different Canarian rift zones, including those of the TVC, outlines notable common characteristics. Rifts are recurrent features that show cyclic

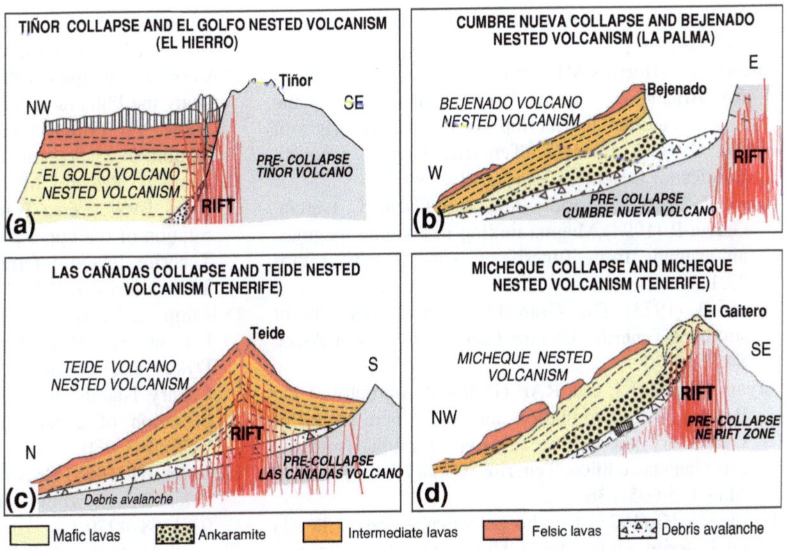

Fig. 4.17 Examples in the Canary Islands of rifts and associated landslides with subsequent nested differentiated volcanism. Note that progressive magmatic differentiation takes place in the sequences filling the landslide scars (from Carracedo et al. 2007)

patterns of growth, instability, flank collapse, nested volcanism, and eruptive decline and dispersion (Carracedo et al. 2011).

Variations in magma composition appear to occur in response to lateral collapses in the Canaries (Fig. 4.17). A collapse implies disruption of an established feeding system of a rift, which allows dense mafic magmas to ascend to the surface by edifice unloading (Manconi et al. 2009). The result is the concentration of progressively centralized eruptions focusing in the interior of the landslide basin, thus progressively filling up the collapse scar (Carracedo et al. 2007, 2011; Longpré et al. 2009). The emplacement of magma at increasingly shallower depths within this nested volcanic edifice will allow for extensive modification of magma and will lead to progressively more differentiated eruptions, commonly reaching felsic compositions (trachytes, phonolites) that become more and more dominant due to the progressive increase in height of the volcanoes nested inside the landslide embayments.

Although felsic volcanic complexes in the Canaries may originate from a variety of processes (Wolff 1983; Pérez-Torrado et al. 1995; Troll and Schmincke 2002; Paris et al. 2005; Longpré et al. 2009), a considerable volume of differentiated volcanism in the Canaries appears to be associated with rift flank collapses that are followed by abundant and prolonged nested volcanism. Regularly, these eruptions evolve from initially mafic to terminally felsic compositions. Lateral collapses may consequently be considered to represent a major cause for structural and petrological variability in ocean islands (Carracedo et al. 2007, 2011; Longpré et al. 2009; Manconi et al. 2009), Teide being a prime example of this feature.

References

Abdel-Monem A, Watkins ND, Gast PW (1971) Potassium-argon ages, volcanic stratigraphy, and geomagnetic polarity history of the Canary Islands; Lanzarote, Fuerteventura, Gran Canaria, and La Gomera. Am J Sci 271:490–521

Abdel-Monem A, Watkins ND, Gast PW (1972) Potassium-argon ages, volcanic stratigraphy, and geomagnetic polarity history of the Canary Islands; Tenerife, La Palma and Hierro. Am J Sci 272(9):805–825

Ablay GJ, Martí J (2000) Stratigraphy, structure and volcanic evolution of the Pico Teide-Pico Viejo formation, Tenerife, Canary Islands. J Volcanol Geotherm Res 103:175–208

Ablay GJ, Carroll MR, Palmer MR, Martí J, Sparks RSJ (1998) Basanite-phonolite lineages of the Teide-Pico Viejo Volcanic Complex, Tenerife, Canary Islands. J Petrol 39:905–936

Ancochea E, Fúster J, Ibarrola E, Cendrero A, Coello J, Hernan F, Cantagrel JM, Jamond C (1990) Volcanic evolution of the island of Tenerife (Canary Islands) in the light of new K-Ar data. J Volcanol Geotherm Res 44:231–249

Ancochea E, Huertas MJ, Cantagrel JM, Coello J, Fúster JM, Arnaud N, Ibarrola E (1999) Evolution of the Canadas edifice and its implications for the origin of the Canadas Caldera (Tenerife, Canary Islands). J Volcanol Geotherm Res 88:177–199

Araña V, Martí J, Aparicio A, García-Cacho L, García-García R (1994) Magma mixing in alkaline magmas: an example from Tenerife, Canary Islands. Lithos 32:1–19

Booth B (1973) The Granadilla pumice deposits of southern Tenerife, Canary Islands. Proc Geol Assoc 84:353–370

Bryan SE, Martí J, Cas RAF (1998) Stratigraphy of the Bandas del Sur formation: an extracaldera record of Quaternary phonolitic explosive eruptions from the Las Canadas edifice, Tenerife (Canary Islands). Geol Mag 135:605–636

Carracedo JC (1975) Estudio paleomagnético de la isla de Tenerife. Ph.D. Thesis, Universidad Complutense, Madrid

Carracedo JC (1979) Paleomagnetismo e historia volcánica de Tenerife. Aula Cultura Cabildo Insular de Tenerife, Santa Cruz de Tenerife, p 81

Carracedo JC (1994) The Canary-Islands—an example of structural control on the growth of large oceanic-island volcanoes. J Volcanol Geotherm Res 60:225–241

Carracedo JC (1996) A simple model for the genesis of large gravitational landslide hazards in the Canary Islands. Geol Soc Spec Publ 110:125–135

Carracedo JC (1999) Growth, structure, instability and collapse of Canarian volcanoes and comparisons with Hawaiian volcanoes. J Volcanol Geotherm Res 94:1–19

Carracedo JC, Rodríguez Badiola E, Soler V (1992) The 1730–1736 eruption of Lanzarote, Canary Islands: a long, high-magnitude basaltic fissure eruption. J Volcanol Geotherm Res 53:239–250

Carracedo JC, Day SJ, Guillou H, Rodríguez Badiola E, Canas JA, Pérez Torrado FJ (1998) Hotspot volcanism close to a passive continental margin: The Canary Islands. Geol Mag 135:591–604

Carracedo JC, Day SJ, Guillou H, Pérez Torrado FJ (1999) Giant Quaternary landslides in the evolution of La Palma and El Hierro, Canary Islands. J Volcanol Geotherm Res 94:169–190

Carracedo JC, Rodríguez Badiola E, Guillou H, De La Nuez J, Pérez Torrado FJ (2001) Geology and volcanology of La Palma and El Hierro (Canary Islands). Estud Geol 57:175–273

Carracedo JC, Rodríguez Badiola E, Guillou H, Paterne M, Scaillet S, Pérez Torrado FJ, Paris R, Fra-Paleo U, Hansen A (2007) Eruptive and structural history of Teide Volcano and Rift Zones of Tenerife, Canary Islands. Geol Soc Am Bull 119:1027–1051

Carracedo JC, Guillou H, Nomade S, Rodríguez-Badiola E, Pérez-Torrado FJ, Rodriguez-González A, Paris R, Troll VR, Wiesmaier S, Delcamp A, Fernández-Turiel JL (2011) Evolution of ocean-island rifts: the northeast rift zone of Tenerife, Canary Islands. Geol Soc Am Bull 123:562–584

D'Albore F, Luongo G (2009) Using analogue models to study the Phlegraean Fields caldera. In: GEOITALIA 2009, VII Forum Italiano di Scienze della Terra, Rimini, 9–11 September 2009, pp 112–113

Delcamp A, Petronis MS, Troll VR, Carracedo JC, de Vries BvW, Pérez-Torrado FJ (2010) Vertical axis rotation of the upper portions of the north-east rift of Tenerife Island inferred from paleomagnetic data. Tectonophysics 492:40–59

Delcamp A, Troll VR, de Vries BvW, Carracedo JC, Petronis MS, Pérez-Torrado FJ, Deegan FM (2012) Dykes and structures of the NE rift of Tenerife, Canary Islands: a record of stabilisation and destabilisation of ocean island rift zones. B Volcanol 74(5):963–980

Dieterich JH (1988) Growth and persistence of Hawaiian volcanic rift zones. J Geophys Res-Solid Earth 93:4258–4270

Elsworth D, Voight B (1995) Dike intrusion as a trigger for large earthquakes and the failure of volcano flanks. J Geophys Res-Solid Earth 100:6005–6024

Elsworth D, Voight B (1996) Evaluation of volcano flank instability triggered by dyke intrusion. In: McGuire WJJAPNJ (ed) Volcano instability on the Earth and other planets. Geological Society Special Publications, vol 110, pp 45–53

Fiske RS, Jackson ED (1972) Orientation and growth of Hawaiian volcanic rifts—effect of regional structure and gravitational stresses. Proc R Soc Lond Ser A 329:299–326

Gee MJR, Masson DG, Watts AB, Mitchell NC (2001) Offshore continuation of volcanic rift zones, El Hierro, Canary Islands. J Volcanol Geotherm Res 105:107–119

Geldmacher J, Hoernle K, van den Bogaard P, Zankl G, Garbe-Schönberg D (2001) Earlier history of the ≥70-Ma-old Canary hotspot based on the temporal and geochemical evolution of the Selvagen Archipelago and neighboring seamounts in the eastern North Atlantic. J Volcanol Geotherm Res 111:55–87

Geyer A, Martí J (2010) The distribution of basaltic volcanism on Tenerife, Canary Islands: implications on the origin and dynamics of the rift systems. Tectonophysics 483:310–326

Gudmundsson A, Marinoni LB, Marti J (1999) Injection and arrest of dykes: implications for volcanic hazards. J Volcanol Geotherm Res 88:1–13

Guillou H, Carracedo JC, Pérez Torrado F, Rodriguez Badiola E (1996) K-Ar ages and magnetic stratigraphy of a hotspot-induced, fast grown oceanic island: El Hierro, Canary Islands. J Volcanol Geotherm Res 73:141–155

Guillou H, Carracedo JC, Paris R, Perez Torrado FJ (2004) Implications for the early shield-stage evolution of Tenerife from K/Ar ages and magnetic stratigraphy. Earth Planet Sci Lett 222(2):599–614

Ibarrola E, Ancochea E, Fuster JM, Cantagrel JM, Coello J, Snelling NJ, Huertas MJ (1993) Cronoestratigrafía

del Macizo de Tigaiga: Evolución de un sector del edificio Cañadas (Tenerife, Islas Canarias). Boletín de la Real Sociedad Española de Historia Natural (Sec Geol) 88(1–4):57–72
Iverson RM (1995) Can magma-injection and groundwater forces cause massive landslides on Hawaiian volcanoes? J Volcanol Geotherm Res 66:295–308
Longpré M-A, Troll VR, Walter TR, Hansteen TH (2009) Volcanic and geochemical evolution of the Teno massif, Tenerife, Canary Islands: some repercussions of giant landslides on ocean island magmatism. Geochem Geophys Geosyst 10:Q12017. doi: 10.1029/2009gc002892
Luongo G, Cubellis E, Obrizzo F, Petrazzuoli SM (1991) A physical model for the origin of volcanism of the Tyrrhenian margin: the case of the Neapolitan area. J Volcanol Geotherm Res 48:173–185
Macdonald GA (1965) Hawaiian calderas. Pacific Sci 19:320–334
Macdonald GA (1972) Volcanoes. Prentice-Hall, Englewood Cliffs
MacFarlane DJ, Ridley WI (1968) An interpretation of gravity data for Tenerife, Canary Islands. Earth Planet Sci Lett 4:481–486
Manconi A, Longpré M-A, Walter TR, Troll VR, Hansteen TH (2009) The effects of flank collapses on volcano plumbing systems. Geology 37:1099–1102
Márquez A, López I, Herrera R, Martín-González F, Izquierdo T, Carreño F (2008) Spreading and potential instability of Teide volcano, Tenerife, Canary Islands. Geophys Res Lett 35:L05305. doi: 10.1029/2007gl032625
Martí J, Mitjavila J, Araña V (1994) Stratigraphy, structure and geochronology of the Las Cañadas Caldera (Tenerife, Canary Islands). Geol Mag 131:715–727
Masson DG, Watts AB, Gee MJR, Urgeles R, Mitchell NC, Le Bas TP, Canals M (2002) Slope failures on the flanks of the western Canary Islands. Earth Sci Rev 57:1–35
McGuire WJ, Pullen AD, Saunders SJ (1990) Recent dyke-induced large-scale block movement at Mount Etna and potential slope failure. Nature 343:357–359
Münn S, Walter TR, Klügel A (2006) Gravitational spreading controls rift zones and flank instability on El Hierro, Canary Islands. Geol Mag 143:257–268
Navarro Latorre JM, Farrujia I (1989) Plan Hidrológico Insular de Tenerife. Zonificación hidrogeológica aspectos geológicos e hidrogeológicos. Cabildo Insular de Tenerife
Navarro Latorre JM, Coello J (1989) Depressions originated by landslide processes in Tenerife. In: ESF meeting on Canarian Volcanism, Lanzarote. ESF, Strasbourg, pp 150–152
Paris R, Guillou H, Carracedo JC, Perez-Torrado FJ (2005) Volcanic and morphological evolution of La Gomera (Canary Islands), based on new K-Ar ages and magnetic stratigraphy: implications for oceanic island evolution. J Geol Soc 162:501–512
Pérez-Torrado FJ, Carracedo JC, Mangas J (1995) Geochronology and stratigraphy of the Roque Nublo Cycle, Gran Canaria, Canary Islands. J Geol Soc 152:807–818
Porter SC (1972) Distribution, morphology, and size frequency of cinder cones on Mauna Kea volcano, Hawaii. Geol Soc Am Bull 83:3607–3612
Ridley WI (1971) The field relations of the Las Cañadas Volcanoes, Tenerife, Canary Islands. Bull Volcanol 35:318–334
Stillman CJ (1987) A Canary Islands dyke swarm: implications for the formation of oceanic islands by extensional fissural volcanism. In: Halls HC, Fahrig WF (eds) Mafic dyke swarms. GA C Spec Pap, vol 34, pp 243–255
Swanson DA, Duffield WA, Fiske RS (1976) Displacement of the south flank of Kilauea Volcano; the result of forceful intrusion of magma into the rift zones, U.S. Geol Surv Prof Pap, vol 963, p 39
Thirlwall MF, Singer BS, Marriner GF (2000) ^{39}Ar-^{40}Ar ages and geochemistry of the basaltic shield stage of Tenerife, Canary Islands, Spain. J Volcanol Geotherm Res 103(1–4):247–297
Troll VR, Schmincke HU (2002) Magma mixing and crustal recycling recorded in Ternary Feldspar from Compositionally Zoned Peralkaline Ignimbrite ‘A’, Gran Canaria, Canary Islands. J Petrol 43:243–270
Urgeles R, Canals M, Baraza J, Alonso B, Masson D (1997) The most recent megalandslides of the Canary Islands: El Golfo debris avalanche and Canary debris flow, west El Hierro Island. J Geophys Res Solid Earth 102:20305–20323
Urgeles R, Masson DG, Canals M, Watts AB, Le Bas T (1999) Recurrent large-scale landsliding on the west flank of La Palma, Canary Islands. J Geophys Res-Solid Earth 104:25331–25348
Vogt PR, Smoot NC (1984) The Geisha Guyots—multibeam bathymetry and morphometric interpretation. J Geophys Res 89:1085–1107
Walker GPL (1986) Koolau dike complex, Oahu—intensity and origin of a sheeted-dike complex high in a Hawaiian volcanic edifice. Geology 14:310–313
Walker GPL (1987) The dike complex of Koolau Volcano, Oahu; internal structure of a Hawaiian rift zone. In: Decker RW, Wright TL, Stauffer PH (eds) Volcanism in Hawaii. The dike complex of Koolau volcano, Oahu: internal structure of a Hawaiian rift zone. U.S. Geological Survey, Prof Pap, vol 1350, pp 962–996
Walker GPL (1992) Coherent intrusion complexes in large basaltic volcanoes—a new structural model. J Volcanol Geotherm Res 50:41–54
Walter TR (2003) Buttressing and fractional spreading of Tenerife, an experimental approach on the formation of rift zones. Geophys Res Lett 30:1296. doi: 10.1029/2002GL016610
Walter TR, Schmincke H-U (2002) Rifting, recurrent landsliding and Miocene structural reorganization on NW-Tenerife (Canary Islands). Int J Earth Sci 91:615–628

Walter TR, Troll VR (2001) Formation of caldera periphery faults: an experimental study. Bull Volcanol 63:191–203
Walter TR, Troll VR (2003) Experiments on rift zone evolution in unstable volcanic edifices. J Volcanol Geotherm Res 127:107–120
Walter TR, Troll VR, Cailleau B, Belousov A, Schmincke H-U, Amelung F, Bogaard PVD (2005) Rift zone reorganization through flank instability in ocean island volcanoes: an example from Tenerife, Canary Islands. Bull Volcanol 67:281–291
Watts AB, Masson DG (1995) A giant landslide on the north flank of Tenerife, Canary Islands. J Geophys Res 100:24487–24498
Watts AB, Masson DG (2001) New sonar evidence for recent catastrophic collapses of the north flank of Tenerife, Canary Islands. Bull Volcanol 63:8–19
Wiesmaier S, Deegan F, Troll V, Carracedo J, Chadwick J, Chew D (2011) Magma mixing in the 1100 AD Montaña Reventada composite lava flow, Tenerife, Canary Islands: interaction between rift zone and central volcano plumbing systems. Contrib Mineral Petrol 162:651–669
Wolff JA (1983) Petrology of Quaternary pyroclastic deposits from Tenerife, Canary Islands. Ph.D. Thesis, University of London, London
Wyss M (1980) Hawaiian rifts and recent Icelandic volcanism—expressions of plume generated radial stress-fields. J Geophys 47:19–22

Pre-Teide Volcanic Activity on the Northeast Volcanic Rift Zone

5

Valentin R. Troll, Frances M. Deegan, Audray Delcamp, Juan Carlos Carracedo, Chris Harris, Benjamin van Wyk de Vries, Michael S. Petronis, Francisco J. Perez-Torrado, Jane P. Chadwick, Abigail K. Barker, and Sebastian Wiesmaier

Abstract

The northeast rift zone of Tenerife (NERZ) presents a partially eroded volcanic rift that offers a superb opportunity to study the structure and evolution of oceanic rift zones. Field data, structural observations,

V. R. Troll (✉) · F. M. Deegan · A. K. Barker
Department of Earth Sciences (CEMPEG), Uppsala University, Uppsala 75236, Sweden
e-mail: Valentin.Troll@geo.uu.se

F. M. Deegan
Laboratory for Isotope Geology, Swedish Museum of Natural History, Stockholm, Sweden

A. Delcamp
Department of Geography, Vrije Universiteit Brussels, Brussels 1050, Belgium

J. C. Carracedo · F. J. Perez-Torrado
Departamento de Física (GEOVOL), Universidad de Las Palmas de Gran Canaria, Las Palmas de Gran Canaria, Canary, Islands, Spain

C. Harris
Department of Geological Sciences, University of Cape Town, Rondebosch 7701, South Africa

B. van Wyk de Vries
Laboratoire Magmas et Volcans, Université Blaise Pascal, Clermont, Ferrand, France

M. S. Petronis
Environmental Geology Natural Resource Management Department, New Mexico Highlands University, Las Vegas, New Mexico 87701, U.S.A

J. P. Chadwick
Science Gallery, Trinity College Dublin, Dublin 2, Ireland

S. Wiesmaier
Department of Earth and Environmental Sciences, Ludwig-Maximilians Universität (LMU), Munich, Germany

J. C. Carracedo and V. R. Troll (eds.), *Teide Volcano*, Active Volcanoes of the World,
DOI: 10.1007/978-3-642-25893-0_5,

isotopic dating, magnetic stratigraphy, and isotope geochemistry have recently become available for this rift and provide a reliable temporal framework for understanding the structural and petrological evolution of the entire rift zone. The NERZ appears to have formed in several major pulses of activity with a particularly high production rate in the Pleistocene (ca. 0.99 and 0.56 Ma). The rift underwent several episodes of flank creep and eventual catastrophic collapses driven by intense intrusive activity and gravitational adjustment. Petrologically, a variety of mafic rock types, including crystal-rich ankaramites, have been documented, with most samples isotopically typical of the "Tenerife signal". Some of the NERZ magmas also bear witness to contamination by hydrothermally altered components of the island edifice and/or sediments. Isotope geochemistry furthermore points to the generation of the NERZ magmas from an upwelling column of mantle plume material mixed with upper asthenospheric mantle. Finally, persistent isotopic similarity through time between the NERZ and the older central edifices on Tenerife provides strong evidence for a genetic link between Tenerife's principal volcanic episodes.

5.1 Ocean Island Rift Zones

Rift zones on ocean islands such as the Canary, Hawaiian, and Cape Verde archipelagos are major volcanic surface alignments associated with intense dyke intrusions at depth. Volcanic rift zones are extremely important in terms of Ocean Island growth and evolution as discussed by Carracedo and Troll in Chap. 4. Firstly, rift zones generally form prominent, major topographical features on ocean islands as they concentrate volcanic activity, and thus control the distribution of both volcanic hazards and natural resources. On Tenerife (Fig. 5.1), for example, access to fresh groundwater is aided by water tunnels (*galerías*) constructed into the rift zone's interior (Navarro and Farrujia 1989). Finally, and perhaps most importantly, Ocean Island rift zones control the structure of a growing volcanic edifice, perhaps even from their initial stages of growth, and thus define the location of large-scale flank collapses, which are particularly prominent on Tenerife (e.g., Carracedo 1994; Walter and Troll 2003; Carracedo et al. 2011; see also Fig. 5.1).

Ocean Island rift zones were initially recognised and described on the Hawaiian Islands (Fiske and Jackson 1972; Swanson et al. 1976; Walker 1986, 1987, 1992; Dieterich 1988). Perhaps the most significant progress in understanding Ocean Island rift zone genesis and structural development has been made through their study in the Canary Islands (e.g., Carracedo 1975, 1979, 1994, 1996, 1999; Carracedo et al. 1992, 1998, 2001, 2007, 2011; Guillou et al. 1996; Walter and Schmincke 2002; Walter and Troll 2003; Walter et al. 2005; Hansen 2009; Delcamp et al. 2010, 2012; Carracedo and Troll, this volume). The Canary Islands as a whole are especially valuable to the study of rift zone development due to the fact that the rifts are generally long-lived, dynamic, and conspicuously large structures. The combination of relatively low plume activity feeding the Canaries, low velocity of the African Plate in comparison to, e.g., the Pacific Plate that underlies the Hawaiian Islands, and a prolonged subaerial volcanic history with the absence of late subsidence, gives rise to long-lived ridges on the Canary Islands that are prone to frequent recurring volcanic activity and also to recurring volcanic failure (see Chap. 4 for further discussion). Hence, the spatial circumstances particular to the Canary archipelago provide us with an outstanding and unique opportunity to investigate rift processes in immense detail and with as yet unparalleled temporal resolution.

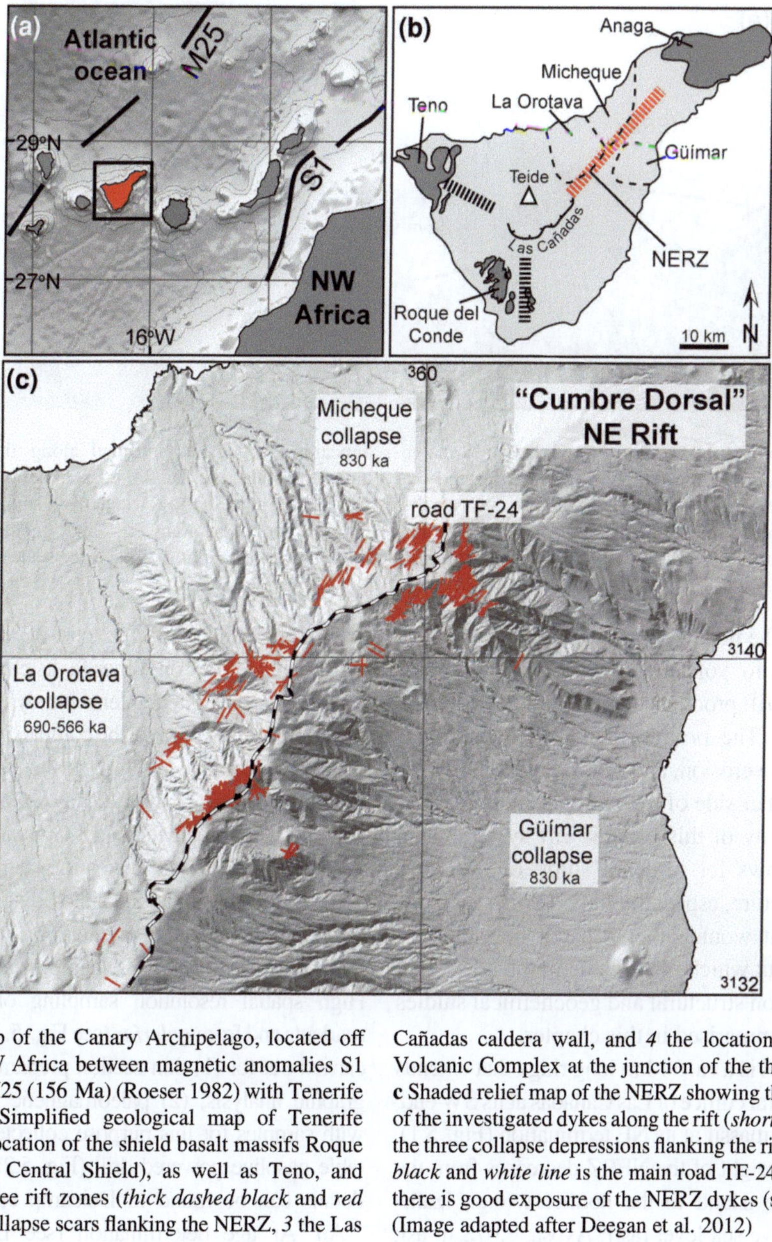

Fig. 5.1 **a** Map of the Canary Archipelago, located off the coast of NW Africa between magnetic anomalies S1 (175 Ma) and M25 (156 Ma) (Roeser 1982) with Tenerife highlighted. **b** Simplified geological map of Tenerife showing *1* the location of the shield basalt massifs Roque del Conde (the Central Shield), as well as Teno, and Anaga, *2* the three rift zones (*thick dashed black* and *red lines*) and the collapse scars flanking the NERZ, *3* the Las Cañadas caldera wall, and *4* the location of the Teide Volcanic Complex at the junction of the three rift zones. **c** Shaded relief map of the NERZ showing the distribution of the investigated dykes along the rift (*short red lines*) and the three collapse depressions flanking the ridge. The *thick black* and *white line* is the main road TF-24, along which there is good exposure of the NERZ dykes (see Fig. 5.2b). (Image adapted after Deegan et al. 2012)

5.2 Geology of the NERZ and Research Developments

Various intervals of rift zone development are represented in the Canary Islands. For instance, the rifts of the younger western islands El Hierro and La Palma are characteristic of the shield development stage. The NW and NE rifts on Tenerife, on the other hand, represent part of the more advanced, post-erosional (rejuvenation) stage of island growth (see Carracedo and Pérez-Torrado, Chap. 2). The NERZ (Fig. 5.1) has been largely inactive for thousands of years along most of its length with only a single known historic eruption in 1704–1705 A.D., which produced relatively small volumes of lava (<3.5 × 10^6 m; Carracedo et al. 2006). The products of this eruption are

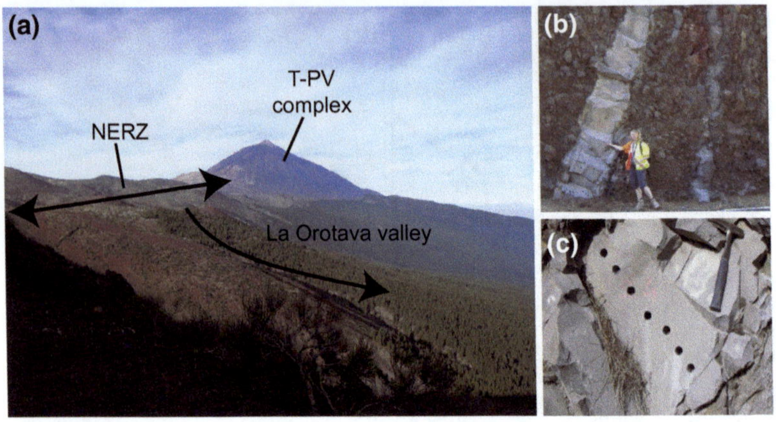

Fig. 5.2 **a** Photograph showing the roughly linear ridge of the Northeast rift zone (NERZ) of Tenerife, and the Teide-Pico Viejo (T-PV) complex, which sits at the south-westernmost projection of the NERZ. La Orotava collapse scar can be seen to the NW of the rift. **b** Field appearance of dykes intruded along the rift exposed along the main road TF-24. **c** Example of a dyke outcrop that has been drilled to retrieve fresh samples of the dyke interior for palaeomagnetic and geochemical studies (hammer to *top right* of image for scale)

restricted to the headwall of the Güímar landslide scar (the Arafo volcano), and probably do not reflect direct rift products themselves (cf. Longpré et al. 2009). The protracted inactivity, coupled with extensive erosion, and several giant landslide events on either side of the rift, has created a dissected anatomy of this oceanic rift system. This situation allows for in-depth study of the rift's internal structure, especially the complex network of dykes that would otherwise not be generally accessible, but which is available in the NERZ for high resolution structural and geochemical studies (Fig. 5.2) summarised in this chapter.

The NERZ is a ca. 30 km long ridge that extends from the central edifice of Las Cañadas at its SW end, to the Anaga massif at its NE termination (Fig. 5.1). The overall height of the NERZ decreases from the centre of the island to the northeast (e.g., Izaña, 2,386 m above sea level [asl]; Ayosa, 2,078 m asl; Joco, 1,956 m asl; Gaitero, 1,748 m asl). The ridge-like topography of the NERZ has been shaped by three successive mass wasting events along its flanks. From aerial view, these landslide scars can be seen as depressions on both sides of the ridge (Fig. 5.1). The Micheque and Güímar landslides were roughly coeval, taking place at ca. 0.83 Ma, while the more recent La Orotava landslide occurred at 0.56 Ma (Figs. 5.1 and 5.2; Carracedo et al. 2011 and references therein).

Although the NERZ is one of the pronounced geological features on Tenerife, its challenging complexity has prevented intense study so far. Recent research efforts to unravel the complexity of the NERZ involved systematic and in-depth mapping of the rift; structural analysis of over 400 dykes, including dykes exposed within the Micheque, Güímar, and La Orotava collapse scars and dykes in water tunnels (*galerías*); and investigations into dyke petrography, morphology, thickness, orientation, and their internal features (e.g., Fig. 5.3; Delcamp et al. 2010, 2012). High spatial resolution sampling of dykes along roadcuts and from *galerías* (see Fig. 5.1) has also been carried out for (1) thin-section preparation and petrographic analysis; (2) paleomagnetic measurements, with samples for this purpose collected using a portable gasoline powered drill (Fig. 5.2; full analytical details can be found in Delcamp et al. 2010); (3) $^{40}Ar/^{39}Ar$ age determination (see Delcamp et al. 2010); and (4) geochemical analysis, including major and trace elements of over 80 dyke samples following the method in Abratis et al. (2002). A sub-set of these 80 samples was analyzed for stable (oxygen) isotopes following the procedure given in Vennemann and Smith (1990) and Fagereng et al. (2008). Finally, a selection of these was further analysed for combined trace and rare earth element (REE) and Sr–Nd–Pb isotopes. Full analytical details, including measurements of internal and external standards, and the entire

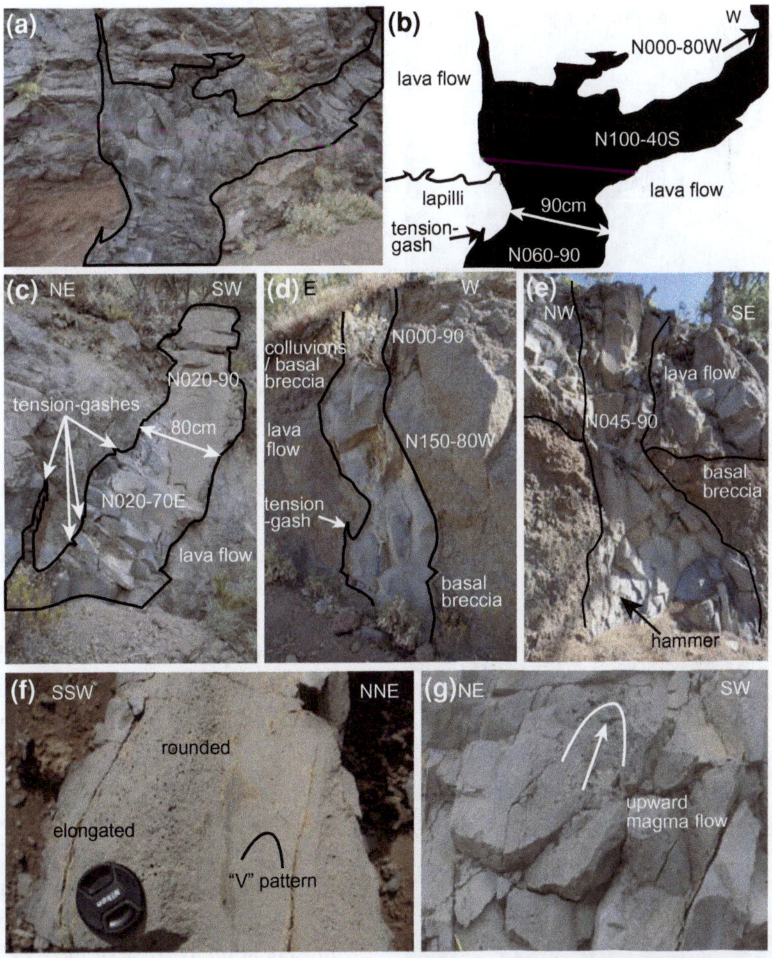

Fig. 5.3 Photographs of the field appearance of representative dykes of the NERZ with variations in orientation (strike, dip or both), thickness, and texture observed on the single intrusion scale. Images from Delcamp et al. (2012). **a**, **b** Dyke orientation is seen to change at the transition between surrounding lapilli and a lava flow. **c** Dyke showing a change in orientation within a lava flow unit. **d** Dyke showing a change in orientation at the transition between surrounding basal breccia and a lava flow. **e** Variation in dyke thickness due to a change in the host-rock lithology. The dyke is thicker within the basal breccia (lower part, low competence layer) than within the lava flow (upper part, high competence layer). **f** Example of vesicle types and distribution in a dyke interior. **g** Example showing the direction of magma flow as recorded by vesicle orientation

geochemical data set can be found in Deegan et al. (2012).

The results of this multi-disciplinary effort are synthesised in this chapter and provide insights into (1) the structural development of the NERZ, (2) the magmas feeding the NERZ plumbing system, and (3) the underlying mantle sources to the NERZ. This research approach thus provides a perspective from source processes to surface expression for oceanic rifts.

5.3 Field Occurrence and Petrography of the Dykes

The dykes exhibit a large degree of variability in terms of their field occurrence, even on the scale of a single intrusion. Thickness, orientation, and texture are frequently found to change, related to, e.g., changes in the competence of the host lithology (cf. Gudmundsson 2002) as can be seen

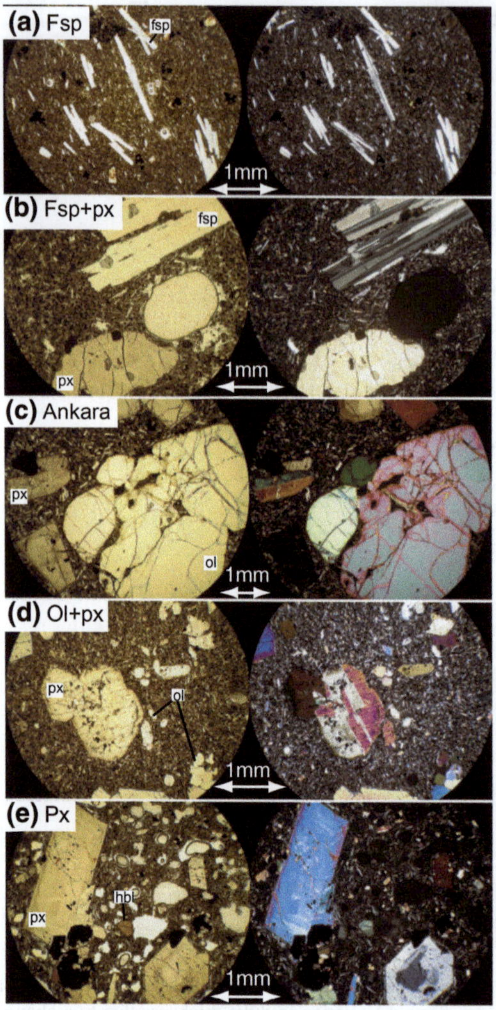

Fig. 5.4 Representative photomicrographs of the petrographic dyke groups (aphyric types not shown). **a** Feldspar-phyric group, **b** Feldspar and pyroxene-phyric group, **c** Ankaramite group, **d** Olivine and pyroxene-phyric group, and **e** Pyroxene-phyric group. Plane polarized view is shown on the left and cross polarized view on the right. Abbreviations: *Fsp* = feldspar, *px* = pyroxene, *ol* = olivine, *ankara* = ankaramite, *hbl* = hornblende. Figure modified from Delcamp et al. (2012)

in Fig. 5.3. In particular, where there are two host lithologies in contact, the intruding dyke undergoes a structural change at the contact point (Fig. 5.3). Other notable variations on the intrusion scale include occasionally observed vesicular and brecciated dyke interiors, locally concentrated flow lobes, folding, and small-scale intrusive "fingers" into non-cohesive country rock (cf. Mathieu et al. 2008; Delcamp et al. 2012).

The mineralogy and petrography of the dykes, established both in the field, based on hand specimen scale observations, and using thin section microscopy (Fig. 5.4; Deegan et al. 2012; Delcamp et al. 2012) revealed recurring textures and the mineralogy of the dykes was used to characterise several petrographic groups:

1. *Aphyric group*. For the most part phenocryst free or only weakly-phyric. Occasional small crystals (<1 mm) of pyroxene and feldspar are observed in a micro- to cryptocrystalline, sometimes glassy, groundmass. Microphenocrysts generally constitute less than 5 % of the rock volume.
2. *Feldspar-rich group* ("fsp" dykes, Fig. 5.4a). Feldspar is the dominant phenocryst phase, occupying up to 50 % of the rock volume. Rare clinopyroxene, olivine, and Fe–Ti oxide crystals may be present with each constituting less than 10 % of the rock volume.
3. *Feldspar- and pyroxene-rich group* ("fsp + px" dykes, Fig. 5.4b). Feldspar and pyroxene are the main phenocryst phases, occupying up to 50 % of the rock volume, with feldspar generally more abundant than pyroxene. Minor olivine and Fe–Ti oxides may be present at up to 10 % of the rock volume.
4. *Ankaramite group* ("ankara", Fig. 5.4c). Olivine and pyroxene phenocrysts (up to 2 cm across) occupying between 40 and 60 % of the rock volume. Minor phenocryst phases also include amphibole and Fe–Ti oxides. Ankaramite groundmass ranges from cryptocrystalline to Fe–Ti oxide-rich.
5. *Olivine- and pyroxene-rich group* ("ol + px" dykes, Fig. 5.4d). Composed of a similar mineral assemblage to the ankaramite group, but with crystal contents ranging from 5 to 35 % of the rock volume. Phenocrysts are a few mm in size, making them substantially smaller than those in the ankaramites. The groundmass is generally rich in plagioclase and Fe–Ti oxides.
6. *Pyroxene-rich group* ("px" dykes, Fig. 5.4e). Clinopyroxenes are the major phenocryst forming phase occupying between 5 and 30 %

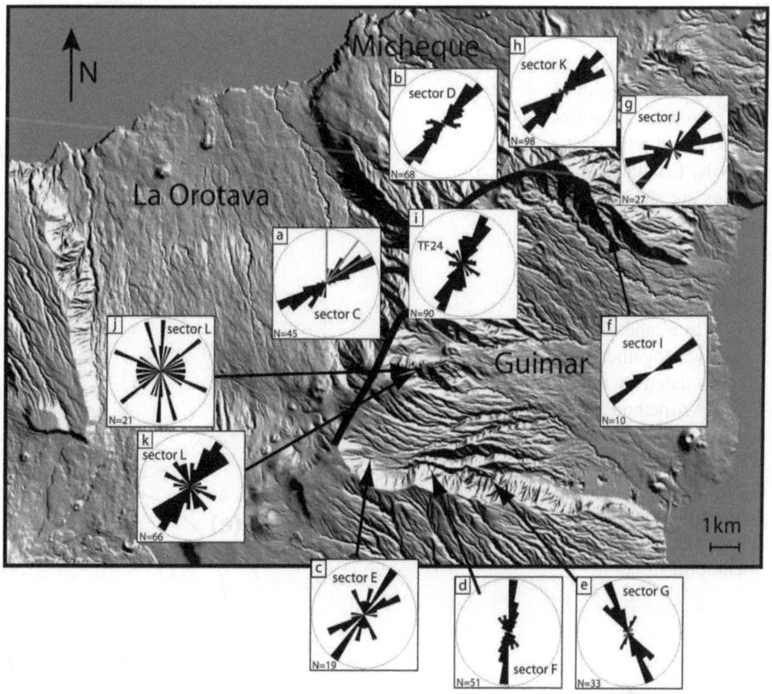

Fig. 5.5 *Rose* diagrams showing a large range of dyke orientations observed in the various portions of the NERZ. Note that dyke orientation along the rift is for most intrusions non-parallel to the rift axis. See text for details

of the rock volume. Minor phenocryst phases in this group include amphibole and Fe–Ti oxides at usually less than 10 % of the rock volume.

5.4 Structural Evolution of the NERZ

Extensive field work has helped to gain insight into the structural development and organisation of the NERZ, in particular the link between dynamic rift-zone development and giant collapse events. In this context, several models have previously been discussed to explain rift zone organisation, as outlined in Chap. 4. The major finding arising from the field study presented in Delcamp et al. (2012) is that dyke orientation in the NERZ is not simply parallel to the rift axis or the collapse embayments but is frequently oblique to the walls of the main collapse scars (Fig. 5.5). This contrasts most previous observations from nature and experiments that document dykes parallel to the rift axis and the walls of collapse scars (e.g., Acocella and Neri 2009). The oblique dyke geometry of the NERZ is interpreted as being caused by flank spreading and associated creep during successive pulses of emplacement of dykes and other shallow intrusions. Flank spreading would initially stabilise a rift, but after a critical amount of magma supply a collapse event would be triggered by continued intrusive activity. A small, but detectable change in dyke orientation on the rift axis seems to be also associated with the major landslides, implying that rift zones do indeed change dynamics and orientation due to external forces such as gravity-driven flank collapses (cf. Walter and Troll 2003; Walter et al. 2005; Carracedo and Troll, this volume). This result serves to change our perception of Ocean Island rifts from simple parallel alignments of intrusions to a complex and dynamic feeder system that develops in response to internal as well as external influences.

5.5 Magnetic Studies and Ages of the Dykes

Paleomagnetic studies of the NERZ were carried out by Carracedo et al. (2011) and Delcamp et al. (2010) to add temporal constraints to the

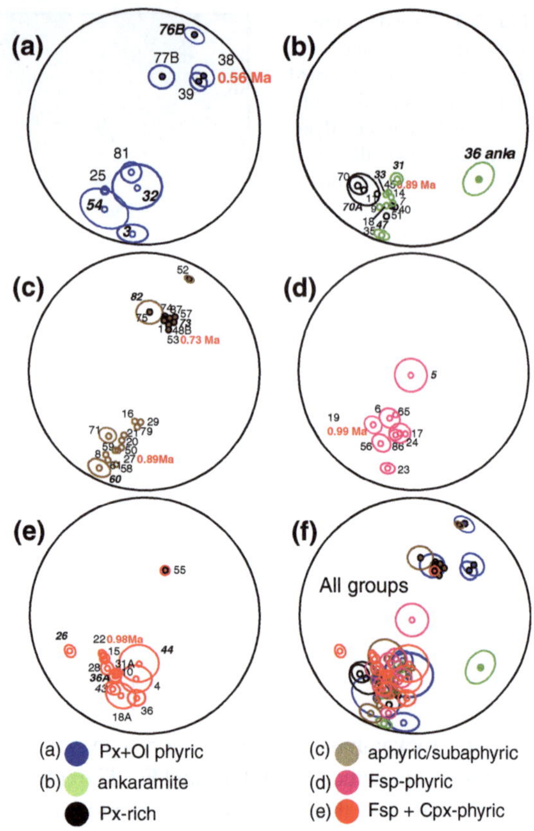

Fig. 5.6 Equal area projection of mean paleomagnetic results for each site (Delcamp et al. 2010). Normal polarities are indicated with open symbols. Confidence ellipses are shown for some samples while others are excluded for clarity. Individual petrographic groups are represented in plots **a** to **e**. and all groups are shown together in plot. **f** Dykes dated by the $^{40}Ar/^{39}Ar$ method are labelled with their age in Ma (see Fig. 5.7 for more detail). Note that petrographic groups do not correspond to individual orientation trends

evolutionary history of the rift. The NERZ shows a range in magnetic polarities, with the dykes spanning at least two polarity intervals, i.e., the Matuyama (reverse) and Bruhnes (normal) chrons (Figs. 5.6 and 5.7). The paleomagnetic data hence points to the NERZ as being a relatively long-lived system with, at times, a considerable magma supply.

New $^{40}Ar/^{39}Ar$ ages reported in Carracedo et al. (2011) and Delcamp et al. (2010) show that intrusive activity on the NERZ is characterised by a semi-continuous magma supply but with swarms of dykes of various petrographic types being intruded close in time, i.e., in pulses. A peak in magma supply during the mid-Pleistocene is thought to have led to flank deformation and the subsequent collapses of the Micheque, Güímar, and La Orotava edifices (Carracedo et al. 2011). The paleomagnetic data also record a 26° clockwise vertical-axis rotation of the sampled rift core. Delcamp et al. (2010) interpret this as a result of a local volcano-tectonic response to strike-slip movements that occurred successively on either side of the rift axis due to flank instabilities. Central areas on the rift may thus experience near surface rotation—a feature also reported from other Canary Islands, such as El Hierro (e.g., Széréméta et al. 1999). It is noteworthy that the petrographic groupings of dykes do not correspond to specific polarities nor are they spatially segregated (Fig. 5.6), which suggests that the various petrographic groups of dykes were not intruded in distinct packages, but instead semi-simultaneously over the lifetime of the rift. A scenario involving a number of smaller storage and feeder reservoirs is hence likely to have fed the near-surface dyke intrusions, causing distinct petrographic types to occur along the rift system at the same time. The NERZ thus represents a highly dynamic and changeable geological and geo-morphological environment. Volcanic rift

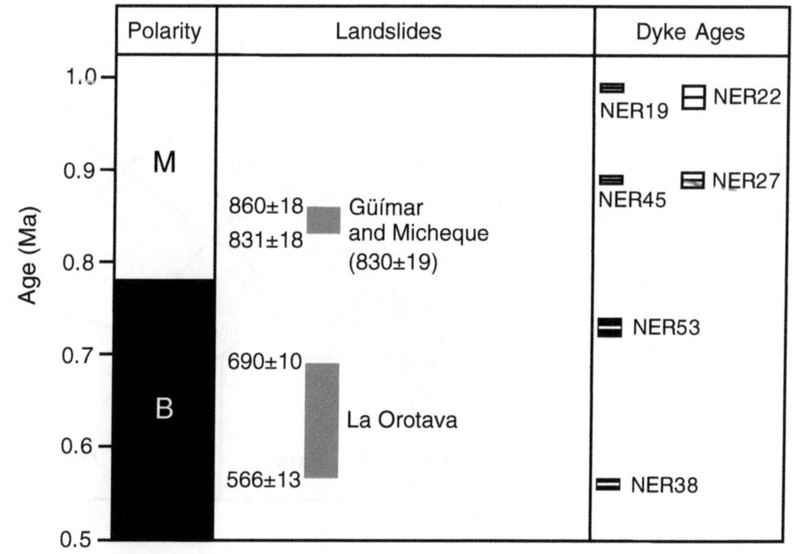

Fig. 5.7 A comparison of the polarity scale (*left*, *M* = Matuyama, *B* = Brunhes), the timing of major collapse events on the NERZ (*centre*, Carracedo et al. 2011), and representative NERZ dyke ages (*right*) after Delcamp et al. (2010). Representative dyke ages are shown as rectangles whose vertical extent corresponds to the analytical uncertainty

zones such as the NERZ on Tenerife are thus not static structural features. In fact, they are characterised by concentrated pulses of active growth and giant destructions that result in measurable geometric re-arrangements.

5.6 Petrogenesis of the NERZ Magmas

The full geochemical and isotope data available for the dykes of the NERZ is reported in Deegan et al. (2012), with an overview provided here. The NERZ dykes can be classified as alkali magmas, similar to the other magma suites on Tenerife, such as the Teide-Pico Viejo complex (Fig. 5.8). Although the dykes of the NERZ display wide petrographic variability, most classify as basanites based on their major element composition (Fig. 5.8). The exceptions are crystal-rich ankaramite dykes, which plot as extremely mafic magma types due to their high content by volume of Mg-rich olivine and pyroxene (cf. Longpré et al. 2009).

Major element variation diagrams for the complete dyke suite display linear trends that are characteristic of fractional crystallisation of a mineral assemblage including olivine, clinopyroxene, plagioclase, Fe–Ti oxides, and apatite (Fig. 5.9). Magnetite has previously been identified as the major Fe–Ti oxide phase in the dykes using reflected light microscopy (Fig. 5.9; Delcamp et al. 2010), which explains the kink in TiO_2 with increasing degrees of differentiation as magnetite fractionates. Trace element variations in the dykes are also broadly consistent with fractional crystallisation. Decreasing Sc, Ni, and Co with decreasing MgO reflects crystallisation of olivine and pyroxene (see Deegan et al. 2012; Fig. 5.9e, f). Fractional crystallisation was quantified by applying least squares minimisation to the aphyric samples, which most closely represent liquid compositions. The broad trends in the major element data can be readily explained by removal of plagioclase, clinopyroxene, Fe–Ti oxides, olivine, and apatite in the proportions 41:25:18:11:5, respectively (Fig. 5.9).

While the major and trace element variability in the dykes is largely controlled by fractional crystallisation, the isotope data suggest additional magmatic processes to have taken place. One key process which could account for the isotope variations in the dykes is crustal contamination of some of the rift zone magmas by hydrothermally altered components of the island edifice and/or oceanic sediments (Fig. 5.10).

Crustal contamination of Ocean Island magmas has traditionally been thought of as minimal. This is due to the relatively thin underlying oceanic crust and the assumption that magma injected into basaltic crust would be of similar

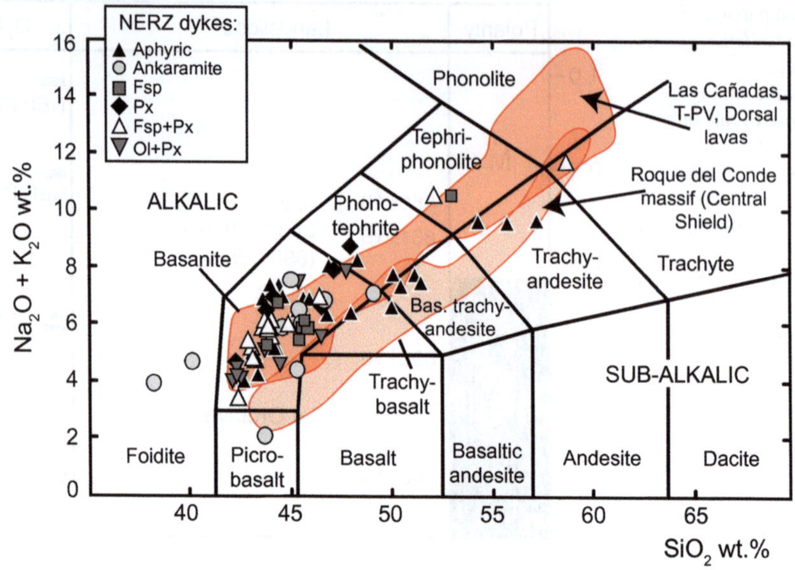

Fig. 5.8 Total alkali versus silica (*TAS*) plot for dykes of the Northeast rift zone (*NERZ*) with boundaries drawn after Irvine and Baragar (1971) and nomenclature after Le Maitre et al. (1989). Symbols correspond to the petrographic groupings of the NERZ. Shaded fields are drawn using data for the Central Shield (Roque del Conde) after Thirlwall et al. (2000) and for the Las Cañadas volcano, the Teide-Pico Viejo complex (T-PV), and lavas from the NERZ after Ablay et al. (1998) and Neumann et al. (1999). Figure modified after Deegan et al. (2012)

enough composition so that contamination would in effect be absent. However, in practice, crustal contamination by various components of the ocean crust, overlying sediments, and the island edifice itself is becoming increasingly recognized in Ocean Island settings (e.g., Clague et al. 1995; Bohrson and Reid 1997; Garcia et al. 1998; Kent et al. 1999; Harris et al. 2000; Wolff et al. 2000; Gurenko et al. 2001; Troll and Schmincke 2002; Hansteen and Troll 2003; Gaffney et al. 2005; Legendre et al. 2005; Wiesmaier 2010). On Tenerife, the most likely contaminants are the plutonic rocks of the island core and variably altered shallow felsic materials, which display a large compositional diversity (e.g., Palacz and Wolff 1989; Wolff et al. 2000; Wiesmaier et al. 2012).

Oxygen isotope analysis of the dykes of the NERZ indicates that some dyke intrusions bear witness to low temperature alteration, i.e., those with coupled high $\delta^{18}O$ and H_2O wt. % values (Deegan et al. 2012). A number of dykes with regular H_2O values still show $\delta^{18}O$ values that are elevated relative to the mantle range, which cannot be readily explained by closed-system fractional crystallisation or post-eruptive alteration (see the Rayleigh fractionation curve on Fig. 5.10; and also Sheppard and Harris 1985). Mixing models suggest that some of the variability in the NERZ data may be explained by uptake of ^{18}O from a combination of oceanic sediment and hydrothermally altered material from the island edifice (Fig. 5.10). Similarly, minor variation in $^{87}Sr/^{86}Sr$ and $^{143}Nd/^{144}Nd$ values, that trend away from the upper mantle range and towards more crustal values, can be explained by minor interaction between some batches of NERZ magma with hydrothermally altered island edifice material (Fig. 5.11; Deegan et al. 2012). This is plausible, given that assimilation of such hydrothermally altered felsic material has been recognized previously for Tenerife (e.g., Palacz and Wolff 1989; Wolff et al. 2000; Wiesmaier 2010). Altered felsic rocks probably occupy a substantial portion of the subvolcanic pile in which the NERZ magmas resided and into which the NERZ dykes were intruded and would hence have been readily available for

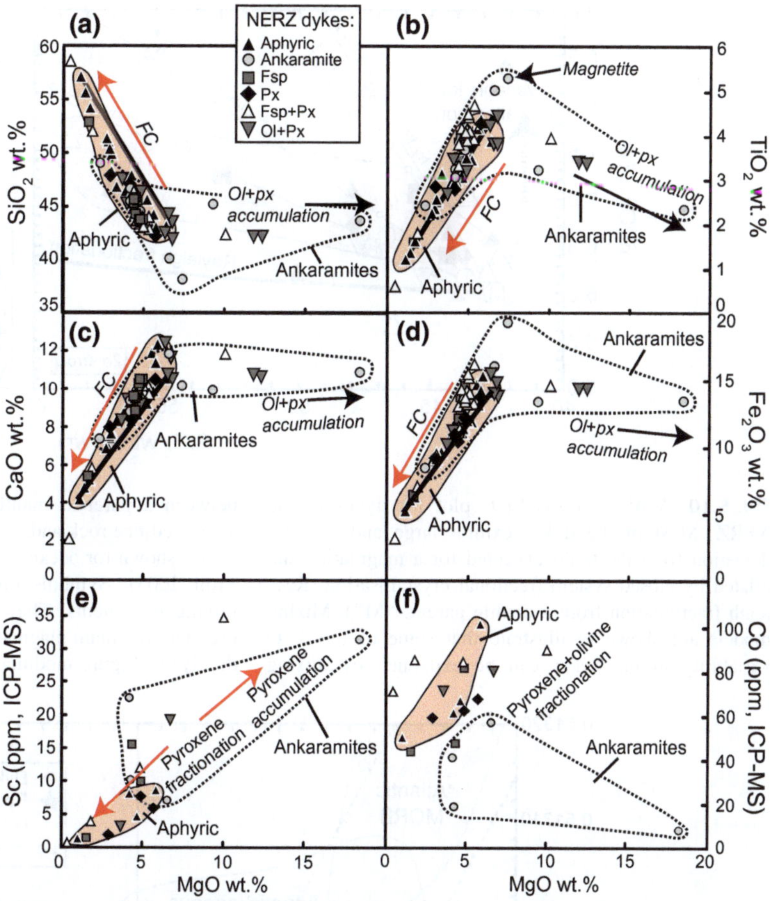

Fig. 5.9 Major and trace element variation diagrams for dykes of the NERZ. Aphyric samples are shown as *black triangles* and, being largely phenocryst-free, are representative of true liquid compositions. *Solid black lines* are fractional crystallisation trajectories for a model involving crystallisation of plagioclase, clinopyroxene, Fe–Ti oxides, olivine, and apatite in the proportions 41:25:18:11:5, respectively (see text). *FC* = fractional crystallisation. Figure modified after Deegan et al. (2012)

interaction. Note that assimilation of ocean sediments and/or hydrothermally altered ocean crust and island edifice has been discussed in the context of several of the Canary Islands, not only Tenerife (e.g., Marcantonio et al. 1995; Thirwall et al. 1997; Gurenko et al. 2001; Troll and Schmincke 2002; Hansteen and Troll 2003; Aparicio et al. 2006, 2010; Troll et al. 2012). The effects of these late stage processes need to therefore be accounted for before attempting to define the mantle processes giving rise to the primary magmas of the NERZ.

Concerning magma generation beneath the NERZ, the Pb isotope data presented in Deegan et al. (2012) provide insights into the nature of the mantle source and, in combination with magnetic polarity data (Delcamp et al. 2010), the genetic ties between various stages in island growth. Lead isotopes are not expected to be overly sensitive to contamination by hydrothermally altered island edifice since Pb does not appear to undergo isotopic exchange during alteration (Cousens et al. 1993; Gaffney et al. 2005) and hence they are particularly useful for the NERZ. Most of the Pb isotope data for the NERZ plot below but roughly parallel to the Northern Hemisphere Reference Line (NHRL, Hart 1984) which represents a mixture between depleted upper mantle (DMM) and a mantle that involves an ancient subducted component (HIMU) (Fig. 5.12). In this context, the NERZ Pb isotope data are consistent with derivation from a young HIMU source, in which the subducted ocean crust component has resided less than 1.5 Ga in the source region (Thirlwall 1997; Geldmacher and Hoernle 2000, 2001; Simonsen et al. 2000; Gurenko et al. 2009). Two of the dykes plot above the NHRL on Fig. 5.12, which, in combination with their elevated $\delta^{18}O$

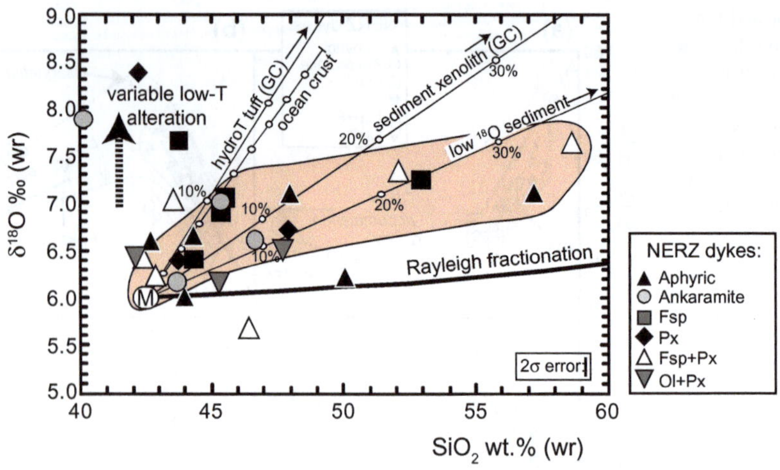

Fig. 5.10 A $\delta^{18}O$ versus SiO_2 plot for dykes of the NERZ. Most of the dykes exhibit large and variable deviation from the trend expected for a magmatic suite related by closed system fractional crystallisation (Rayleigh fractionation from a mantle parent, "M"). Mixing models are shown to illustrate that some of the $\delta^{18}O$ variability in the dykes can be explained by mixing between a parental mantle source and hydrothermally altered edifice rock and/or ocean sediment. A mixing line is also shown for ocean crust (layer 2 from Hansteen and Troll 2003). Symbols on the *mixing lines* are 10 % mixing increments. The *orange shaded* field serves to highlight the main magmatic contamination trend among the dykes. Figure modified after Deegan et al. (2012)

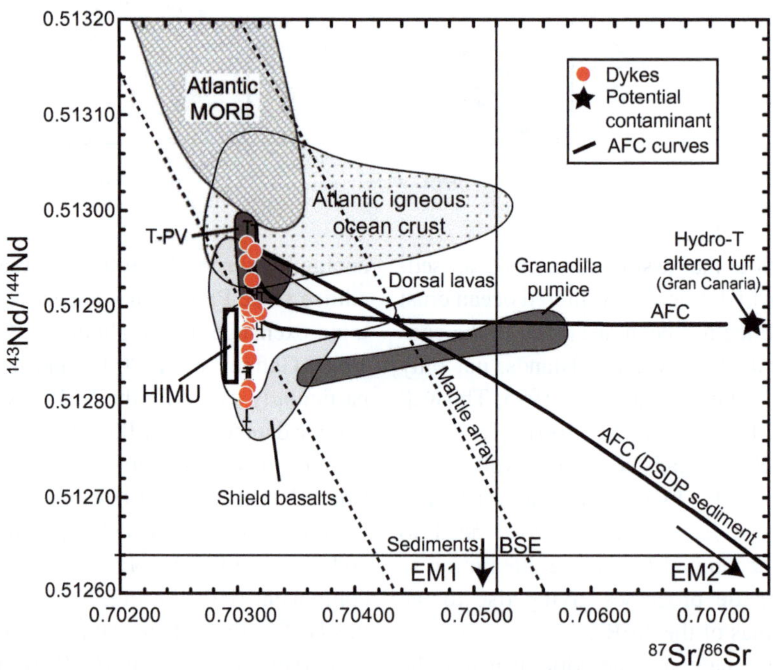

Fig. 5.11 $^{87}Sr/^{86}Sr$—$^{143}Nd/^{144}Nd$ isotope variation diagram for dykes of the NERZ, and some reference fields including: Atlantic MORB (Ito et al. 1987); Atlantic igneous ocean crust (Hoernle 1998); Tenerife shield basalts (Simonsen et al. 2000; Gurenko et al. 2006); the Teide-Pico Viejo (T-PV) complex (Wiesmaier et al. 2010); NERZ lavas (Simonsen et al. 2000); and the Granadilla pumice (Palacz and Wolff 1989). A hydrothermally altered tuff from Gran Canaria is also plotted (Troll 2001). The dykes of the NERZ form a roughly linear array between the fields for HIMU-type mantle and Atlantic MORB. A number of dykes are slightly offset from the array. Their offset can be explained by incorporation of small amounts of hydrothermally altered material during magma storage in the island edifice (heavy black lines are assimilation-fractional crystallisation (AFC) curves)

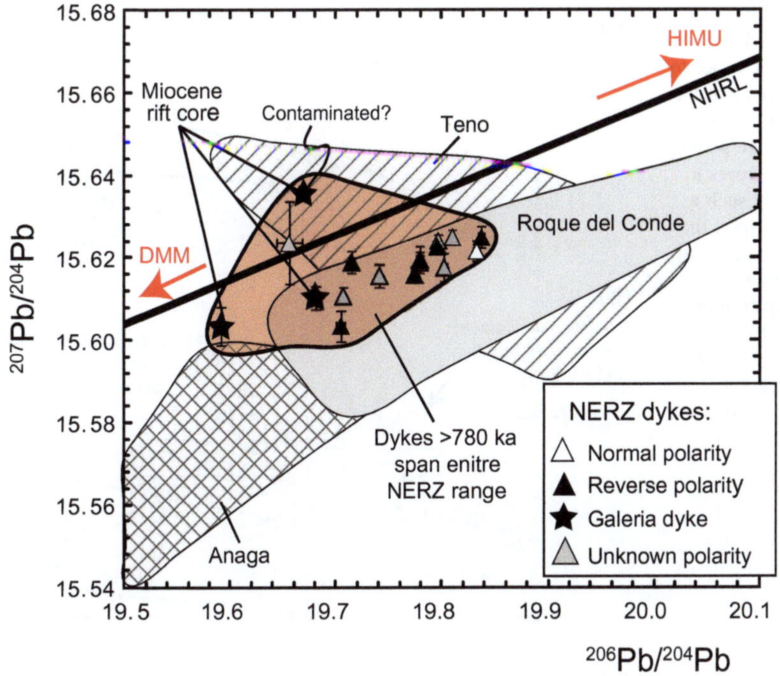

Fig. 5.12 $^{206}Pb/^{204}Pb$—$^{207}Pb/^{204}Pb$ isotope variation diagram for dykes of the NERZ, with symbols corresponding to the ages of the dykes. Magnetic polarities are taken from Delcamp et al. (2010): reverse polarity dykes are ≥780 ka old; normal polarity dyke is ≤780 ka old. *Black stars* are *galería* samples taken from the core of the rift and hence represent an older, likely Pliocene, part of the system (Carracedo et al. 2011)

values, points to more intense sediment contamination of these two samples. Discarding samples that show evidence for crustal contamination, Strontium-Neodymium isotopes are consistent with DMM + HIMU mantle mixing (Deegan et al. 2012).

Using magnetic polarity data obtained for some of the dykes (Delcamp et al. 2010), we can place temporal constraints on the isotope data. As can be seen in Fig. 5.12, the Pb isotope compositional range of the NERZ has been largely unchanged through most of its evolution, i.e., the old *galería* dykes and the dykes of normal, reverse, and unknown polarities span the entire NERZ range, and do not form isotopically distinguishable groups. This observation suggests that the mantle source feeding the NERZ has been isotopically constant through its lifetime. Moreover, of the three shield-stage volcanoes on Tenerife (Teno, Anaga, and Roque del Conde), the NERZ is isotopically most similar to the old central edifice of Roque del Conde (Fig. 5.12). It therefore appears likely that the NERZ had a different "mantle highway" than both Teno and Anaga, and instead formed as an extension of the Central Shield, gradually growing toward what is now Anaga in the NE (Fig. 5.1, see also Guillou et al. 2004; Carracedo et al. 2011).

The Pb-isotope signature of the NERZ is also similar to the Las Cañadas volcano (Simonsen et al. 2000) suggesting that they both shared a similar parental source too. This is not surprising as the Pleistocene NERZ was emplaced and erupted coeval to the central Las Cañadas volcano (e.g., Ancochea et al. 1990; Bryan et al. 1998; Edgar et al. 2002; Carracedo et al. 2011). The relatively young Teide-Pico Viejo (T-PV) complex, however, extends towards less radiogenic $^{206}Pb/^{204}Pb$—$^{207}Pb/^{204}Pb$ values, which, coupled with high $^{143}Nd/^{144}Nd$ values, suggests that this most recent phase of magmatism on Tenerife had

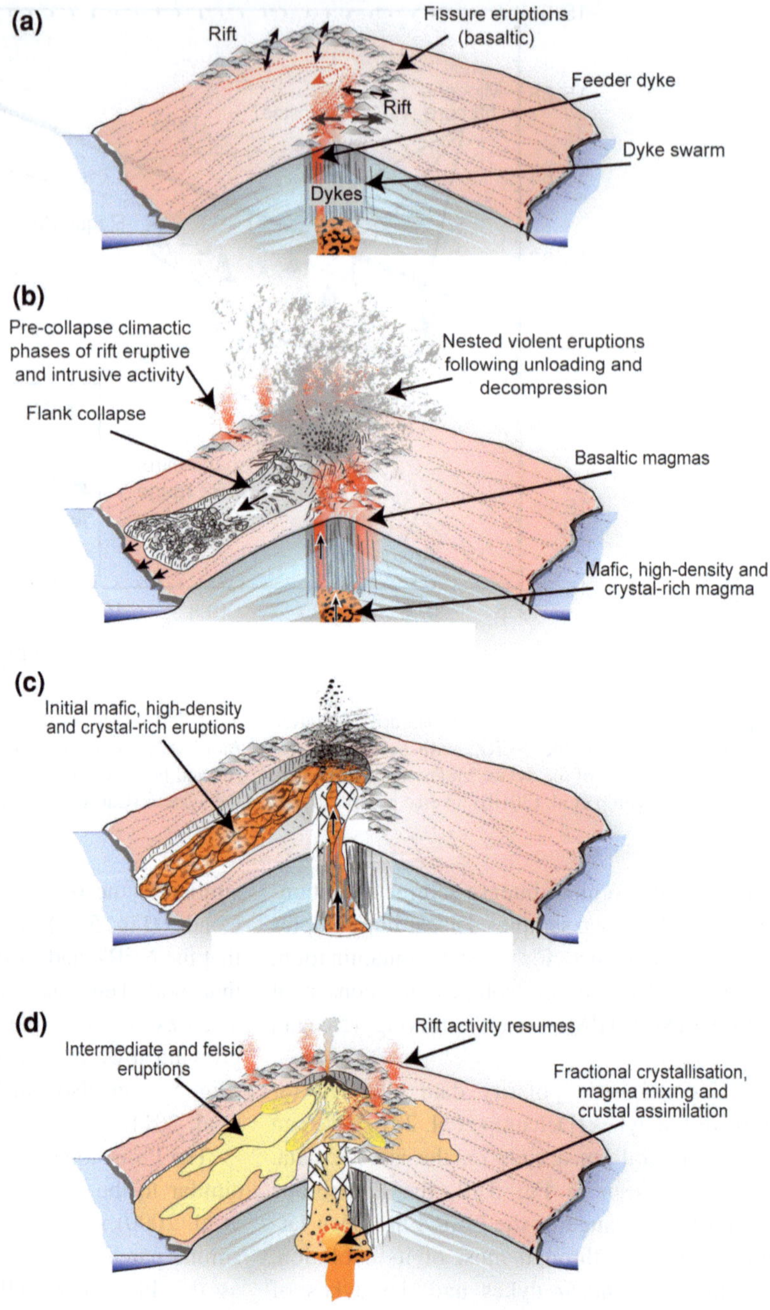

Fig. 5.13 Schematic model of Canary Island rift evolution (after Carracedo et al. 2011). Note that the rift evolution proceeds until high intrusive activity causes flank collapses to occur. Following such a catastrophic collapse, the plumbing system needs to readjust, leading to structural and petrologic modifications of the rift (cf. Longpré et al. 2009; Manconi et al. 2009; Delcamp et al. 2010, 2012; Carracedo et al. 2011)

a slightly different mantle source with a greater input of DMM material (i.e., a lower proportion of a deep HIMU-plume component).

The persistence of a HIMU-dominated mantle source throughout the lifetime of the NERZ can be interpreted in the context of the "blob" model for the Canary Islands (Hoernle et al. 1991; Hoernle and Schmincke 1993). This model involves formation of a zone of discrete HIMU mantle blobs at the top of a mantle plume beneath the Canaries. These blobs become entrained with upper asthenospheric mantle (DMM), giving rise to the mixed DMM + HIMU mantle isotope signal of the Canaries. We suggest that a large column of HIMU-type mantle, or a quick succession of compositionally similar HIMU-blobs, occupied the melting zone beneath central Tenerife from the Miocene to Pleistocene. This was possibly accommodated for such a long period of time by slow plate movement beneath the Canaries (~ 10 mm yr^{-1}; Hoernle and Schmincke 1993 and references therein). The recent T-PV complex appears to reflect a lesser influence of the HIMU blob and greater incorporation of DMM material, possibly due to an increase of edge-DMM entrainment into the melting zone and thinning of the blob(s) with time (see Deegan et al. 2012).

5.7 Unravelling the NERZ from Source to Surface

The recent intense and multi-disciplinary research efforts to unravel the structural and petrogenetic history of the NERZ underscores the premise that ocean island rift zones are long-lived, complex, and dynamically evolving systems. They can neither be thought of as a simple parallel arrangement of dykes and volcanic vents along fractures, nor as purely reflecting long-lived ocean crust fractures. Indeed, it has been shown that the NERZ is a dynamic morphological environment, which developed from a series of intrusive pulses interspersed with periods of relative quiescence, but also with flank creep, and particularly with collapse events that changed the structural arrangement of dyke intrusions and eruptive sites repeatedly (e.g., Delcamp et al. 2010, 2012; Carracedo et al. 2011; and Carracedo and Troll, this volume; see also Fig. 5.13 for a summary).

In terms of petrogenesis, the history of the NERZ began with magma generation from a long-lived HIMU-dominated mantle source. Following segregation from the mantle region, many of the NERZ magmas underwent fractional crystallisation, and some underwent assimilation of ocean sediments and hydrothermally altered island edifice material. Isotope geochemistry places the initiation and growth of the NERZ into the wider geological context of Tenerife. The data discussed in this chapter, in conjunction with geochronological evidence (Carracedo et al. 2011), support a model of the NERZ magmas being related to the magma source of the central edifice (Roque del Conde and Las Cañadas volcanoes), which implies that the NERZ likely originated from the central part of Tenerife to eventually link up with the Anaga edifice in the NE.

References

Ablay GJ, Carroll MR, Palmer MR, Martí J, Sparks RSJ (1998) Basanite-phonolite lineages of the teide-pico viejo volcanic complex, tenerife, Canary Islands. J Petrol 39:905–936

Abratis M, Schmincke H-U, Hansteen TH (2002) Composition and evolution of submarine volcanic rocks from the central and western Canary Islands. Int J Earth Sci 91:562–582

Acocella V, Neri M (2009) Dike propagation in volcanic edifices: overview and possible developments. Tectonophysics 471:67–77

Ancochea E, Fuster JE, Ibarrola E, Cendrero A, Coello J, Hernan F, Cantagrel JM, Jamond C (1990) Volcanic evolution of the island of tenerife (Canary Islands) in the light of new K-Ar data. J Volcanol Geotherm Res 44:231–249

Aparicio A, Bustillo MA, Garcia R, Araña V (2006) Metasedimentary xenoliths in the lavas of the timanfaya eruption (1730–1736, Lanzarote, Canary Islands): metamorphism and contamination processes. Geol Mag 143:181–193

Aparicio A, Tassinari CCG, García R, Araña V (2010) Sr and Nd isotope composition of the metamorphic, sedimentary and ultramafic xenoliths of lanzarote (Canary Islands): implications for magma sources. J Volcanol Geotherm Res 189:143–150

Bohrson WA, Reid MR (1997) Genesis of silicic peralkaline volcanic rocks in an ocean island setting by crustal melting and open system processes: Socorro Island, Mexico. J Petrol 38:1137–1166

Bryan SE, Martí J, Cas RAF (1998) Stratigraphy of the bandas del sur formation: an extracaldera record of quaternary phonolitic explosive eruptions from the las cañadas edifice, tenerife (Canary Islands). Geol Mag 135:605–636

Carracedo JC (1975) Estudio Paleomagnético de la Isla de Tenerife. Dissertation, Universidad Complutense de Madrid

Carracedo JC (1979) Paleomagnetismo e Historia Volcánica de Tenerife: Tenerife, Aula de Cultura del Cabildo de Tenerife, p 81

Carracedo JC (1994) The Canary Islands: an example of structural control on the growth of large oceanic island volcanoes. J Volcanol Geotherm Res 60:225–242

Carracedo JC (1996) Morphological and structural evolution of the western Canary Islands: hotspot induced three-armed rifts or regional tectonic trends? J Volcanol Geotherm Res 72:151–162
Carracedo JC (1999) Growth, structure, instability and collapse of Canarian volcanoes and comparisons with Hawaiian volcanoes. J Volcanol Geotherm Res 94:1–19
Carracedo JC, Rodríguez Badiola E, Soler V (1992) The 1730–1736 eruption of Lanzarote: an unusually long, high-magnitude fissural basaltic eruption in the recent volcanism of the Canary Islands. J Volcanol Geotherm Res 53:239–250
Carracedo JC, Day S, Guillou H, Rodríguez Badiola E, Canas JA, Pérez Torrado FJ (1998) Hotspot volcanism close to a passive continental margin: the Canary Islands. Geol Mag 135:591–604
Carracedo JC, Rodríguez Badiola E, Guillou H, De La Nuez J, Pérez Torrado FJ (2001) Geology and volcanology of La Palma and El Hierro, Western Canaries. Estud Geol 57:175–273
Carracedo JC, Rodríguez Badiola E, Guillou H, Paterne M, Scaillet S, Pérez Torrado FJ, Paris R, Criado C, Hansen A, Arnay M, González Reimers E, Fra-Paleo U, González Pérez R (2006) Los volcanes del Parque Nacional del Teide: El Teide, Pico Viejo y las dorsales activas de Tenerife. In: Carracedo JC (ed) Los Volcanes del Parque Nacional del Teide, Serie Técnica: Madrid, Organismo de Parques Nacionales, Ministerio del Medio Ambiente, Madrid, pp 175-199
Carracedo JC, Rodríguez Badiola E, Guillou H, Paterne M, Scaillet S, Pérez Torrado FJ, Paris R, Fra-Paleo U, Hansen A (2007) Eruptive and structural history of Teide volcano and rift zones of Tenerife, Canary Islands. Geol Soc Am Bull 19:1027–1051
Carracedo JC, Guillou H, Nomade S, Rodríguez Badiola E, Paris R, Troll VR, Wiesmaier S, Delcamp A, Fernández-Turiel JL (2011) Evolution of ocean-island rifts: The northeast rift zone of Tenerife, Canary Islands. Geol Soc Am Bull 123:562–584
Carracedo JC, Troll VR (2012) Structural and geological elements of the Teide Volcanic complex: rift zones and gravitational collapses (this volume)
Clague DA, Moore JG, Dixon JE, Friesen WB (1995) Petrology of submarine lavas from Kilauea's Puna Ridge. Hawaii. J Petrol 36:299–349
Cousens BL, Spera FJ, Dobson PF (1993) Post-eruptive alteration of silicic ignimbrites and lavas, Gran Canaria, Canary Islands: strontium, neodymium, lead, and oxygen isotopic evidence. Geochim Cosmochim Acta 57:631–640
Deegan FM, Troll VR, Barker AK, Chadwick JP, Harris C, Delcamp A, Carracedo JC (2012) Crustal versus source processes recorded in dykes of the Northeast volcanic rift zone of Tenerife, Canary Islands. Chem Geol 334:324–344
Delcamp A, Petronis MS, Troll VR, Carracedo JC, van Wyk de Vries B, Perez-Torrado FJ (2010) Vertical axis rotation of the upper portions of the north-east rift of Tenerife Island inferred from paleomagnetic data. Tectonophysics 492:40–59
Delcamp A, Troll VR, van Wyk de Vries B, Carracedo JC, Petronis MS, Pérez-Torrado FJ, Deegan FM (2012) Stabilisation and destabilization of ocean island rift-zones: the NE-rift of Tenerife, Canary Islands. Bull Volcanol 74:963–980
Dieterich JH (1988) Growth and persistence of Hawaiian volcanic rift zones. J Geophys Res 93:4258–4270
Edgar CJ, Wolff JA, Nichols HJ, Cas RAF, Martí J (2002) A complex Quaternary ignimbrite-forming phonolitic eruption: the Poris member of the Diego Hernández Formation (Tenerife, Canary Islands). J Volcanol Geotherm Res 118:99–130
Fagereng A, Harris C, LaGrange M, Stevens G (2008) Stable isotope study of the Archean rocks of the Vredefort impact structure, central Kaapvaal Craton, South Africa. Contrib Mineral Petrol 155:63–78
Fiske R, Jackson ED (1972) Orientation and growth of Hawaiian volcanic rifts: the effect of regional structure and gravitational stresses. Proc R Soc London 329:299–326
Gaffney AM, Nelson BK, Reisberg L, Eiler J (2005) Oxygen-osmium isotope systematics of West Maui lavas: a record of shallow-level magmatic processes. Earth Planet Sci Lett 239:122–139
Garcia MO, Ito E, Eiler JM, Pietruszka AJ (1998) Crustal contamination of Kilauea volcano magmas revealed by oxygen isotope analyses of glass and olivine from Puu Oo eruption lavas. J Petrol 39:803–817
Geldmacher J, Hoernle K (2000) The 72 Ma geochemical evolution of the Madeira hotspot (eastern North Atlantic): recycling of Paleozoic ($\leq$500 Ma) oceanic lithosphere. Earth Planet Sci Lett 183:73–92
Geldmacher J, Hoernle K (2001) Corrigendum to: 'The 72 Ma geochemical evolution of the Madeira hotspot (eastern North Atlantic): recycling of Paleozoic ($\leq$ 500 Ma) oceanic lithosphere' (Earth Planet Sci Lett 2000 183:73–92). Earth Planet Sci Lett 186:333
Gudmundsson A (2002) Emplacement and arrest of sheets and dykes in central volcanoes. J Volcanol Geotherm Res 116:279–298
Guillou H, Carracedo JC, Pérez Torrado F, Rodríguez Badiola E (1996) K-Ar ages and magnetic stratigraphy of a hotspot-induced, fast grown oceanic island: El Hierro, Canary Islands. J Volcanol Geotherm Res 73:141–155
Guillou H, Carracedo JC, Paris R, Pérez Torrado FJ (2004) K/Ar ages and magnetic stratigraphy of the miocene-pliocene shield volcanoes of tenerife, canary islands: implications for the early evolution of tenerife and the canarian hotspot age progression. Earth Planet Sci Lett 222:599–614
Gurenko AA, Chaussidon M, Schmincke H-U (2001) Magma ascent and contamination beneath one intraplate volcano: evidence from S and O isotopes in glass inclusions and their host clinopyroxenes from Miocene basaltic hyaloclastites southwest of Gran Canaria (Canary Islands). Geochimica et Cosmochimica Acta 65:4359–4374
Gurenko AA, Hoernle KA, Hauff F, Schmincke H-U, Han D, Miura YN, Kaneoka I (2006) Major, trace

element and Nd-Sr-Pb-O-He-Ar isotope signatures of shield stage lavas from the central and western Canary Islands: insights into mantle and crustal processes. Chem Geol 233:75–112

Gurenko AA, Sobolev AV, Hoernlae KA, Hauff F, Schmincke H-U (2009) Enriched, HIMU-type peri dotite and depleted recycled pyroxenite in the canary plume: a mixed-up mantle. Earth Planet Sci Lett 277:514–524

Hansen A (2009) Volcanología y Geomorfología de la Etapa de Rejuvenecimiento Plio-Pleistocena de Gran Canaria (Islas Canarias) Ph.D. thesis, Las Palmas, Gran Canaria, University of Las Palmas

Hansteen TH, Troll VR (2003) Oxygen isotope composition of xenoliths from the oceanic crust and volcanic edifice beneath Gran Canaria (Canary Islands): consequences for crustal contamination of ascending magmas. Chem Geol 193:181–193

Harris C, Smith HS, le Roex AP (2000) Oxygen isotope composition of phenocrysts from Tristan da Cunha and Gough Island lavas: variation with fractional crystallization and evidence for assimilation. Contrib Mineral Petrol 138:164–175

Hart SR (1984) A large scale isotope anomaly in the southern hemisphere mantle. Nature 309:753–757

Hoernle K (1998) Geochemistry of Jurassic ocean crust beneath Gran Canaria (Canary Islands): implications for crustal recycling and assimilation. J Petrol 39:859–880

Hoernle K, Schmincke H-U (1993) The role of partial melting in the 15-Ma geochemical evolution of Gran Canaria: a blob model for the Canary hotspot. J Petrol 34:599–626

Hoernle K, Tilton G, Schmincke H-U (1991) Sr-Nd-Pb isotopic evolution of Gran Canaria: evidence for shallow enriched mantle beneath the Canary Islands. Earth Planet Sci Lett 106:44–63

Irvine TN, Baragar WRA (1971) A guide to the chemical classification of the common volcanic rocks. Can J Earth Sci 8:523–548

Ito E, White WM, Göpel C (1987) The O, Sr, Nd and Pb isotope geochemistry of MORB. Chem Geol 62:157–176

Kent AJR, Norman MD, Hutcheon ID, Stolper EM (1999) Assimilation of seawater-derived components in an oceanic volcano: evidence from matrix glasses and glass inclusions from Loihi Seamount, Hawaii. Chem Geol 156:299–319

Legendre C, Maury RC, Caroff M, Guillou H, Cotton J, Chauvel C, Bollinger C, Hémond C, Guille G, Blais S, Rossi P, Savanier D (2005) Origin of exceptionally abundant phonolites on Ua Pou Island (Marquesas, French Polynesia): Partial melting of basanites followed by crustal contamination. J Petrol 46:1925–1962

Le Maitre RW, Bateman P, Dudek A, Keller J, Lameyre J, Le Bas MJ, Sabine PA, Schmid R, Sorensen H, Streckeisen A, Woolley AR, Zanettin B (1989) A classification of igneous rocks and glossary of terms. Blackwell, Oxford

Longpré M-A, Troll VR, Walter TR, Hansteen TH (2009) Volcanic and geochemical evolution of the Teno massif, Tenerife, Canary Islands: some repercussions of giant landslides on ocean island magmatism. Geochem Geophys Geosy 10:Q12017. doi: 10.1029/2009GC002892

Manconi A, Longpré M-A, Walter TR, Troll VR, Hansteen TH (2009) The effects of flank collapses on volcano plumbing systems. Geology 37:1099–1102

Marcantonio F, Zindler A, Elliott T, Staudigel H (1995) Os isotope systematics of La Palma, Canary Islands: evidence for recycled crust in the mantle source of HIMU ocean islands. Earth Planet Sci Lett 133:197–410

Mathieu L, van Wyk de Vries B, Troll VR, Holohan EP (2008) Dykes, cups, saucers and sills: Analogue experiments on magma intrusion into brittle rocks. Earth Planet Sci Lett 271:1–13

Navarro JM, Farrujia I (1989) Plan Hidrológico Insular de Tenerife: Zonificación Hidrológica, Aspectos Geológicos e Hidrogeológicos: Cabildo Insular de Tenerife, Santa Cruz de Tenerife, p 133

Neumann E-R, Wulff-Pedersen E, Simonsen SL, Pearson NJ, Martí J, Mitjavila J (1999) Evidence for fractional crystallization of periodically refilled magma chambers in Tenerife, Canary Islands. J Petrol 40:1089–1123

Palacz ZA, Wolff JA (1989) Strontium, neodymium and lead isotope characteristics of the Granadilla Pumice, Tenerife: a study of the causes of strontium isotope disequilibrium in felsic pyroclastic deposits. In: Saunders AD, Norry MJ (eds) Magmatism in the ocean basins. Geol Soc Spec Publ 42: 147–159

Roeser HA (1982) Magnetic anomalies in the magnetic quiet zone off Morocco. In: Rad U, Hinz K, Sarnthein M, Seibold E (eds) Geology of the northwest african continental margin. Springer, Berlin, pp 61–68

Sheppard SMF, Harris C (1985) Hydrogen and oxygen isotope geochemistry of Ascension Island lavas and granites: variation with crystal fractionation and interaction with seawater. Contrib Mineral Petrol 91:74–81

Simonsen SL, Neumann E-R, Seim K (2000) Sr-Nd-Pb isotope and trace-element geochemistry evidence for a young HIMU source and assimilation at Tenerife (Canary Island). J Volcanol Geotherm Res 103:299–312

Swanson DA, Duffield WA, Fiske RS (1976) Displacements of the south flank of Kilauea Volcano: the result of forceful intrusions of magma into the rift zones. US Geol Surv Prof Paper 963:39

Szérémèta N, Laj C, Guillou H, Kissel C, Mazaud A, Carracedo JC (1999) Geomagnetic paleosecular variation in the brunhes period, from the island of El Hierro (Canary Islands). Earth Planet Sci Lett 165:241–253

Thirlwall MF (1997) Pb isotopic evidence for OIB derivation from young HIMU mantle. Chem Geol 139:51–74

Thirlwall MF, Singer BS, Marriner GF (2000) ^{39}Ar-^{40}Ar ages and geochemistry of the basaltic shield stage of

Tenerife, Canary Islands, Spain. J Volcanol Geotherm Res 103:247–297
Troll VR (2001) Evolution of large peralkaline silicic magma chambers and associated caldera systems: a case study from Gran Canaria, Canary Islands. Ph. D. thesis Christian-Albrechts-Universität zu Kiel
Troll VR, Schmincke H-U (2002) Alkali-feldspar in compositionally zoned peralkaline rhyolite/trachyte ignimbrite "A", Gran Canaria: implications for magma mixing and crustal recycling. J Petrol 43:243–270
Troll VR, Klügel A, Longpré M-A, Burchardt S, Deegan FM, Carracedo JC, Wiesmaier S, Kueppers U, Dahren B, Blythe LS, Hansteen TH, Freda C, Budd DA, Jolis EM, Jonsson E, Meade FC, Berg SE, Mancini L, Polacci M, Pedroza K (2012) Floating stones off El Hierro, Canary Islands: xenoliths of pre-island sedimentary origin in the early products of the October 2011 eruption. Solid Earth 3:97–110
Vennemann TW, Smith HS (1990) The rate and temperature of reaction of ClF_3 with silicate minerals, and their relevance to oxygen isotope analysis. Chem Geol 86:83–88
Walker GPL (1986) Koolau dike complex, Oahu: Intensity and origin of a sheeted-dike complex high in a Hawaiian volcanic edifice. Geology 14:310–313
Walker GPL (1987) The dike complex of Koolau Volcano, Oahu: internal structure of a Hawaiian Rift Zone. In: Decker RW, Wright RL, Stauffer PJ (eds) Volcanism in Hawaii, vol 2. US Geol Surv Prof Pap 1350:961–993
Walker GPL (1992) Coherent intrusion complexes in large basaltic volcanoes; a new structural model. In: Harris PG, Cox KG, Baker PE (eds) Essays on magmas and other earth fluids, vol 50. Elsevier, The Netherland, pp 41–54
Walter TR, Schmincke H-U (2002) Rifting, recurrent landsliding and Miocene structural reorganization on NW-Tenerife (Canary Islands). Int J Earth Sci 91:615–628
Walter TR, Troll VR (2003) Experiments on rift zone evolution in unstable volcanic edifices. J Volcanol Geotherm Res 127:20–107
Walter TR, Troll VR, Cailleau B, Belousov A, Schmincke H-U, Amelung F, v.d.Bogaard P (2005) Rift zone reorganisation through flank instability in ocean island volcanoes: an example from Tenerife, Canary Islands. Bull Volcanol 67:281–291
Wiesmaier S (2010) Magmatic differentiation and bimodality in oceanic island settings—implications for the petrogenesis of magma in Tenerife, Spain. PhD thesis, University of Dublin, Trinity College
Wiesmaier S, Troll VR, Carracedo JC, Ellam RM, Bindeman I, Wolff JA (2012) Bimodality of lavas in the Teide–Pico Viejo succession in Tenerife—the role of crustal melting in the origin of recent phonolites. J Petrol 53:2465–2495
Wolff JA, Grandy JS, Larson PB (2000) Interaction of mantle-derived magma with island crust? Trace element and oxygen isotope data from the Diego Hernandez Formation, Las Cañadas, Tenerife. J Volcanol Geotherm Res 130:343–366

Dating the Teide Volcanic Complex: Radiometric and Palaeomagnetic Methods

6

Hervé Guillou, Catherine Kissel, Carlo Laj, and Juan Carlos Carracedo

Abstract

This chapter describes the integration of radiometric dating and palaeomagnetic investigations to decipher the spatial and temporal evolution of volcanic edifices of Tenerife Island. Accurate ages are crucial to reconstruct the recent eruptive history of Tenerife (specifically Teide Volcano and the North West and North East Rift Zones). Samples have been dated using both the K–Ar and the $^{40}Ar/^{39}Ar$ method in order to assess the reliability of the ages obtained. When the two methods yielded similar results and precision, accurate pooled ages were calculated. The correlation of these ages with the geomagnetic polarity of the lavas (referred to as the geomagnetic and astronomical polarity time scales) has recently been successfully applied to establish the magnetic stratigraphy of volcanoes in the Canary Islands and to constrain the main geological units. Moreover, this well-constrained and high resolution geochronological framework is of prime interest to track and study geomagnetic reversals and excursions. As an example, results are presented from three lava flows in Tenerife from the Mono Lake geomagnetic excursion, the youngest in the documented geological record.

6.1 Introduction

Progress in the study of volcanic areas has been, in many aspects, parallel to the development of methods for dating volcanic rocks. Since the pioneering work of McDougall (1963, 1964) in the Hawaiian Islands, radiometric ages (K/Ar and $^{40}Ar/^{39}Ar$) have been determined for many volcanic regions, particularly the Hawaiian and Canary Islands (Fig. 6.1). Consequently, the geochronology, stratigraphy and volcanic history of these archipelagos are presently among the best known in the world.

H. Guillou (✉) · C. Kissel · C. Laj
Laboratoire des Sciences du Climat et de L'Environnement/IPSL (CEA-CNRS-UVSQ), Gif sur Yvette, France

J. C. Carracedo
Departamento de Física (GEOVOL), Universidad de Las Palmas de Gran Canaria, Las Palmas de Gran Canaria, Spain

J. C. Carracedo and V. R. Troll (eds.), *Teide Volcano*, Active Volcanoes of the World,
DOI: 10.1007/978-3-642-25893-0_6, © Springer-Verlag Berlin Heidelberg 2013

Before 2007, although several authors attempted the geological study of the Teide Volcanic Complex (TVC), only limited progress had been made on the reconstruction of the latest (postcaldera) eruptive history of Tenerife. This was mainly due to the lack of geochronological information, restricted at that time to a single age by Ablay et al. (1995) from Montaña Blanca (Fig. 6.2). Without abundant radiometric ages, it was in fact very difficult to distinguish the recent volcanic formations from one another.

Until recently, researchers stated that a more precise reconstruction of the recent eruptive period of Tenerife (Teide Volcano and the North West and North East Rift Zones) was not achievable because of the inapplicability of radiometric dating techniques in this geological context (Araña et al. 2000). The reason invoked was that lavas were too young for K/Ar and $^{40}Ar/^{39}Ar$ dating, and that suitable organic material (charcoal) for radiocarbon dating was absent.

However, since the first K/Ar ages were obtained from the Canary Islands by Abdel Monem et al. (1971, 1972), and in particular during the last decade, considerable efforts have been made to extend the K/Ar and $^{40}Ar/^{39}Ar$ chronology towards younger and younger ages. Based on these techniques, a recent intensive dating program was implemented and abundant K/Ar and $^{40}Ar/^{39}Ar$ ages are now available from lavas of the different Canary Islands.

As far as radiocarbon ages are concerned, in contrast to the Hawaiian Islands, where abundant radiocarbon ages have been determined from modern (present-day carbon) to lavas older than 38 ka (Rubin et al. 1987), radiocarbon dating in the Canaries has, for a long time, been mainly conducted as part of archaeological research, and as such determined on organic remains (shells, roots, etc.).

Only a few dates were obtained from charcoals derived from Canarian lava flows (e.g., Pellicer 1977; Ablay et al. 1995) prior to the works of Guillou et al. (1998) in La Palma, Carracedo et al. (2007) in Tenerife, Rodriguez-Gonzalez et al. (2009) in Gran Canaria, and Pérez Torrado et al. (2011) in El Hierro. These new radiocarbon dates have considerably improved the reconstruction of the recent volcanic history of the central and western Canaries.

Fig. 6.1 Sampling one of the phonolite flows of Teide inside the Orotava Valley that gave radiometric ages and palaeomagnetic data corresponding to the Mono Lake excursion. Pooled K/Ar ages and $^{40}Ar/^{39}Ar$ plateau ages produce the best age estimates for these flows and proved to be very successful to date key events such as geomagnetic field excursions. Conversely, comparison of radiometric and palaeomagnetic data ensures the reliability of the dating methods used to reconstruct the volcanic stratigraphy and history of the Teide Volcanic Complex

Finally, a combination of palaeomagnetic results and radiometric dating on a number of lavas that have recorded characteristic changes in the Earth's magnetic field (e.g., geomagnetic reversals and excursions) was employed to test the precision of radiometric ages because the duration of some of these geomagnetic reversal events is shorter than the intrinsic error of radiometric dating. These investigations focused both on methodological (testing the instrumental capability to measure increasingly lower percentages of $^{40}Ar^*$) and geological objectives (refining the geochronology and volcanic stratigraphy and reconstructing the volcanic history of the TVC). In turn, decoupled from the history of the Teide volcano but of great interest for the scientific community, the combined palaeomagnetic and radiometric investigations of geomagnetic instabilities were used to constrain tie points in the magnetostratigraphic time scale used in sediments, in particular for palaeoclimatic applications.

This chapter describes attempts to extend the bracketed time range for which these dating methods are currently applied towards younger

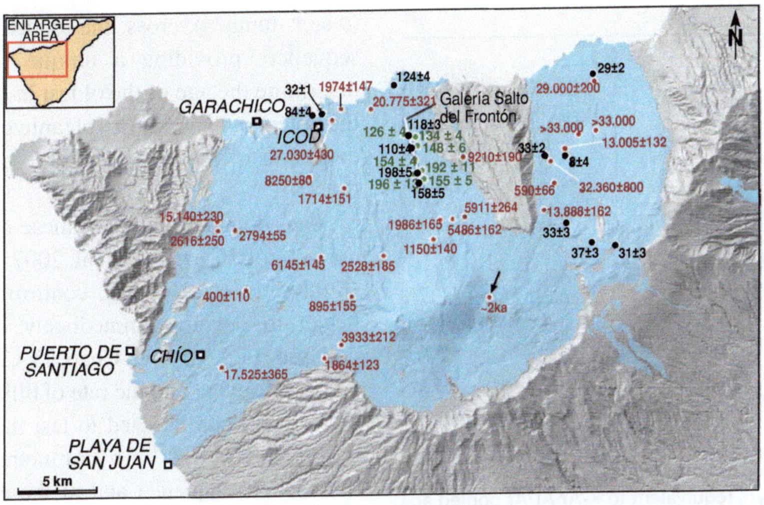

Fig. 6.2 Map of the TVC showing published radiometric ages (Carracedo et al. 2007). Before this work, the only chronological data reported for the TVC was the ~2 ka age dating Montaña Blanca (Ablay et al. 1995), marked with an arrow. Black circles: K/Ar ages, in ka; red circles: ^{14}C ages, in calibrated year B.P.; green circles: ^{40}Ar/^{39}Ar, in ka

ages using lavas from TVC, comparing the results obtained from three different methods (^{14}C, K/Ar and ^{40}Ar/^{39}Ar) and combining them with palaeomagnetic investigations for some past geomagnetic instabilities.

6.2 Testing Dating Methods in the TVC

The suitability of the K/Ar and ^{40}Ar/^{39}Ar dating methods on increasingly younger ages was tested on two phonolitic lava flows from the TVC, one inside the Orotava Valley (CITF-98) and the other from Playa San Marcos (CITF-301) (see Figs. 8.15 and 8.20), corresponding to the latest stage of evolution of Teide. The results show a remarkable agreement between the two dating methods. Unspiked K/Ar analysis (Charbit et al. 1998) at the Laboratoire des Sciences du Climat et de L'Environnement (LSCE) gave ages of 33.1 ± 1.8 and 31.6 ± 1.9 ka, respectively. Samples with equivalent groundmass to the K/Ar experiments were analysed at the new ^{40}Ar/^{39}Ar facility developed at the LSCE. The ^{40}Ar/^{39}Ar ages obtained for the two samples, calculated from three independent experiments, are 32.4 ± 1.8 and 31.4 ± 1.7 ka (2σ) (Guillou et al. 2011). Within error, the reported ^{40}Ar/^{39}Ar ages agree with the K/Ar ages, and are of similar precision. Therefore, this study demonstrates that precise ages can be obtained from young volcanic rocks using the new ^{40}Ar/^{39}Ar method and confirms the accuracy and precision of the K/Ar unspiked method to date the TVC.

This approach is also relevant to check the accuracy of ^{14}C ages used to date the TVC. A charcoal sample, suitable for ^{14}C dating, from within the basal scoriae of phonolitic flow CITF-98 gave a precise pooled K/Ar and ^{40}Ar/^{39}Ar age of 32.2 ± 1.2 ka. Using this date and the available calibration curve INTCAL09 14C (Reimer et al. 2009) a ^{14}C age of about 28.2 ka for this flow would be expected (Fig. 6.3). This age is approximately 4 ka younger than the radiocarbon age of 32.36 ± 800 years BP (2σ) which was actually obtained from the charcoal using the AMS technique (Carracedo et al. 2007). The K/Ar and ^{40}Ar/^{39}Ar age of this sample at 32.2 ± 1.2 ka is retained as the reliable calendar age for this flow.

There are two main interpretations for the discrepancy between the ^{14}C and K/Ar clock derived ages. The first one would be to question the radiocarbon calibration. Given the stringent criteria adopted these days to update the

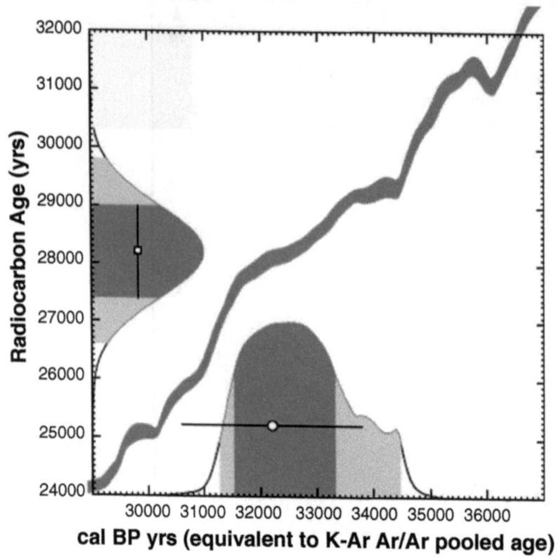

Fig. 6.3 Radiocarbon Age vs. Calibrated Age diagram established using the Radiocarbon calibration program: CALIB REV6.0.0. (Copyright M. Stuiver and P.J. Reimer). White circle: Pooled K/Ar $^{40}Ar/^{39}Ar$ age used as reference to recalculate; *white square* radiocarbon age (from Guillou et al. 2011)

calibration curve between 0 and 50,000 years, we discard this first hypothesis. Errors in ^{14}C dates and their possible sources were already documented over 30 years ago and are evident in several volcanic areas such as the Eifel (Bruns et al. 1980), the provinces of Grosseto and Siena (Saupé et al. 1980), and in the Azores (Pasquier-Cardin et al. 1999). In these areas, significant to large ^{14}C depletions may occur in many plants, due to assimilation of ^{14}C endogenous CO_2, the consequence of which, as demonstrated by the study of modern plants, is an error in excess of the radiocarbon ages that can reach some ka. We suggest that the apparent older radiocarbon age discussed here results from the fact that the analysed charcoal probably derived from a plant which grew close to active volcanic fumaroles, which are sources emitting ^{14}C-free CO_2.

6.3 Dating Old Teide

The oldest sequences of the TVC, outcropping in the northern coastal cliffs, have been dated between 84 ± 4 and 124 ± 4 ka (see Fig. 6.2; Carracedo et al. 2007). However, several *galerías* (water tunnels) cross the entire post-collapse sequence, providing a unique opportunity to determine the age of the oldest lava sequences of the volcano, the rates of volcanic growth, and the evolution of magmas along the construction of the TVC.

Ages obtained in two of these *galerías*—Salto del Frontón (Carracedo et al. 2007) and La Gotera (Boulesteix et al. 2012), confirm that the TVC began to develop immediately after the Icod collapse (Fig. 6.4) (see Chap. 7). Constraining the age of this event and the rate of filling the collapse embayment can be used to test the suitability of unspiked K/Ar and $^{40}Ar/^{39}Ar$ methods in this type of volcanic sequence and at these emplacement conditions. In both *galerías*, the K/Ar ages obtained for the earliest post-collapse lavas from two different laboratories and using the same unspiked method (161 ± 5 and 158 ± 5 ka) are indistinguishable within the range of analytical error. However, a flow in the middle of the *galería* Salto del Frontón gave considerably older ages (about 40 ka older), with consistent and indistinguishable results from the two different laboratories, again both using the unspiked K/Ar and $^{40}Ar/^{39}Ar$ techniques (Fig. 6.4a).

A plausible explanation for these apparently contradictory ages may lie in the different depositional contexts along the *galería*. A homogeneous filling of the entire collapse embayment is very unlikely. In fact, post-collapse instability, particularly in the walls of the embayment, and vigorous dissection by erosion of the relatively soft debris result in a very irregular and changing topography in the embayment to which the successive flows have to adjust, lava flows tending to follow previous incisions (see Chap. 3). Lateral changes, even over very short distances, are common, and it is improbable that the oldest possible lava will be identified, to give a true age of the collapse (Fig. 6.4b).

A second potential explanation concerns post-collapse effusive activity, which resumed immediately with very high eruptive rates and frequencies, as indicated by the lack of any sign of discontinuity (interbedded pyroclasts, soils, dykes, etc.). These conditions of intensive volcanism may account for the hydrothermal

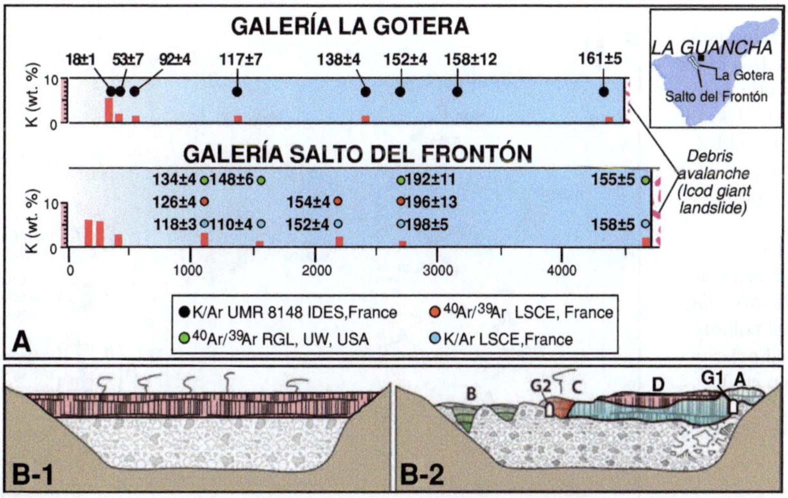

Fig. 6.4 **a** K/Ar and $^{40}Ar/^{39}Ar$ ages obtained by different laboratories from a sequence of lavas in a *galería* on the northern flank of Teide Volcano. The sequence comprises the bulk of the growth of the TVC, from the debris avalanche associated with the ~200 ka giant landslide to the late peripheral phonolitic lava domes. Ages obtained at the end of the *galería* consistently give younger values than the equally consistent values from lavas near the central part of the *galería*. The discrepancy may be explained by hydrothermal alteration of the initial post-collapse sequence and very rapid growth at the early stage of filling of the collapse embayment (based on Carracedo et al. 2007 and Boulesteix et al. 2012). **b** Different conceptions of volcanic filling of the collapse embayment: *B-1* Assuming that the fill of the collapse scar is homogeneous (Boulesteix et al. 2012). *B-2* Post-collapse vigorous dissection by erosion results in a very irregular and changing topography in the embayment to which the successive flows have to adjust (Carracedo et al. 2007). *Galerías* G1 and G2 show the contact of lava flows with the debris avalanche deposits, but the age of the lavas can be very different. Because of erosion and the successive eruptive events, the "minimum" age for the collapse given by G2 would be "younger" than the one provided by G1. The first model (*B-1*) seems geologically unrealistic

alteration observed in the lower part of the lava sequence, at the end of the *galería*. Alteration and exposure to high temperatures (through reheating) can allow radiogenic argon ($^{40}Ar^*$) to be released, causing the calculated K/Ar age to become younger than the "true" age of the dated lava. Similarly, reheating events and diffusion of argon can result in lower $^{40}Ar/^{39}Ar$ ages. Therefore, the internally consistent ages of about 190 ka obtained in the middle of the *galería*, from lavas free of any sign of alteration or reheating (erupted at significantly lower rates and frequencies), may represent a minimum age for the onset of volcanism after the Icod giant landslide.

6.4 Geomagnetic Instabilities in Volcanic Formations of the TVC and the NERZ: Dynamics of the Volcanic Edifices, Mapping and Correlation and Chronological Tie Points

6.4.1 Geomagnetic Reversals

Geomagnetic reversals have been studied to date and correlate volcanic formations in the Canaries since the early 1970s (Carracedo 1975, 1979; Guillou et al. 1996, 2001, 2004a; Carracedo

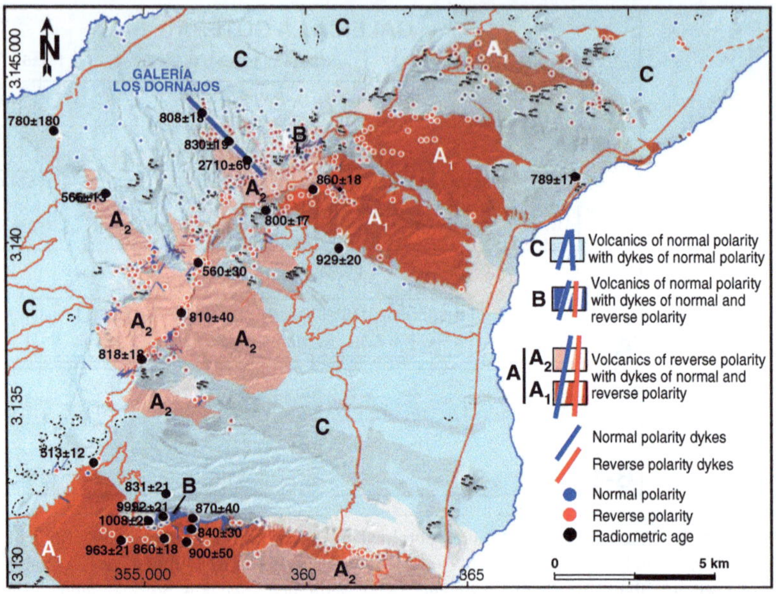

Fig. 6.5 Main magnetostratigraphic units defined as a function of the polarity of 415 oriented cores of lavas and dykes in the NE Rift Zone NERZ. These units have proven to be extremely useful in correlating and reconstructing the successive eruptions that have formed the rift. *Blue* indicates normal polarity, and *red* reversed polarity (from Carracedo et al. 2011)

et al. 2001, 2011). Combined use of palaeomagnetic and isotopic dating (geomagnetic inversions and radiometric dating) also helps to evaluate the reliability of K/Ar ages (Guillou et al. 2004b). This combination of dating techniques (K/Ar and $^{40}Ar/^{39}Ar$) was therefore used to identify and define palaeomagnetic events in the Canaries (Guillou et al. 1996; Singer et al. 2002).

The comparison of the ages and geomagnetic polarity of the lavas with the geomagnetic and astronomical polarity time scales (GPTS and APTS) has been successfully applied to establish the magnetic stratigraphy of volcanoes in the Canary Islands, the NERZ being a good example of this (Fig. 6.5; Carracedo et al. 2011). This combined stratigraphic, isotopic, and palaeomagnetic work has been conducted not only on outcrops (e.g., the Pared de Güímar section, Fig. 6.6), but also in *galerías* from the NERZ, where significant stratigraphic discordances allowing the recognition of a northbound lateral collapse were recognised (Fig. 6.7). These collapses would have been otherwise unnoticeable since post-collapse volcanism filled this basin, extending beyond the coastline to conceal the scar and the avalanche breccia. This approach of integrating geomagnetic reversals and radiometric dating significantly contributed to outlining the spatial and temporal evolution of the NE Rift Zone of Tenerife (NERZ), especially the duration of its main stages of growth and the timing of the catastrophic lateral collapses forming the Valleys of La Orotava and Güímar.

Similarly to many other rift zones, the NERZ evolved very rapidly with high eruption rates, which apparently persisted between 1 Ma and 840 ka. Effusive rates up to 2.5 km^3/k.y. and volcanic growth of 3.5–4 m/k.y have been measured (Carracedo et al. 2011). Such rapid growth, implying high frequency of lava flow emission, allows detailed recording of changes in the geomagnetic field. In fact, the ages obtained for the NERZ activity indicate that this place is probably the most favourable setting to identify a significant part of the normal Jaramillo subchron. Particularly suitable for this purpose is the southern wall of Valley de Güímar (Fig. 6.6a), a 500 m-thick sequence of basaltic flows, and the scar of a $\sim$47 km^3, 10 $\times$ 10 km lateral collapse that occurred between 860 and 830 ka ago (respectively the age of the top of the collapse wall and of early lavas filling the embayment; Carracedo et al. 2011). Radiometric (K/Ar) ages constrain the upper part of the sequence between

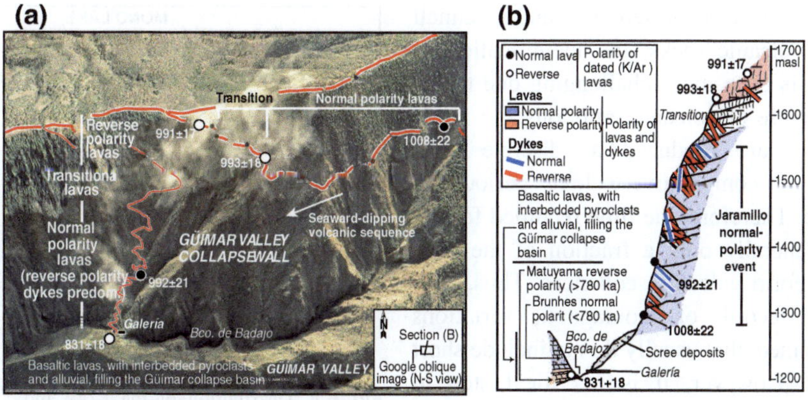

Fig. 6.6 Oblique view (Google Earth) from the north of the southern wall of the Güímar giant landslide scar (the Pared de Güímar). **a** Ages dating the pre- and postcollapse formation. **b** Stratigraphy and magnetostratigraphy of the Pared de Güímar (from Carracedo et al. 2011)

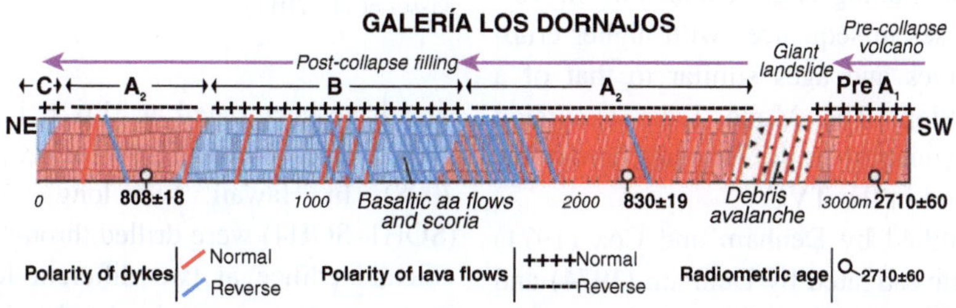

Fig. 6.7 Geomagnetic polarity of the volcanic formations found in the Los Dornajos *galería*, on the north flank of the NERZ. The magnetostratigraphic units shown in Fig. 6.5 are crossed in this *galería*. A debris avalanche deposit and an older formation intersected at the end of the *galería* represent a giant landslide and the pre-collapse deep structure of the rift zone, not visible at the surface (from Carracedo et al. 2011)

1008 ± 22 and 860 ± 18 ka, yielding an eruptive growth of about 3.7 m/ky (equivalent to 2–3 flows/ky). The polarity of this lava sequence shifts from normal polarity at the lower part of the sequence (dated at 1008 ± 22 ka) to reverse polarity at the top (dated at 860 ± 18 ka), indicating that the lower part of the sequence with normal polarity corresponds to the Jaramillo subchron (987–1052 ka; Singer et al. 2004), and the upper part, of reverse polarity, to the Matuyama chron (Fig. 6.6b).

6.4.2 The Mono Lake Excursion

Geomagnetic excursions have attracted increasing interest in the scientific community because, due to their short time constant, they may be used as precise time markers in various geological records. Excursions are relatively brief geomagnetic instabilities characterised by a decrease in intensity associated with large directional shifts from the dipolar field direction (larger than secular variation), immediately followed by a return to the pre-excursional state (see review in Laj and Channell 2007). Geomagnetic excursions are difficult to identify in geological sections. In sediments, palaeomagnetists have recently accessed a number of very suitable marine sequences thanks to palaeoclimatic interest in high sediment accumulation areas coupled with newly developed marine coring facilities. This has greatly increased the number of high-resolution reconstructions of past geomagnetic field changes in which excursions can be identified (Laj and Channell 2007). In volcanic rocks, the identification of excursions is even more challenging due to the sporadic nature of volcanic eruptions.

Eruptions are produced as discrete units separated by comparatively long periods of quiescence. Therefore, the time recorded for the flows is generally only a fraction of the time elapsed to form volcanic sequences. Thus, only incomplete records of geomagnetic variations can be obtained, that usually do not include short excursions. However, it is a critical step to achieve even this as lavas are the only geological archive yielding the absolute dating and absolute palaeointensity data used to characterise the excursional earth magnetic field. The probability of finding short excursions in volcanics increases in sequences with higher eruptive frequencies and ages similar to that of a given excursion. The Mono Lake excursion (MLE), the youngest in the Brunhes chron, has been identified in the TVC.

First identified by Denham and Cox (1971) and further investigated by Denham (1974) and Liddicoat and Coe (1979) at Mono Lake, western USA, the MLE has been initially dated at about 24–25 ka B.P. based on ^{14}C ages on ostracods, ash layers and the assumption of uniform sedimentation rates. Following this pioneering study, other data were obtained from various continental sections (lacustrine and loess deposits) suggesting the global character of the MLE. However, their ages were scattered and imprecise, mainly resulting from difficulties in obtaining reliable ages from continental sedimentary sections (e.g., Kissel et al. 2011). Unfortunately, the most recent $^{40}Ar/^{39}Ar$ investigation of the interbedded ash layers in the Mono Lake type section did not allow the problem to be resolved because of the absence of juvenile eruptive crystals (Cassata et al. 2010).

New absolute age data were therefore needed in order to anchor this excursion to the geomagnetic instability time scale (GITS, Singer et al. 2002). Volcanic records of the MLE were until very recently limited to two volcanic provinces, New Zealand and Hawaii. In the Auckland volcanic field (New Zealand), three

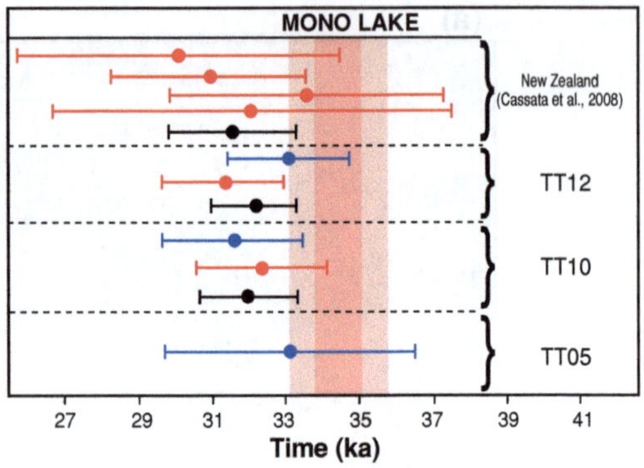

Fig. 6.8 Distribution of the K/Ar and $^{40}Ar/^{39}Ar$ ages obtained from volcanic rocks recording the Mono Lake excursion. *Blue* dots represent K/Ar determinations, *red* dots $^{40}Ar/^{39}Ar$ determinations and *black dots* are for pooled ages (all at 2σ uncertainty relative to ACRs at 1.193 Ma equivalent to FCs at 28.02 Ma). Modified from Kissel et al. (2011)

sites have been dated at 31.6 ± 1.8 ka (plateau pooled age) using $^{40}Ar/^{39}Ar$ (Cassata et al. 2008). In Hawaii, two long volcanic cores (SOH1–SOH4) were drilled through the Kilauea volcano edifice at two different locations, but could not be accurately dated using $^{40}Ar/^{39}Ar$ and/or K/Ar due to the very low radiogenic argon contents of the lavas.

Lavas of Teide Volcano, erupted during the narrow time interval covered by this excursion (Fig. 6.8), provide an opportunity to better refine the age and confirm the global distribution of the MLE (Kissel et al. 2011). Palaeomagnetic data were analysed from three lava flows with appropriate K/Ar ages. All the palaeomagnetic directions are significantly anomalous with respect to the geocentred axial dipole (GAD) field direction at this location ($D = 0°$; $I = 47.2°$). Inclinations for two sites TT-05 and TT-12 are shallower than the GAD value ($36.9° \pm 4.6°$ and $22.5° \pm 4.5°$ respectively) while the third site (TT-10) has an inclination similar to the GAD value but the declination is strongly deviated at 67.9° ($\alpha_{95} = 4.7°$) (Kissel et al. 2011).

These directional deviations from the GAD field are associated with low intensity values determined using the Thellier and Thellier (1959) procedure and PICRIT-03 set of criteria (see Kissel et al. 2011 for details). The average

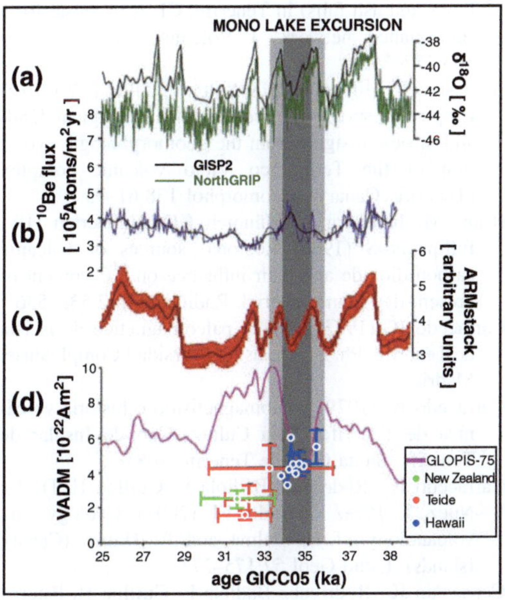

Fig. 6.9 Climatic and geomagnetic records at the time of the Mono Lake excursion. **a** Oxygen isotope records from Greenland ice with GISP2 in *black* (Grootes et al. 1993) and NGRIP in *green* (North Greenland Ice Core Project Members 2004) illustrating changes in atmospheric temperature over Greenland at the time of the Mono Lake geomagnetic excursion. **b** ^{10}Be flux record initially obtained from GRIP core (Muscheler et al. 2004). **c** Sedimentary drift deposit stack as a tracer for changes in the deep circulation strength in the North Atlantic showing similar characteristics and age to those of oxygen isotopes in ice (Kissel et al. 2008). **d** GLOPIS-75 continuous record in *pink* (Laj et al. 2004) and volcanic data reported as VADM values with ages as in Fig. 6.2 (modified from Kissel et al. 2011)

intensities obtained for the three flows are 21.4 ± 6.6 μT (TT-05), 7.8 ± 1.4 μT (TT-10) and 12.4 ± 3.9 μT (TT-12) corresponding to the virtual axial dipole moment (VADM) values of 4.3 ± 1.3 × 1022 Am_2; 1.6 ± 0.3 × 1022 Am_2; 2.5 ± 0.8 × 1022 Am_2 respectively, significantly lower than the present value of 8 × 1022 Am_2.

While for TT10, the angular difference with the directions expected for a GAD field (43.3°) is large, it falls in the range of secular variation (11° and 24.6°, respectively) for TT-05 and TT-12. However, when the intensity of the field prevailing at the moment of emplacement is considered, then even the last two sites cannot be considered as reflecting the usual secular variation. It therefore appears that these flows have recorded an excursion of the geomagnetic field.

The three lavas had been dated using the K/Ar method before the palaeomagnetic sampling. While $^{40}Ar/^{39}Ar$ dating was not attempted on sample TT-05 because of low radiogenic ^{40}Ar ($^{40}Ar^*$) content, the two other lava flows (TT-10, TT-12) were dated using both unspiked K/Ar and $^{40}Ar/^{39}Ar$ methods. As shown above, the two methods yielded similar precision and allowed accurate pooled ages to be obtained (samples TT-10 and TT-12 are labelled CITF-301 and CITF-98 respectively in Carracedo et al. (2007) and Guillou et al. (2011).

Given their radiometric age, this is clearly identified as the Mono Lake Excursion, the only one in this age range. The three ages from the Teide flows largely overlap with those from New Zealand (Cassata et al. 2008), confirming the brief duration of the excursion, which is shorter than the uncertainty associated with the radiometric ages. Such a brief duration has already been proposed and evaluated at about 1,500 years for the Laschamp excursion (Laj et al. 2000, 2004).

The new results reported in Fig. 6.9, together with those from the other volcanic localities and from other archives, suggest that the magnetic field intensity during the MLE may have been more reduced than previously believed. At TT-10, it is about 20 % of the present-day field, which is half of the value measured in New Zealand and in Hawaii.

Figure 6.9 shows, in addition, palaeomagnetic and climatic records from different geological archives versus the most recent Greenland ice core age model (GICC05 age model, Andersen et al. 2006). The ^{10}Be peak, around 34 kyr (width at mid-height of about 0.6–0.8 kyr), resulting from the weakening in the Earth's magnetic field intensity at the MLE is significant with respect to the background curve, and it is coeval with the rapid cold stadial within the millennial climatic fluctuation 7 (Dansgaard-Oeschger cycle). The age of about 34 kyr for this fluctuation is defined in the Greenland ice core by annual layer counting.

When integrated into this large dataset, it appears that although the radiometric ages from Tenerife and New Zealand are not statistically different from the ages from ice cores, the mean radiometric age for each formation is systematically younger than the stratigraphic age from the ice cores. However, given the challenge that constitutes obtaining accurate K/Ar and Ar/Ar ages from such young lava flows, this study is extremely encouraging for discovering additional volcanic lavas in which both palaeomagnetic and dating approaches can be combined in order to constrain the age of the MLE.

These new data are, with those from Hawaii, the only two volcanic records of the Mono Lake from the northern hemisphere. This study, together with those conducted in New Zealand and Hawaii, is very promising because they show that significant information can be retrieved from geological sequences recording very brief geomagnetic features and which, therefore, constitute precise tie points for stratigraphic correlations. On the other hand, this is also a test of the precision and accuracy of the radiometric dating methods used to date the Teide Volcanic Complex (Carracedo et al. 2007).

References

Abdel-Monem A, Watkins ND, Gast PW (1971) Potassium-argon ages, volcanic stratigraphy and geomagnetic polarity history of the Canary Islands: Lanzarote, Fuerteventura, Gran Canaria, and La Gomera. Am J Sci 271:490–521

Abdel-Monem A, Watkins ND, Gast PW (1972) Potassium -argon ages, volcanic stratigraphy and geomagnetic polarity history of the Canary Islands: Tenerife, La Palma and Hierro. Am J Sci 272:805–825

Ablay GJ, Ernst GGJ, Marti J, Sparks RSJ (1995) The 2 ka subplinian eruption of Mña. Blanca, Tenerife. Bull Volcanol 57:337–355

Andersen KK, Svensson A, Johnsen SJ, Rasmussen SU, Bigler M, Röthlisberger R, Ruth U, Siggaard-Andersen ML, Steffensen JP, Dahl-Jensen D, Vinther BM, Clausen HB (2006) The greenland ice core chronology 2005, 15–42 ka. Part 1: constructing the time scale. Quatern Sci Rev 25:3246–3257

Araña V, Felpeto A, Astiz M, García A, Ortiz R, Abella R (2000) Zonation of the main volcanic hazards (lava flows and ash falls) in Tenerife, CI. A proposal for a surveillance network. J Volcanol Geotherm Res 103:377–391

Boulesteix T, Hildenbrand A, Soler V, Gillot P-Y (2012) Eruptive response of oceanic islands to giant landslides: new insights from the geomorphologic evolution of the Teide-Pico Viejo volcanic complex (Tenerife, Canary). Geomorphol 138:61–73

Bruns M, Ingeborg L, Münnich KO, Hubberten HW, Fillipakis S (1980) Regional sources of volcanic carbon dioxide and their influence on ^{14}C content of present-day plant material. Radiocarbon 2:532–536

Carracedo JC (1975) Estudio paleomagnético de la isla de Tenerife. Ph. D. Thesis, Universidad Complutense, Madrid

Carracedo JC (1979) Paleomagnetismo e historia volcánica de Tenerife. Aula Cultura Cabildo Insular de Tenerife, Santa Cruz de Tenerife, p 81

Carracedo JC, Rodríguez Badiola E, Guillou H, De La Nuez J, Pérez Torrado FJ (2001) Geology and volcanology of La Palma and El Hierro (Canary Islands). Estud Geol 57:175–273

Carracedo JC, Rodríguez Badiola E, Guillou H, Paterne M, Scaillet S, Pérez Torrado FJ, Paris R, Fra-Paleo U, Hansen A (2007) Eruptive and structural history of Teide volcano and rift zones of Tenerife, Canary Islands. Geol Soc Am Bull 119:1027–1051

Carracedo JC, Guillou H, Nomade S, Rodríguez-Badiola E, Pérez-Torrado FJ, Rodríguez-González A, Paris R, Troll VR, Wiesmaier S, Delcamp A, Fernández-Turiel JL (2011) Evolution of ocean island rifts: the Northeast rift zone of Tenerife, Canary Islands. Geol Soc Am Bull 123:562–584

Cassata WS, Singer BS, Cassidy J (2008) Laschamp and Mono Lake geomagnetic excursions recorded in New Zealand. Earth Planet Sci Lett 268:76–88

Cassata WS, Singer BS, Liddicoat JC, Coe RS (2010) Reconciling discrepant chronologies for the geomagnetic excursion in the Mono Basin, California: insights from new $^{40}Ar/^{39}Ar$ dating experiments and a revised relative paleointensity correlation. Quatern Geochronology 5:533–543

Charbit S, Guillou H, Turpin L (1998) Cross calibration of K-Ar standard minerals using an unspiked Ar measurement technique. Chem Geol 150:147–159

Denham CR (1974) Counter-clockwise motion of palaeomagnetic directions 24,000 years ago at Mono Lake, California. J Geomag Geoelec 26:487–498

Denham CR, Cox A (1971) Evidence that the Laschamp polarity event did not occur 13,300–34,000 years ago. Earth Planet Sci Lett 13:181–190

Grootes PM, Stuiver M, White JWC, Johnsen S, Jouzel J (1993) Comparison of oxygen isotope records from the GISP2 and GRIP Greenland ice cores. Nat 366:552–554

Guillou H, Carracedo JC, Pérez Torrado F, Rodríguez Badiola E (1996) K-Ar ages and magnetic stratigraphy of a hotspot-induced, fast grown oceanic island: El Hierro, Canary Islands. J Volcanol Geotherm Res 73:141–155

Guillou H, Carracedo JC, Day SJ (1998) Dating of the upper Pleistocene–Holocene volcanic activity of La Palma using the unspiked K-Ar technique. J Volcanol Geotherm Res 86:137–149

Guillou H, Carracedo JC, Duncan R (2001) K-Ar, $^{40}Ar/^{39}Ar$ Ages and magnetostratigraphy of Brunhes and Matuyama lava sequences from La Palma Island. J Volcanol Geotherm Res 106:175–194

Guillou H, Carracedo JC, Paris R, Pérez Torrado FJ (2004a) K/Ar ages and magnetic stratigraphy of the Miocene-Pliocene shield volcanoes of Tenerife, Canary Islands: Implications for the early evolution of Tenerife and the Canarian hotspot age progression. Earth Planet Sci Lett 222:599–614

Guillou H, Pérez Torrado FJ, Hansen Machin AR, Carracedo JC, Gimeno D (2004b) The Plio-Quaternary volcanic evolution of Gran Canaria based on new K-Ar ages and magnetostratigraphy. J Volcanol Geotherm Res 135:221–246

Guillou H, Nomade S, Carracedo JC, Kissel C, Laj C, Wandres C (2011) Effectiveness of combined unspiked K-Ar and $^{40}Ar/^{39}Ar$ dating methods in the ^{14}C age range. Quat Geochronol 6:530–538

Kissel C, Laj C, Piotrowski AM, Goldstein SL, Hemming SR (2008) Millenialscale propagation of Atlantic deep waters to the glacial southern ocean. Paleoceanogr 23:PA2102

Kissel C, Guillou H, Laj C, Carracedo JC, Nomade S, Perez-Torrado F, Wandres C (2011) The Mono Lake excursion recorded in phonolitic lavas from Tenerife (Canary Islands): palaeomagnetic analyses and coupled K/Ar and Ar/Ar dating. Phys Earth Planet Inter 187:232–244

Laj C, Channell JET (2007) Geomagnetic excursions. In: Kono M (ed), Treatise in geophysics. Geomagnetism, Encyclopedia of Geophysics, pp 373–416

Laj C, Kissel C, Mazaud A, Channell JET, Beer J (2000) North Atlantic palaeointensity stack since 75 ka (NAPIS-75) and the duration of the Laschamp event. Philos. Trans. R. Soc. London, Ser A 358:1009–1025

Laj C, Kissel C, Beer J (2004) High Resolution Global Paleointensity Stack since 75 kyrs (GLOPIS-75) calibrated to absolute values. AGU Monograph, ''Timescales of the Geomagnetic Field'' 145:255–265

Liddicoat JC, Coe RS (1979) Mono Lake geomagnetic excursion. J Geophys Res 84:261–271

McDougall I (1963) Potassium-argon ages from western Oahu, Hawaii. Nat 197:344–345

McDougall I (1964) Potassium-argon ages from lavas of the Hawaiian Islands. Geol Soc Am Bull 75:107–128

Muscheler R, Beer J, Wagner G, Laj C, Kissel C, Raisbeck GM, Yiou F, Kubik PW (2004) Changes in the carbon cycle during the last deglaciation as indicated by the comparison of ^{10}Be and ^{14}C records. Earth Planet Sci Lett 219:325–340

North Greenland Ice Core Project Members (2004) High resolution climate record of the northern hemisphere reaching into the last interglacial period. Nat 431:147–151

Pasquier-Cardin A, Allard P, Ferreira T, Hatte C, Coutinho R, Fontugne M, Jaudon M (1999) Magma-derived CO_2 emissions recorded in ^{14}C and ^{13}C contents of plants growing in Furnas caldera, Azores. J Volcanol Geotherm Res 92:195–207

Pellicer MJ (1977) Estudio volcanológico de la isla de El Hierro (Islas Canarias). Estud Geol 33:181–197

Pérez Torrado FJ, Rodríguez González A, Carracedo JC, Fernández Turiel JL, Guillou H, Hansen A, Rodríguez Badiola E (2011) Edades C-14 Del Rift ONO de El Hierro (Islas Canarias). In: XIII Reunión Nacional del Cuaternario, Andorra la Vella, pp 101–104

Reimer PJ, Baillie MGL, Bard E, Bayliss A, Beck JW, Bertrand C, Blackwell PGB, Buck CE, Burr G, Cutler KB, Damon PE, Edwards RLE, Fairbanks RG, Friedrich M, Guilderson TP, Hughen KA, Kromer B, McCormac FG, Manning S, Bronk Ramsey C, Reimer RW, Remmele S, Southon JR, Stuiver M, Talamo S, Taylor FW, van der Plicht J, Weyhenmeyer CE (2009) Intcal09 and Marine 09 radiocarbon age calibration curves, 0–50,000 years - cal BP. Radiocarbon 51:1111–1150

Rodriguez-Gonzalez A, Fernandez-Turiel JL, Perez-Torrado FJ, Hansen A, Aulinas M, Carracedo JC, Gimeno D, Guillou H, Paris R, Paterne M (2009) The Holocene volcanic history of Gran Canaria island: implications for volcanic hazards. J Quatern Sci 24:697–709

Rubin M, Gargulinski LK, McGeehin JP (1987) Hawaiian radiocarbon dates. In: Decker RW, Wright TL, Stauffer PH (eds) Volcanism in Hawaii: Papers to commemorate the 75th anniversary of the founding of the Hawaiian Volcano Observatory vol 1. US Geol Survy Prof Pap, vol 1350, pp 213–242

Saupé F, Strappa O, Coppens R, Guillet B, Jaegy R (1980) A possible source of error in ^{14}C dates: volcanic emanations (examples from the Monte Amaita district, provinces of Grosseto and Sienna, Italy). Radiocarbon 22:525–531

Singer BS, Relle MK, Hoffman KA, Battle A, Laj C, Guillou H, Carracedo JC (2002) Ar/Ar ages from transitionally magnetized lavas on La Palma, Canary Islands, and the geomagnetic instability timescale. J Geophys Res 107:2307

Singer BS, Brown LL, Rabassa JO, Guillou H (2004) $^{40}Ar/^{39}Ar$ chronology of late Pliocene and early Pleistocene geomagnetic and glacial events in southern Argentina. In: Channell JET et al (eds) AGU geophysical monograph series 145: timescales of the palaeomagnetic field. AGU, Washington, pp 175–190

Thellier E, Thellier O (1959) Sur l'intensité du champ magnétique terrestre dans le passé historique et géologique. Ann Geophys 15:285–376

Volcanic History and Stratigraphy of the Teide Volcanic Complex

7

Juan Carlos Carracedo, Hervé Guillou, Francisco J. Perez-Torrado, and Eduardo Rodríguez-Badiola

Abstract

The Teide Volcanic Complex, comprising the Teide and Pico Viejo stratocones and the recent eruptives of the Northwest and Northeast rift zones, are a product of the latest eruptive cycle of the island of Tenerife. This cycle commenced with the lateral collapse of the island's northern flank to form part of the Las Cañadas Caldera and the Icod Valley some 200 ka ago. A large set of recently obtained ages allows the precise reconstruction of the eruptive and structural evolution of this volcanic complex, aided by geological mapping and the establishment of an internally consistent chronostratigraphy. Abundant radiometric ages (K/Ar and $^{40}Ar/^{39}Ar$), obtained not only from the older (shield stage) formations but also from the youngest Teide Volcanic Complex, have allowed, for the first time, a view of the island as an integrated temporal system, with Teide volcano at the culmination of the island's several million year-long geological history.

J. C. Carracedo (✉) · F. J. Perez-Torrado
Departamento de Física (GEOVOL), Universidad de Las Palmas de Gran Canaria, Las Palmas de Gran Canaria, Canary Islands, Spain
e-mail: jcarracedo@proyinves.ulpgc.es

H. Guillou
Laboratoire des Sciences du Climat et de l'Environnement, CCEA-CNRS, Gif-sur-Ivette, France

E. Rodríguez-Badiola
Museo Nacional de Ciencias Naturales, CSIC, Madrid, Spain

7.1 Introduction

The most recent period of volcanic activity of Tenerife had been until a few years ago, insufficiently studied, particularly in regards to the geochronological control of the successive eruptions. Apart from the historical events of the last 500 years, for which period there are written eye-witness accounts, very few eruptions of the last 200 ky have been dated, especially those in the Holocene. Consequently, the greater part of the Late Pliocene-Holocene volcanism of the island was scarcely understood. The volcanic history, the evolution of this latest volcanic phase of Tenerife, and the evaluation of related eruptive hazards, were largely speculative, due

J. C. Carracedo and V. R. Troll (eds.), *Teide Volcano*, Active Volcanoes of the World,
DOI: 10.1007/978-3-642-25893-0_7, © Springer-Verlag Berlin Heidelberg 2013

to a lack of radioisotopic ages, precise mapping and reconstruction of eruptive events.

This situation improved considerably after the work carried out between 2004 and 2007, when 60 new radiometric ages (^{14}C, K/Ar and $^{40}Ar/^{39}Ar$) were obtained (Carracedo et al. 2003, 2007). An important part of this work included the study and correlation of the early phases of filling of the Las Cañadas Caldera using the numerous *galerías* (tunnels excavated for groundwater exploitation), which provided new insights into the origin of Las Cañadas Caldera.

7.2 The Teide Volcanic Complex

It could be readily argued that the current Pu'u 'Ō'ō-Kupaianaha eruption is part of Kilauea Volcano, as are the southwest and east rift zones and the summit Caldera. However, it is not as straightforward to accept that the 1909 Chinyero eruption (10 km distant from the main Teide stratocone) and the northwest and northeast rift zones form part of the Teide volcanic complex (TVC) (Fig. 7.1).

Four main structural features can be distinguished in the central part of Tenerife (Fig. 7.2): (1) The main stratocones (Teide and Pico Viejo), (2) A partially filled depression (the Valley of Icod-La Guancha, (3) The Rift Zones (NW and NE), and (4) The pre-Caldera Las Cañadas Volcano (LCV). Although the first three features are supposed to be stratigraphically separated from the pre-collapse LCV, the TVC is probably just the latest post-collapse stage after several mass destruction events, i.e. the volcanic renewal of the LCV. Therefore, some distinctions may not be diagnostic if based on volcanological considerations alone, since the rifts and the central volcanoes act as an essentially continuous, interconnected system.

The differences between basaltic rift eruptions and the phonolitic eruptions of the central volcano are unambiguous, and represent the extreme ends of a bimodal compositional system. However, these differences are less pronounced in the transitional part of the rifts where lava composition changes gradually. Whether

Fig. 7.1 The 3,718 m high Teide Volcano, nested inside the Las Cañadas Caldera, towers over the island of Tenerife, and represents the latest phase of the volcanic construction of the island

Montaña Blanca should be considered to be related to the proximal activity of the NE Rift, or as an adventive dome of the Teide volcano, is open to debate. This same issue also arises with Pico Viejo volcano, which may simply represent an enlarged polygenetic parasitic vent of Teide volcano at the eastern end of the NW rift zone (NWRZ).

The only ambiguity remaining is which part of the rift zones can be related to post-collapse volcanism and hence to the TVC, since the activity of both rift zones precedes the collapse. In this case the only possible method of separation is geochronological. In fact, the rift zones may be the main factor causing the generation of the TVC, inducing the lateral collapse that produced the depression, the subsequent volcanism which partially filled it, and the magmatic differentiation giving place to the felsic stratovolcanoes.

7.3 The Initial Collapse

General consensus takes the collapse that formed the Las Cañadas Caldera about 170–200 ky ago as the starting point for the construction of the TVC. Although the type of collapse (vertical or lateral) is still under debate,

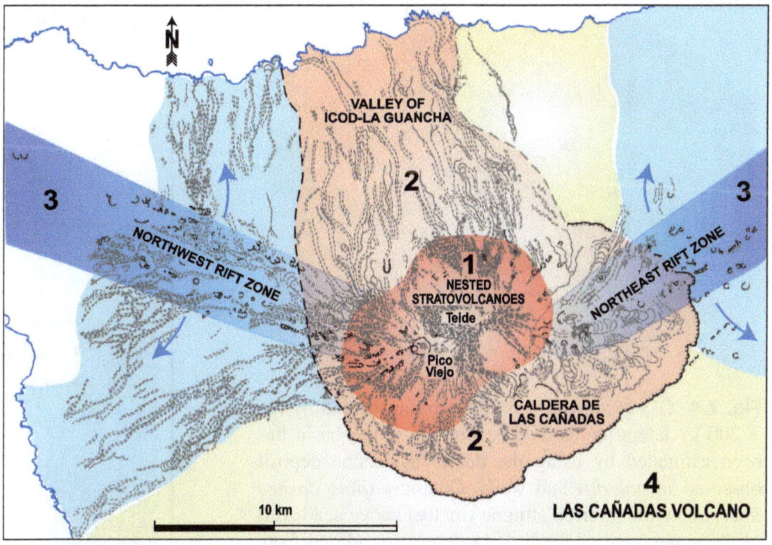

Fig. 7.2 Interactive volcanic features forming the Teide Volcanic Complex: *1* The main stratocones (Teide and Pico Viejo). *2* The partially filled landslide depression. *3* The Rift Zones (NW and NE). *4* The pre-collapse Las Cañadas Volcano (LCV)

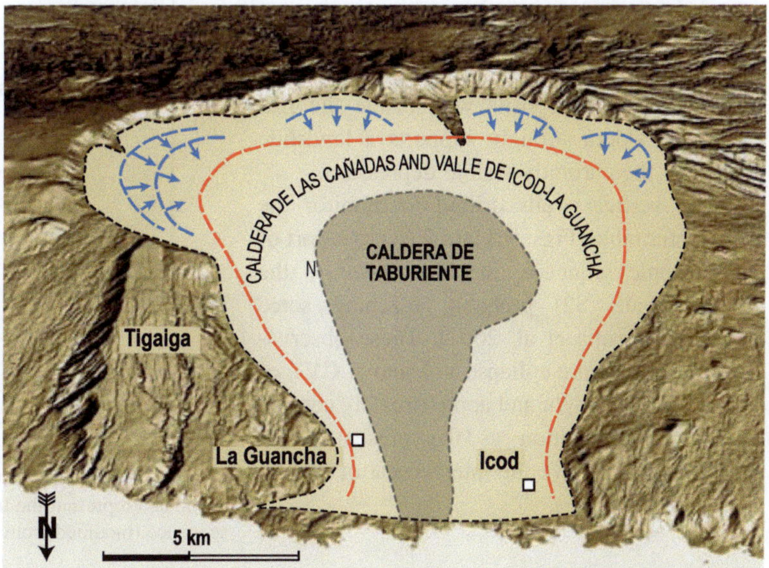

Fig. 7.3 Depression formed by the ~200 ky giant landslide (the Las Cañadas Caldera and the Icod-La Guancha Valley). The Taburiente Caldera (La Palma) is shown for comparison. The dotted red line marks the inferred position of the collapse scarp, while the blue lines with arrows indicate erosive retrogradation and headwalls of the main *barrancos* (canyons)

it seems evident that the well documented 200 ky lateral Icod collapse strongly modified many earlier features (Watts and Masson 1995).

The Icod lateral collapse carved a horseshoe-type depression, the Valley of Icod-La Guancha, which was probably later enlarged by retrograde erosion to configure the 15 km wide Las Cañadas Caldera (Fig. 7.3), that is considerably larger than prototypical lateral collapse embayments like the Taburiente Caldera on La Palma.

Galerías are extremely useful for constraining the deep structure beneath the Icod collapse basin (Fig. 7.4). Certain *galerías* crosscut polymictic breccia underlying the volcanic sequences that fill the depression, which are widely interpreted as a debris avalanche deposit (Navarro Latorre and Coello 1989). The geometry of this TVC basement, as deduced from the *galerías* (Márquez et al. 2008), is a continuous layer that extends inside the Las Cañadas

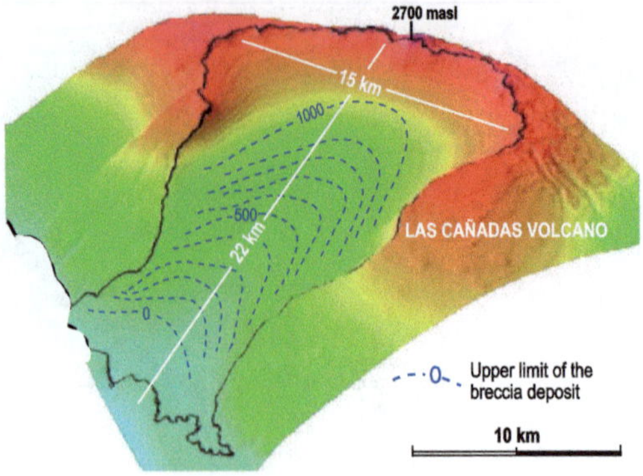

Fig. 7.4 GIS reconstruction of the basin formed by the ~200 ky lateral collapse. The bottom of the basin has been estimated by using the debris avalanche deposit observed in *galerías* and wells. Contours (*blue dashed lines*) show the inferred altitude (metres above sea level) of the upper limit of the breccia deposit (modified from Márquez et al. 2008)

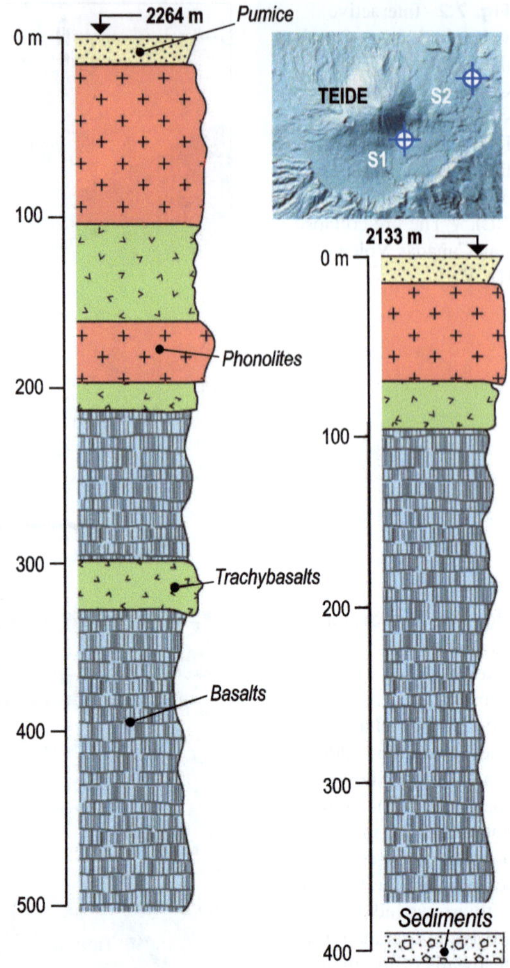

Fig. 7.5 Stratigraphy of the deep boreholes (505 and 404 m) drilled in the Las Cañadas Caldera to monitor the central aquifer of Tenerife. The wells penetrate the volcanic formations that partially filled the collapse basin, and represent the upper part of the Teide Volcano sequence (modified from Farrujia et al. 2001)

Caldera below the present Teide stratocone (dashed lines in Fig. 7.4). The presence of these debris avalanche deposits well inside the Las Cañadas Caldera strongly supports a landslide origin for the currently visible depression.

Deep vertical wells drilled to monitor the local water table (Figs. 7.5 and 7.6) cross part of the volcanic sequence, the one closer to the Caldera wall (S2) probably reaching scree deposits (Farrujia et al. 2001). These observations point to a pre-collapse volcano (LCV), of about 3,200 m height and approximately centred in the same position as the present Teide stratocone, i.e. above the intersection of the rift zones (see Fig. 7.2).

7.4 Geochronology of the Teide Volcanic Complex

Only one eruption in the entire TVC—the Montaña Blanca sequence —was assigned an age date before 2007. The absence of age control is particularly adverse in this context, because without the constraints of radiometric ages, the detailed mapping and the definition of individual eruptive units of the TVC becomes extremely difficult due to the often homogeneous appearance of these volcanics in the field.

Several authors (Araña et al. 2000) stated that a more precise reconstruction of the recent eruptive period of Tenerife (the TVC) was unfeasible, because of the impossibility of applying K/Ar and $^{40}Ar/^{39}Ar$ techniques to rocks from such a young time period coupled with the absence of suitable organic material (charcoal) for radiocarbon dating. However, the work of Ablay et al. (1998) and Ablay and Martí (2000) provided abundant petrological and stratigraphic data for the TVC, and the afore-mentioned age of ~2 ky of Montaña Blanca (Ablay et al. 1995),

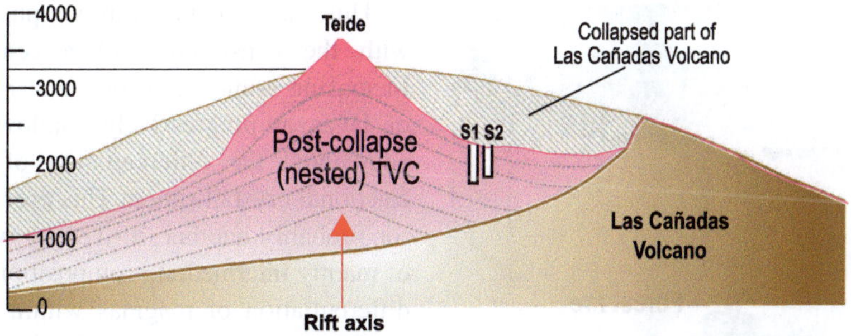

Fig. 7.6 Inferred configuration of the pre-collapse Las Cañadas Volcano and the present profile of the TVC. Both volcanic edifices were probably centred at the convergence of the NW and NE rift zones

in agreement with a previous age of 2.470 ± 110 year (Navarro 1980). In 2004, abundant charcoal remnants were obtained during detailed geological mapping of the volcanic complex, and samples were found particularly concentrated in areas underlying the more silicic lavas (Figs. 7.7 and 7.8) (see Chap. 6). Following stringent criteria for the selection of suitable samples (Fig. 7.8), 26 ^{14}C ages were determined for the NW and NE rift zones, the Teide and Pico Viejo stratovolcanoes and the peripheral lava domes (Table 7.1). To cross-check the reliability of these radiocarbon ages, another 19 unspiked K/Ar ages were obtained from these volcanic units (Table 7.2).

Finally, ages were determined for lavas from one of the deepest *galerías* (the 4,500 m long Salto del Frontón), which crosses the entire volcanic sequence that fills the collapse basin (Table 7.3), using unspiked K/Ar and ^{40}Ar/^{39}Ar (Fig. 7.9) (see Chap. 6 for further detail).

These ages gave crucial constraints for the age of the lateral collapse, the filling of the Icod collapse depression, the construction of the TVC, and the progressive magmatic evolution of the post-collapse volcanics.

7.5 The Main Volcano-Stratigraphic Units

To facilitate description, analysis and discussion, five main volcano-stratigraphic units have been defined in the TVC (Fig. 7.10): (1) Teide Volcano, (2) Pico Viejo Volcano, (3) Peripheral domes and lava domes of the main stratocones, (4) NWRZ volcanism, and (5) NERZ volcanism.

7.5.1 Teide Volcano

Although Teide Volcano is geographically well defined as a stratocone resting on rocks of the Las Cañadas Caldera, from a geological point of view it is a much more complex feature, being the third highest volcanic structure on earth (3,718 m a.s.l., >7 km high) and virtually unique in terms of intraplate volcanism.

The initial growth of Teide Volcano is related to the earliest lavas filling the lateral collapse embayment that forms the horseshoe-shaped Icod-La Guancha Valley. The oldest formations (124–32 ky) are restricted to the interior of the Caldera and vary in composition from basanite to basalt. These sequences outcrop in cliffs on the northern coast (Fig. 7.11). However, still older formations, belonging to the deep core of the volcano, have been reached through *galerías* that cross the entire fill sequence up to the basal debris avalanche (Fig. 7.12). The oldest age obtained (~195 ky, see Figs. 7.11, 7.12 and Tables 7.2 and 7.3) is equivalent to the age of a deposit from an explosive eruption containing syenite fragments which caps the Caldera wall, and is not observed inside the collapse basin. Large volumes of lavas were erupted during the depressurisation that immediately followed the lateral collapse.

The ages obtained for these products are between 173 and 176 ky with the K/Ar method (Mitjavila 1990) and between 179 ± 18 and

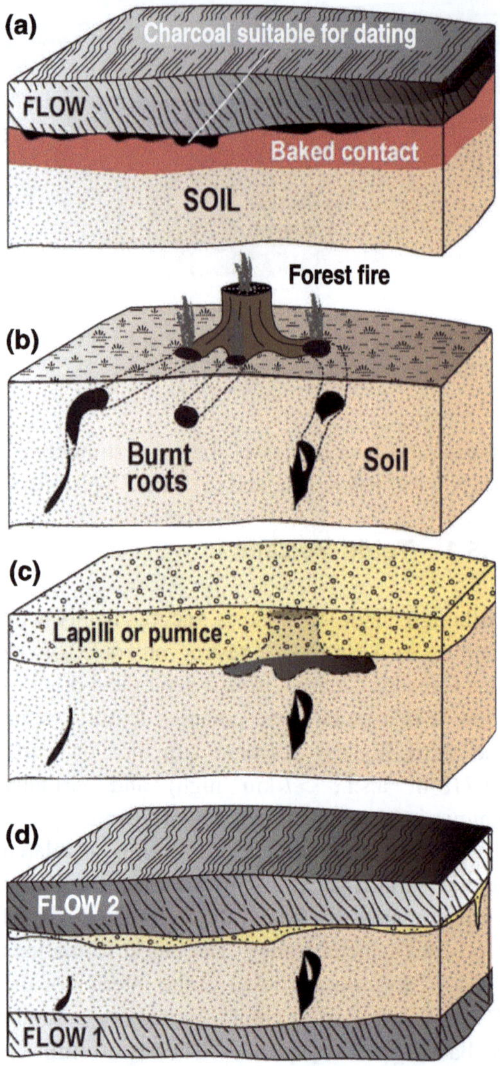

Fig. 7.7 Precautions in the sampling of wood charcoal for radiocarbon dating. Only the charcoal in (**a**) can yield an unambiguous radiocarbon age of the overlying lava flow, while in the remaining scenarios (**b–d**), charcoal fragments may be related to post-eruption forest fires

183 ± 8 ky with the $^{40}Ar/^{39}Ar$ method (Mitjavila and Villa 1993). This explosive deposit was interpreted as being related to the lateral collapse that formed the Las Cañadas Caldera (Ancochea et al. 1999). Radiometric ages show that a sequence of about 350 m of mafic (basanitic and basaltic) lavas accumulated inside the embayment during a period of about only 50 ky, with highest accumulation rates at the junction of the rift arms.

This early mafic eruptive phase continued with the construction of a central volcano. Increasing edifice elevation and residence time of lavas at progressively shallower reservoirs augmented the differentiation of magmas by fractional crystallisation. This process continued for probably another 70–120 ky with eruptions of mainly intermediate composition. Progressive differentiation of magmas within one or more shallow phonolitic magmatic chambers eventually fed the phonolitic eruptions of Teide (Valentin et al. 1990).

The development of the stratovolcano was apparently completed at about 30 ky (Old Teide). Since then, only one eruption of Teide has been observed, which formed the summit crater and the phonolitic flows of the Lavas Negras, at 1150 ± 140 year BP (see Figs. 7.11 and 7.12).

The eruptive activity that built the Teide volcano was coeval with eruptions from the NE and NW rift zones, explaining the sequences of alternating phonolitic and basaltic pyroclastic layers observable in road cuts and in the *galerías* (see photo in Fig. 7.17b).

A noteworthy feature of Teide Volcano is the phreatomagmatic eruption located on the NW flank of the main stratovolcano (Pérez Torrado et al. 2004, 2006). The explosive, surge-type episodes, probably related to snow-melt interacting with the shallow magma, formed thick slabs of indurated, whitish-coloured breccias (Las Calvas del Teide), interbedded with the final sequence of Old Teide lavas, and overlain by lavas of Pico Viejo and the Lavas Negras of the last eruption of Teide volcano itself.

How much higher can the Teide stratocone grow? Obviously a volcano cannot grow indefinitely. As a volcano grows the eruption rate, especially of central felsic volcanoes, frequently decreases. One of the main factors is the distance the magma has to ascend to reach the summit of the volcano. An empiric limit in volcano height of about 3,000 m has been suggested (Davidson and De Silva 2000), above which it is increasingly difficult for lavas, especially the less evolved and thus denser ones, to reach the summit of a large stratovolcano. A

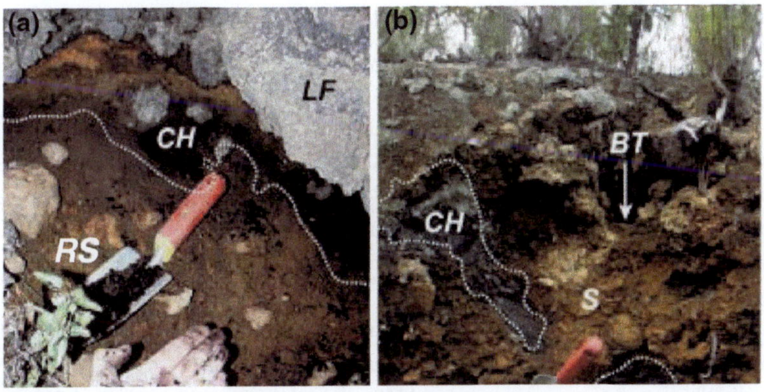

Fig. 7.8 **a** Reliable radiocarbon age from wood charcoal (*CH*) in a red soil (*RS*) overlain by a phonolitic flow (*LF*). **b** Burnt roots (*BT*) from a contemporaneous forest fire running deep into a loose soil (*S*)

"density filter" is therefore established, thus promoting later lavas to be more evolved ones. A second consequence involves the spatial distribution of lavas, with eruptions tending to concentrate around the volcano's basal periphery instead of at the summit (Fig. 7.13).

Teide's eruptive rates and composition seem to adhere to this model. Teide volcano apparently reached the critical height for the transition between basaltic and intermediate composition magmas at about 120 ky, approximately the age of the first evolved lavas (phonotephrites). At about 30 ky the volcano came to a critical height for summit eruptions, with the exception of the mentioned medieval obsidian phonolites (Lavas Negras). From that age onwards, eruptions of Teide volcano itself have been almost exclusively phonolitic lava domes located at the basal perimeter of the stratocone.

7.5.2 Pico Viejo Volcano

Most of the eruptive activity of this stratovolcano occurred between 17 and 27 ky (Fig. 7.14). Matching the evolution of Teide, Pico Viejo evolved from initial basaltic lavas (predominantly pāhoehoe plagioclase basalts) through intermediate composition and finally to phonolitic eruptions.

The elevation of Pico Viejo (3,100 m a.s.l.) is misleading, because the entire edifice forms a relatively thin mantle on the western flank of the older Teide Volcano (see cross section in Fig. 7.14). Such a high elevation is, therefore, mainly due to the location of the vent at about 3,000 m on the flank of the pre-existing Teide edifice. The lava flows of Pico Viejo consistently overlie the Teide Volcano lavas, with the noted exception of the Teide summit eruption.

This suggests that the Pico Viejo may be an adventive (polygenetic) eruptive centre of Teide, probably originating from the flankwards migration of magmas when Teide reached a critical altitude for summit eruptions. In fact, the ages of early basaltic eruptions of Pico Viejo seem to match those of the final phonolitic eruptions of Teide. Pico Viejo lava flows have almost completely resurfaced the western side of the collapse embayment, have overflowed the scarp, and resurfaced the flanks of the NWRZ.

Minor eruptions at the rim and inside the main crater of Pico Viejo Volcano persisted until relatively recent times (Holocene), but it is only a question of definition as to whether these eruptions correspond to Pico Viejo or to the NWRZ. In one of the eruptions, phreatomagmatic explosive phases removed part of the summit of Pico Viejo's edifice forming an explosion crater and mantling the upper flanks of the volcano with brecciated material. As at Teide Volcano, this explosive event was probably driven by snow-melt interacting with magma during the eruption.

Table 7.1 Radiocarbon ages of the Teide volcanic complex (from Carracedo et al. 2011)

Eruption	Analysis[a]	^{14}C age years BP	Calibrated age years BP
NE Rift Zone			
Forest fire[b]	β-counting	580 ± 25	590 ± 66
Volcán del Portillo (upper)[b]	AMS	11,080 ± 80	13,005 ± 132
Volcán del Portillo (lower)[b]	AMS	12,020 ± 80	13,988 ± 163
Montaña Guamasa[b]	β-counting	>33,000	–
Montaña Cerrillar[b]	β-counting	>33.000	–
NW Rift Zone			
Montaña Boca Cangrejo[c]	β-counting	350 ± 60	400 ± 110
Montaña Reventada[c]	β-counting	990 ± 70	895 ± 155
Volcán Los Hornitos[b]	AMS	1,930 ± 80	1,964 ± 123
Volcán El Ciego-1[b]	AMS	2,600 ± 160	2,616 ± 250
Volcán El Ciego-2[c]	AMS	26,600 ± 40	2,794 ± 55
Montaña de Chío[b]	AMS	3,620 ± 70	3,933 ± 212
Volcán Cueva del Ratón[c]	β-counting	5,370 ± 50	6,145 ± 145
Montaña Liferfe[c]	AMS	7,400 ± 40	8,250 ± 80
Montaña del Banco[c]	AMS	12,810 ± 60	–
Teide-Pico Viejo stratovolcanoes			
Teide (latest eruption)[b]	AMS	1240 ± 60	1,150 ± 140
Pico Viejo (South flank lavas)[c]	AMS	14,630 ± 50	17,525 ± 365
Pico Viejo (North flank lavas)[b]	AMS	17,570 ± 150	20,775 ± 321
Pico Viejo (North flank pāhoehoe lavas)[b]	AMS	27,030 ± 430	–
Old Teide (phonolites in the Orotava Valley)**	AMS	32,360 ± 800	–
Teide-Pico Viejo phonolitic peripheral lava domes			
Roques Blancos II[b]	AMS	1,790 ± 120	1,714 ± 151
Roques Blancos I[b]	AMS	2,010 ± 120	1,974 ± 147
Montaña Blanca? (pumice airfall bed?)[b]	AMS	2,020 ± 140	1,996 ± 165
El Boquerón[b]	AMS	24,200 ± 140	2,528 ± 185
La Abejera Baja[b]	AMS	4,790 ± 140	5,496 ± 162
La Abejera Alta[b]	AMS	5,170 ± 110	5,911 ± 264
Montaña Negra-Los Tomillos[b]	AMS	8,220 ± 120	9,210 ± 190

[a] The ß-counting method is based on the detection of radioactive decay of the radiocarbon (^{14} C) atoms. The AMS (Accelerator Mass Spectrometry) method is based on the detection of mass of ^{14} C atoms in the sample (and therefore its ratio of ^{14} C to ^{12} C)

[b] Laboratoire des Sciences du Climat et de l'Environnement, CEA-CNRS, France

[c] Beta Analytic Dating Laboratory, USA

The apparent westward migration of most of the eruptive activity of the TVC from about 30 ky is evident from the abundant Holocene volcanism in the NWRZ. This may be caused by diversion of magma transport coinciding with the overgrowth and abrupt decline of Teide's eruptive activity.

Some of the most characteristic features of Teide and Pico Viejo stratovolcanoes are shown in Fig. 7.15.

7.5.3 The Peripheral Lava Domes

The increasing elevation of both stratocones above a critical level finally restricted eruptions to parasitic phonolitic lava domes around the central volcanoes and basaltic fissure volcanism on the rift zones, mainly in the NW. This late episode of volcanism of the TVC occurred mainly during the Holocene.

Table 7.2 K/Ar ages of the Teide volcanic complex (from Carracedo et al. 2011)

Eruption	Weight molten g	K* wt %	^{40}Ar* %	^{40}Ar* 10^{-12} mol/g ± 1σ	Age ± 2σ ky
NE rift zone					
Volcán del Portillo	1.84000	1.486 ± 0.015	0.267	0.022 ± 0.009	8 ± 4
	3.15460		0.255	0.018 ± 0.006	
Montaña de Enmedio	1.38104	1.577 ± 0.016	0.656	0.083 ± 0.006	31 ± 3
	2.09397		0.917	0.089 ± 0.005	
Montaña de Guamasa	1.98317	1.527 ± 0.015	0.405	0.087 ± 0.005	33 ± 3
	21.14727		0.7	0.088 ± 0.00	
			78	5	
Montaña del Cerrillar	1.77942	1.395 ± 0.014	0.671	0.085 ± 0.005	378 ± 3
	1.90060		0.595	0.093 ± 0.005	
NW rift zone					
Volcán del Palmar	1.05197	0.000 ± 0.013	3.846	0.334 ± 0.010	378 ± 6
	2.10684		5.031	0.361 ± 0.006	
Volcán Teno Alto	1.03751	0.000 ± 0.016	2.323	0.451 ± 0.013	378 ± 6
	1.93700		5.101	0.495 ± 0.007	
Montaña Los Silos	1.27816	0.000 ± 0.016	2.762	0.512 ± 0.013	378 ± 8
	1.08304		3.314	0.547 ± 0.012	
Volcán Tierra del Trigo	1.22597	0.000 ± 0.018	6.333	0.786 ± 0.011	378 ± 7
	1.13251		3.245	0.874 ± 0.011	
Icod Valley (upper flow, western rim	1.11669	0.000 ± 0.016	6.145	1.194 ± 0.011	378 ± 10
	1.44401		4.473	1.152 ± 0.011	
Montaña de Taco	0.95023	0.000 ± 0.037	29.554	4.463 ± 0.026	378 ± 15
	1.44732		43.764	4.598 ± 0.024	
Playa de La Arena (South flank)	1.09534	0.000 ± 0.016	6.779	2.400 ± 0.017	378 ± 20
	1.20229		8.809	2.502 ± 0.016	
Icod Valley (western rim, Las Cañadas Volcano)	1.19247	0.000 ± 0.021	43.291	4.134 ± 0.023	378 ± 25
	1.37110		30.003	4.272 ± 0.023	
Teide-Pico Viejo					
Playa San Marcos (phonolitic flow)	1.71730	4.284 ± 0.043	2.351	0.243 ± 0.009	32 ± 1
	1.65063		1.612	0.221 ± 0.011	
Old Teide (phonolites inside the Orotava Valley)	2.02239	3.777 ± 0.038	3.366	0.230 ± 0.009	33 ± 2
	2.41121		4.395	0.207 ± 0.007	
Old Teide (flows, base of cliff at Playa San Marcos)	1.07373	1.544 ± 0.015	2.090	0.215 ± 0.092	88 ± 4
	1.60741		1.311	0.244 ± 0.006	
Old Teide (flows, base of cliff at Playa Santo Domingo)	1.38494	1.610 ± 0.016	3.241	0.338 ± 0.007	124 ± 4
	20.00200		2.900	0.349 ± 0.005	
Galería Salto del Frontón					
2,700 m	1.65487	1.503 ± 0.017	4.477	0.525 ± 0.008	198 ± 5
	2.33548		3.408	0.513 ± 0.005	
1,500 m	1.43489	1.677 ± 0.017	2.909	0.316 ± 0.007	110 ± 4
	2.22959		2.939	0.323 ± 0.005	
1,100 m	2.14574	3.179 ± 0.032	5.233	0.656 ± 0.006	118 ± 3
	2.18357		4.503	0.644 ± 0.007	

Note Age calculations are based on the decay and abundance constants from Steiger and Jager (1977). The weighted mean ^{40} Ar* values from the replicate experiments are used in the final age calculation

Table 7.3 $^{40}Ar/^{39}Ar$ ages (ky) of lava flows from Galería Salto del Frontón, crossing the entire Teide volcanic sequence

Sample	Depth m	40Ar/39Ar[a] ky	40Ar/39Ar[b] ky
GSF-2	4,700	155 ± 5	–
GSF-10	2,700	192.3 ± 5	195.7 ± 12.7
GSF-12	2,200	–	154 ± 4
GSF-14	1,500	148 ± 5.9	–
GSF-15	1,070	134.2 ± 4	126 ± 4

[a] Rare Gas Laboratory, University of Wisconsin–Madison, USA

[b] Laboratoire des Sciences du Climat et de l'Environnement, CEA-CNRS, France

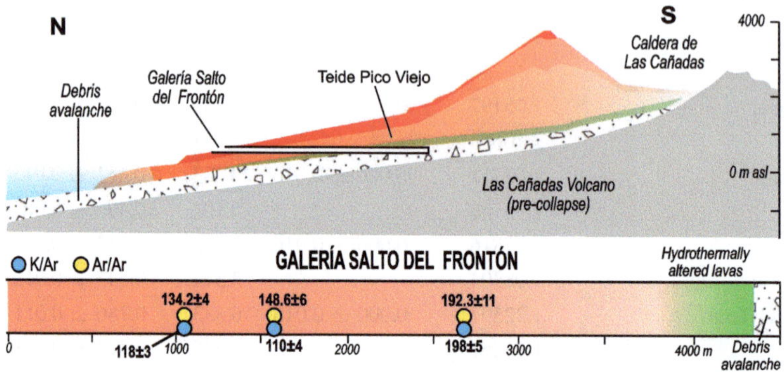

Fig. 7.9 Galería Salto del Frontón crosses the entire post-collapse TVC sequence. The section along the *galería* (*below*) shows the composition and age of the volcanic formations filling the collapse embayment. The deepest part of the section, from ~3,400 m, was unsuitable for dating because of the high degree of hydrothermal alteration of the lavas (modified from Carracedo et al. 2007)

Fig. 7.10 Main volcano-stratigraphic units of the Teide Volcanic Complex: *1* Teide Volcano, *2* Pico Viejo Volcano, *3* Peripheral domes of the main stratocones, *4* NWRZ volcanism, *5* NERZ volcanism

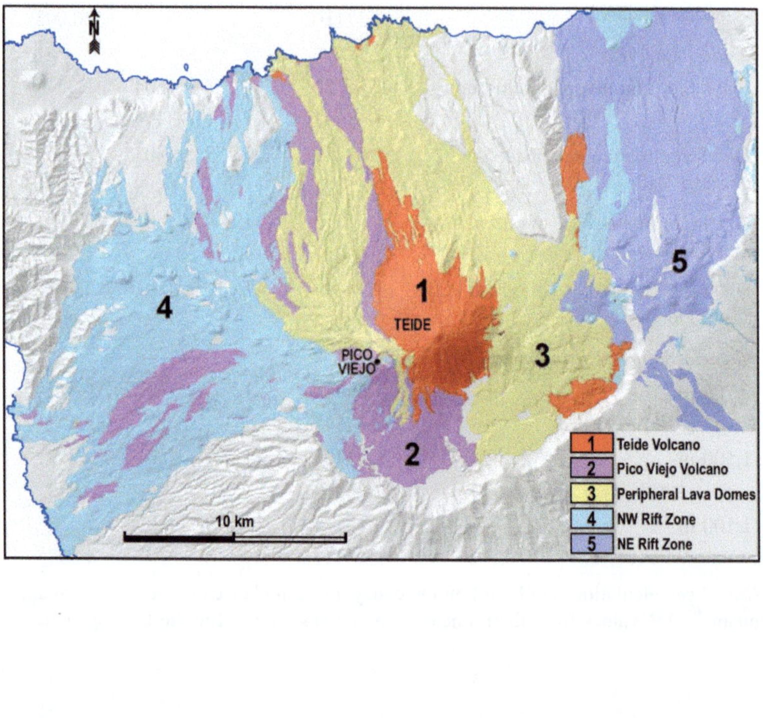

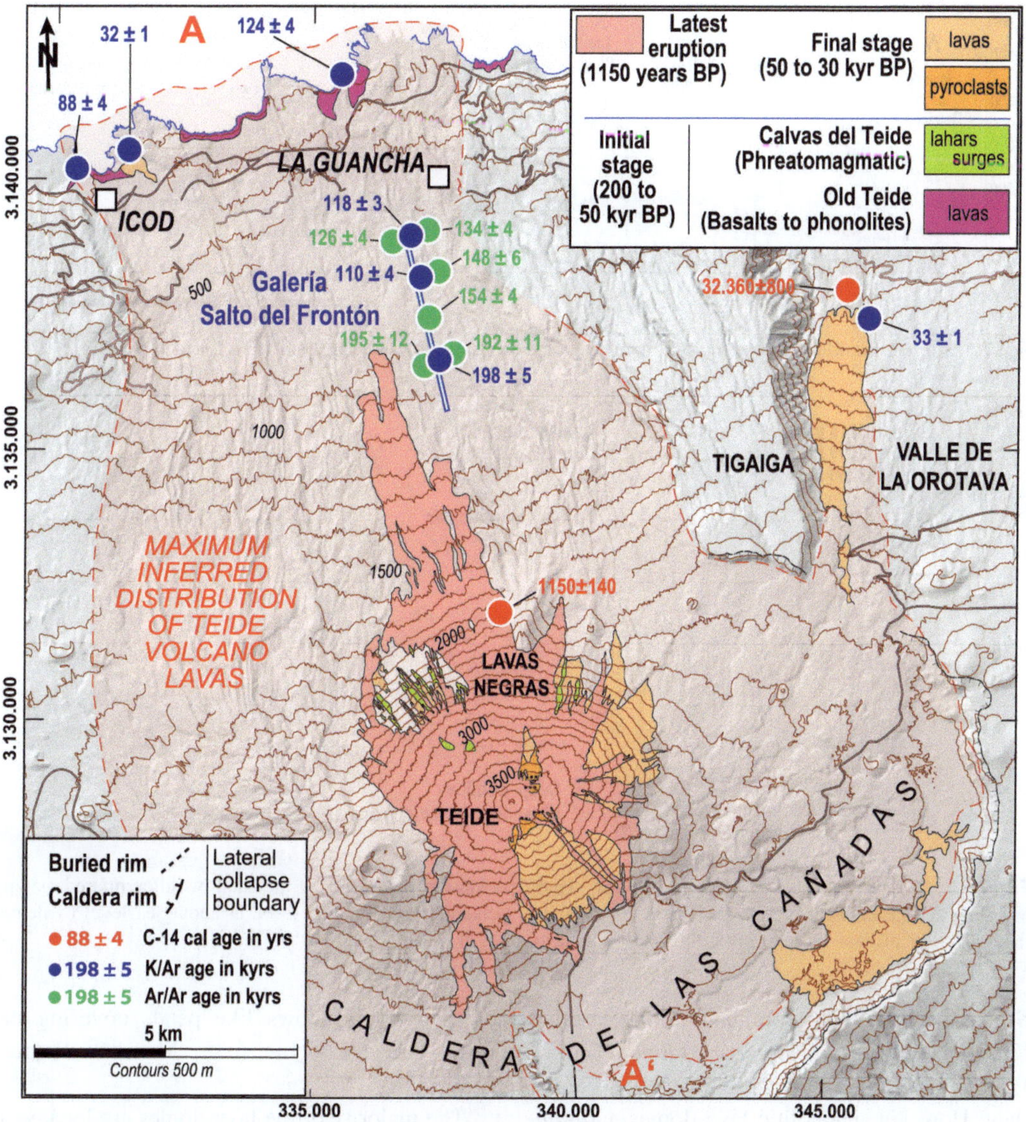

Fig. 7.11 Geological map of the TVC (modified from Carracedo et al. 2007)

As shown in Fig. 7.16, overpressure facilitating lava ascent to the summit of the stratocones is progressively lowered by increasing edifice height. When lateral thrust finally exceeds magma ascent (1 in Fig. 7.16), magma bulging and cryptodomes occur on the flanks (2 in Fig. 7.16) and radial fractures open to form flank vents, the upper ones degassing the system in explosive events, and the effusive basal vents forming bulges and lava domes (3 and 4 in Fig. 7.16).

These phonolitic domes, several forming spectacular steep-fronted blocky flows (*coulées*) reaching the 15 km distant northern coast, are one of the most characteristic features of the TVC (Fig. 7.17). Most of the exposed lava domes and *coulées* have been dated, and their ages range from about 9210 ± 190 to 1714 ± 151 year BP (see Table 7.1), contributing considerably to the relatively high frequency of eruptive events in the central part of Tenerife.

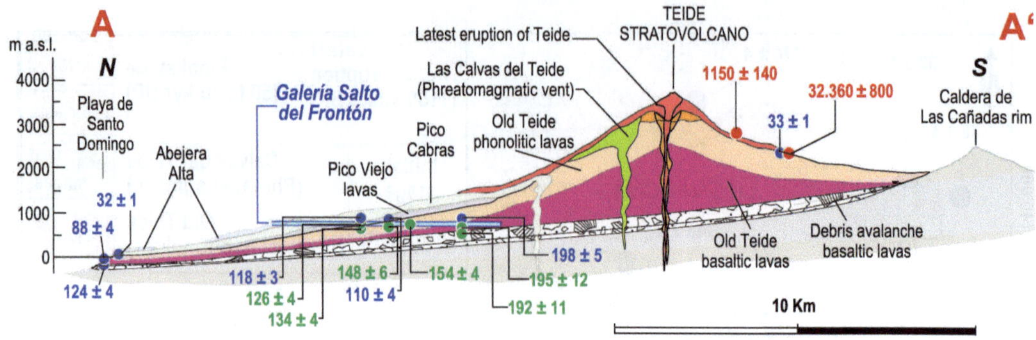

Fig. 7.12 Cross section (N–S) along the main axis of the collapse basin. (modified from Carracedo et al. 2007)

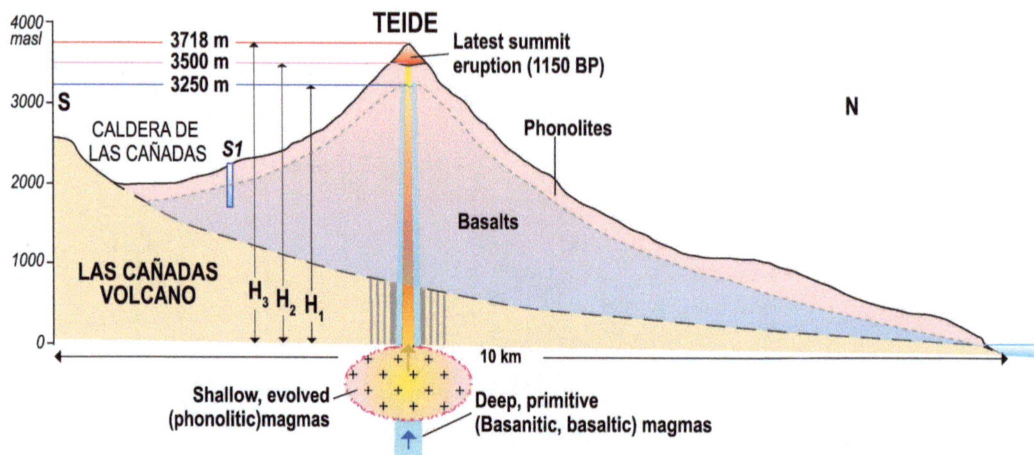

Fig. 7.13 Critical heights in the growth of Teide Volcano: H_1 basalts and intermediate lavas; H_2 evolved lavas. H_3 the present altitude after the latest medieval eruption. The depth of the shallow differentiated magma chamber is estimated to be at about sea level (Valentin et al. 1990)

Silicic, generally andesitic, lava domes in other tectonic environments are very common, and the literature pertaining to these features is abundant. However, phonolitic lava domes and *coulées* of this volume and flow length are rarely found in oceanic islands.

Depending on topography, whether open or confining, the lava domes in the TVC are characterised by two main features: (1) very thick (up to 50–80 m) and long (10-15 km) *coulées*, with conspicuous levees and pressure ridges, generated on open inclined surfaces (e.g., Roques Blancos, Fig. 7.17c); and (2) near symmetrical, hemispherical domes composed of viscous phonolitic glassy short-length lavas formed on flat or low angle confined surfaces, with characteristic "rosette" structures, made up of many small flows like petals covering the surface of the dome (e.g., Montaña Rajada, Fig. 7.17d).

The majority of the lava-domes are located at the basal periphery of the main stratovolcanoes (Figs. 7.18, 7.19, 7.20). Lava-dome eruptions tend to be moderately explosive episodes, forming extensive air-fall deposits of pumice. Collapse of *coulée* fronts also occurs giving rise to hot avalanches and block-and-ash flows or surges. However, no evidence of pyroclastic flows associated with such lava domes has been observed at the TVC. Air-fall pumice deposits are however ubiquitous, and generally composed of thin (cm-scale) layers. A thicker (about 1 m) and more extensive layer of pumice covers the N and NE flanks of the Teide stratocone and the

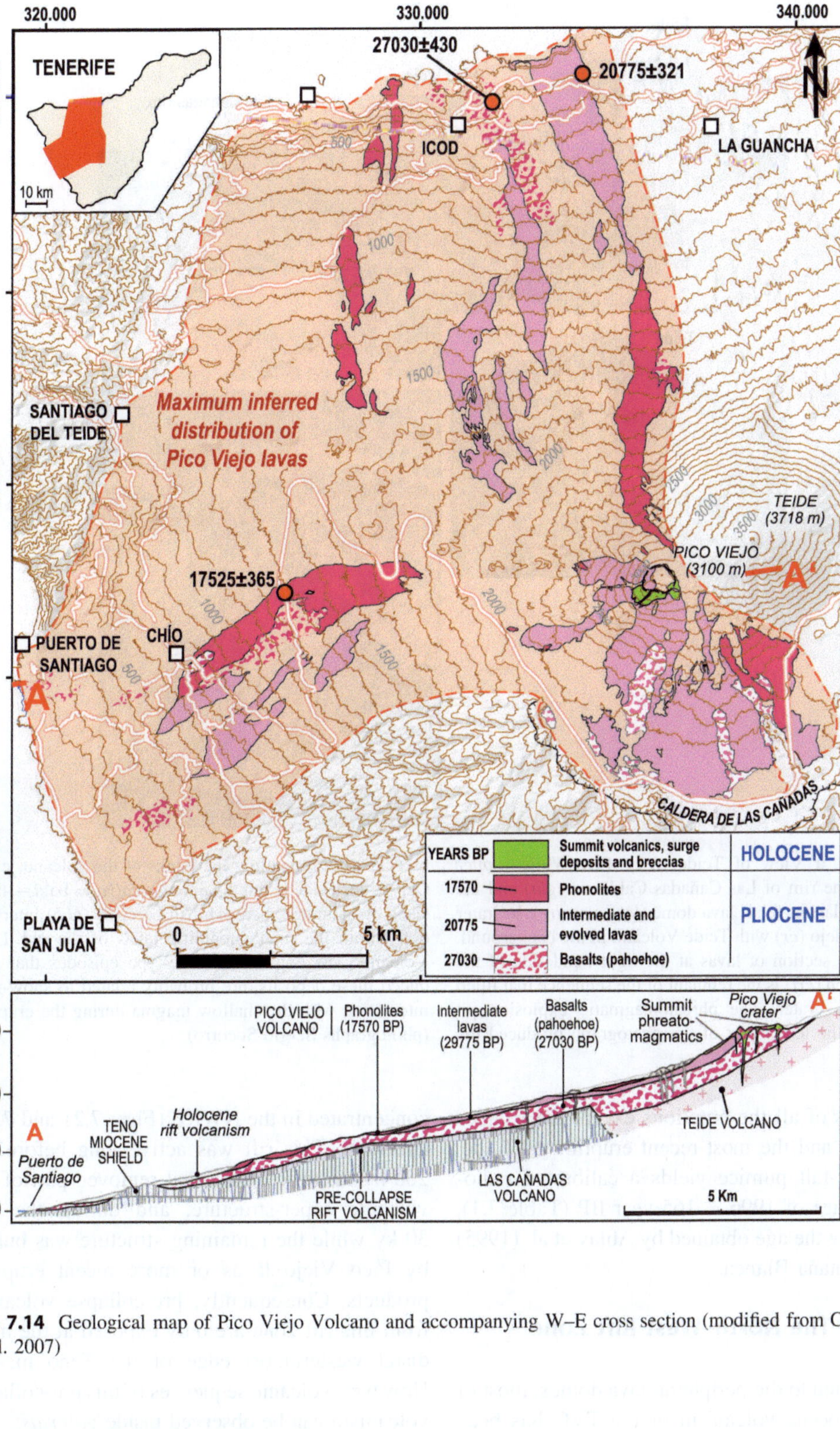

Fig. 7.14 Geological map of Pico Viejo Volcano and accompanying W–E cross section (modified from Carracedo et al. 2007)

Fig. 7.15 **a** View of Teide (*left*), Pico Viejo (*upper centre*), the rim of Las Cañadas Caldera (*right*) and the Roques Blancos felsic lava dome (*lower centre*). **b** Crater of Pico Viejo (*cr*) with Teide Volcano in the background. The thick section of lavas at the right hand side of the photograph (*crf*) is the remnant of the sequence that filled the entire crater. The phreatomagmatic explosion pit (*php*) in the lower part of the photograph produced the breccias (*phb*) mantling the flanks of the volcano. **c** Las Calvas del Teide (from the Spanish *calvo* = bald)—thick slabs of indurated, bare whitish breccias (*phb*) interbedded within the final phonolitic lavas of the old Teide Volcano. The explosive, surge-type episodes that produced these deposits, are probably related to snow-melt interacting with the shallow magma during the eruption (photographs Sergio Socorro)

products of all the eruptions except the Roques Blancos and the most recent eruption of Teide. This air-fall pumice yields a calibrated radiocarbon age of 1996 ± 165 year BP (Table 7.1), matching the age obtained by Ablay et al. (1995) for Montaña Blanca.

7.5.4 The North-West Rift Zone

In addition to the peripheral lava domes, most of the Holocene volcanism of the TVC has been concentrated in the NWRZ (Figs. 7.21 and 7.22; Chap. 4). This rift was active long before the 200 ky lateral collapse that removed part of the volcanic super-structure, and during the last 30 ky while the remaining structure was buried by Pico Viejo lavas or more recent eruptive products. Consequently, pre-collapse volcanics from this rift zone are only exposed at the most distal westernmost edge of the Teno massif. However, volcanic sequences of this pre-collapse volcanism can be observed inside *galerías*.

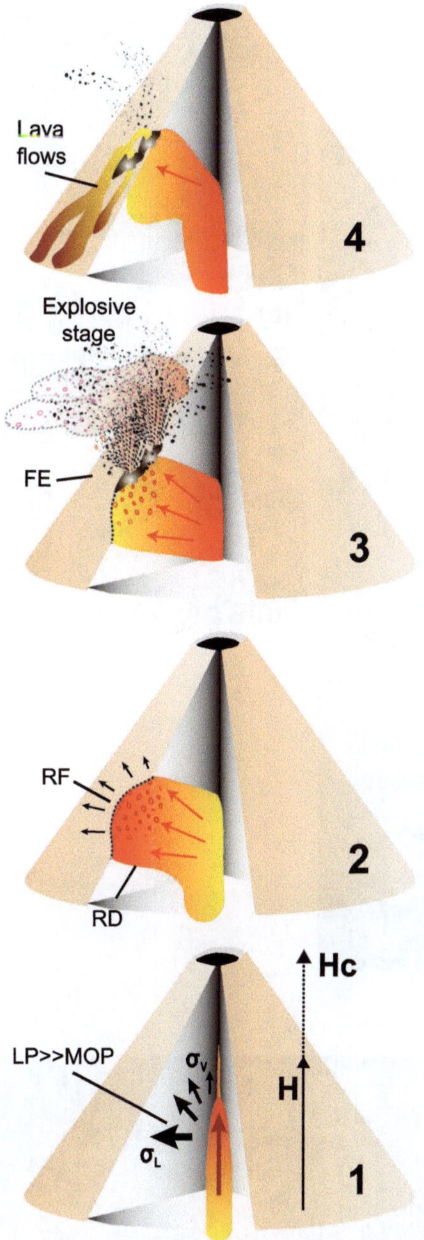

Fig. 7.16 Schematic diagram showing the relationship between the increasing lithostatic pressure (*LP*) of a growing stratocone (e.g. Teide Volcano) and the magmatic overpressure (*MOP*) required for the magma to ascend to the summit of the volcano. *1* When the critical height (*Hc*) is reached, the vertical push of magma is changed to lateral. *2* Bulging may result in radial dykes (*RD*) and related radial fissures (*RF*). *3* Eventually, evolved fissure eruptions (*FE*) from radial fractures begin, with initial explosive phases. *4* Finally, effusive phases produce thick evolved (phonolitic) lava flows and *coulées*

Eruptive vents at the extension of the NW rift into the Teno massif (distal end) are considerably older (see Table 7.2) than those in the central part of the NW rift and near the central stratovolcanoes (proximal end), where the younger eruptions from the NWRZ cap the preceding volcanism (cross-section in Fig. 7.23). Frequent eruptions occurred along this rift during the entire Holocene, including four historical eruptions: Boca Cangrejo (1492), Garachico (1706), Chahorra (1798) and Chinyero (1909).

Interesting examples of magma mixing are found at the proximal end of the NWRZ, near the central volcanoes (see Chap. 11). Here, a lava unit from an eruption of Montaña Reventada shows noteworthy magma mixing textures, containing mingled basaltic and phonolitic lavas (Araña et al. 1994; Rodríguez-Badiola et al. 2006; Wiesmaier et al. 2011). Similar processes may have occurred at the distal end of the NERZ.

The Holocene eruptions of the NWRZ cluster along two distinct volcanic chains, the Chío and the Garachico volcanic fractures, which are separated by a saddle (Fig. 7.22). This feature is crucial in the control of flow direction; the lava flows from vents in the Chío chain consistently spread down the southern flank of the rift, those in the Garachico chain flow down the northern flank, while those located between both structures generally flow along the saddle in a SE-NW direction (e.g., Montaña Reventada, Chinyero, etc.).

7.5.5 The North-East Rift Zone

The NE Rift (see also Chap. 5) extends from the base of the Teide stratovolcano towards the NE of the island, overlying the western flank of the Anaga Mio-Pliocene shield volcano (Fig. 7.24). The age of this rift zone probably extends back to the Pliocene, interacting with the development of the Las Cañadas Volcano (Carracedo et al. 2007, 2011). According to Martí et al. (1994), Cantagrel et al. (1999), and Huertas et al. (2002), basaltic fissure eruptions, with the eruptive centres aligned in a similar pattern to

Fig. 7.17 **a** Montaña de la Arena, a group of basaltic vents of the NERZ attached to the rim of the Las Cañadas Caldera. **b** Interbedded layers of basaltic lapilli (*bl*) and phonolitic pumice (*pm*) from coeval basaltic eruptions of the NERZ and phonolitic events from the peripheral lava domes. **c** Phonolitic lava dome and *coulées* of Roques Blancos attached to the NW flank of the presumed Pico Viejo stratocone. Despite their presumed viscosity, the flows are several kilometres long (the photograph is 7 km from *top* to *bottom*). **d** Characteristic rosette shaped appearance, of very viscous flows from one of the peripheral phonolitic domes of Teide Volcano (Montaña Rajada)

Fig. 7.18 Peripheral lava domes of the Teide-Pico Viejo complex: *1* Roques Blancos, *2* Montaña Blanca, *3* El Boquerón, *4* Los Gemelos, *5* Mancha Ruana, *6* Montaña Majúa, *7* Montaña de La Cruz, *8* Abejera Baja, *9* Abejera Alta, *10* Pico Cabras, *11* Abrunco, *12* Montaña de las Lajas, *13* Montaña de los Conejos, *14* Bocas de Doña María, *15* Montaña Negra

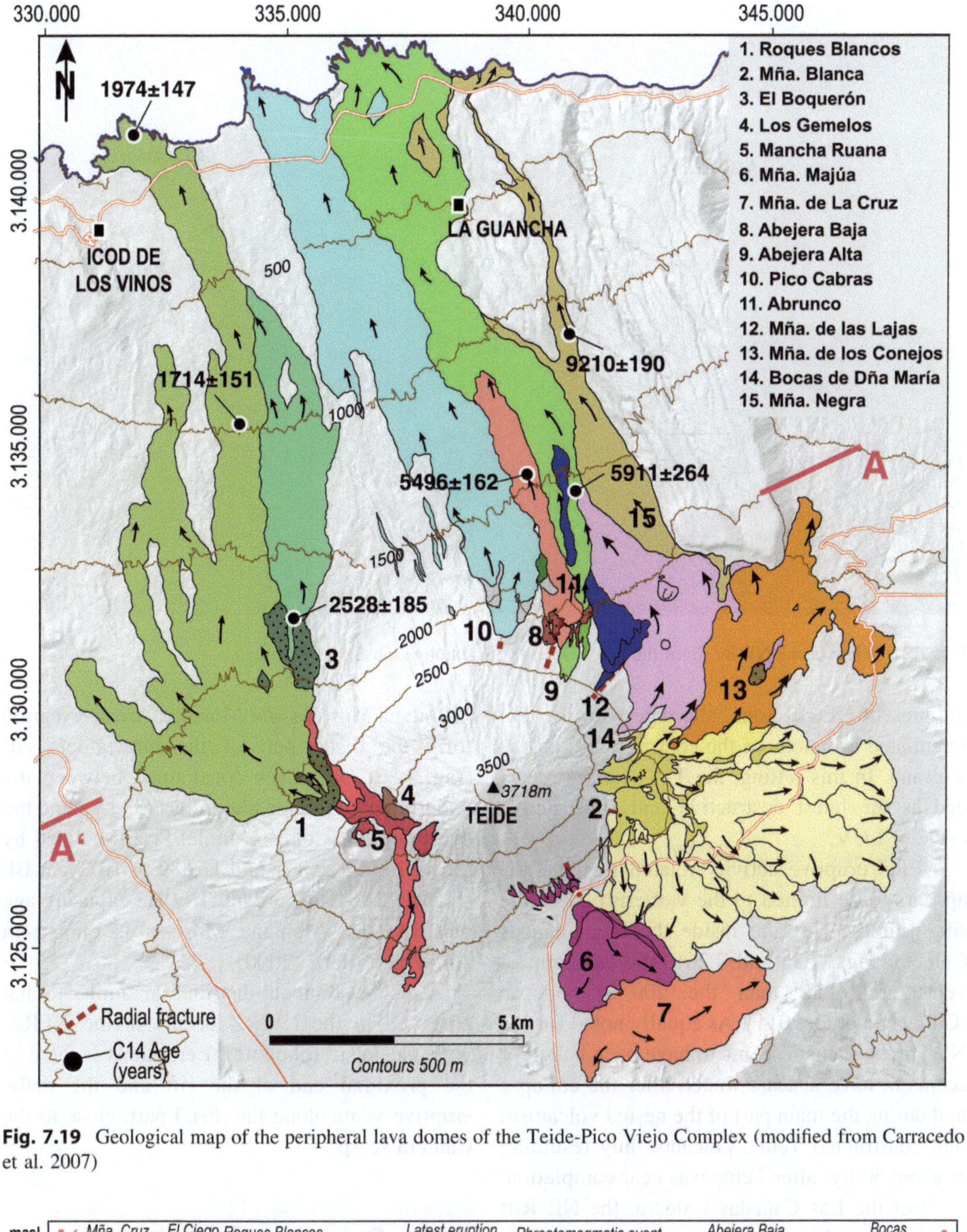

Fig. 7.19 Geological map of the peripheral lava domes of the Teide-Pico Viejo Complex (modified from Carracedo et al. 2007)

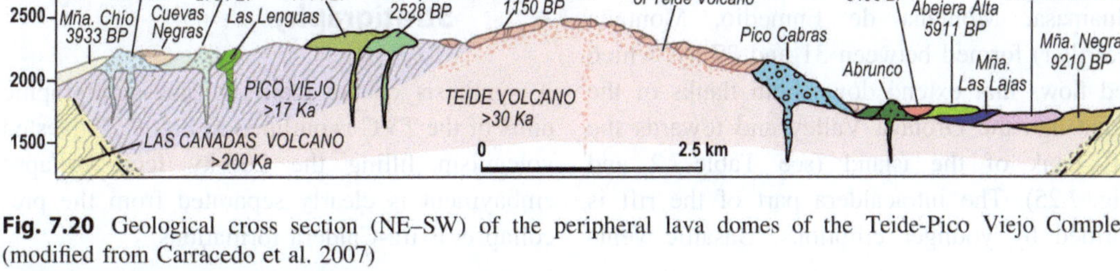

Fig. 7.20 Geological cross section (NE–SW) of the peripheral lava domes of the Teide-Pico Viejo Complex (modified from Carracedo et al. 2007)

Fig. 7.21 View of the NWRZ from the top of Pico Viejo (photograph Sergio Socorro)

the present NE rift zone, alternate with the differentiated eruptions of the central Las Cañadas Volcano. In this setting, the TVC can be regarded as the latest destructive and constructive cycle of LCV.

NERZ eruptive activity related to the TVC appears to be limited to the western part of the rift, near the rim and inside the Las Cañadas Caldera (Figs. 7.25 and 7.26). The easternmost vents are older than the 200 ky collapse (Carracedo et al. 2011). As equally noted for the NWRZ, volcanism seems to have been halted or seems to have at least waned after the collapse and during the main part of the nested volcanism that constructed Teide Volcano, only resuming at about 30 ky, after Teide was near completion.

Near the Las Cañadas Caldera, the NE Rift comprises a cluster of basaltic vents (Montaña Guamasa, Montaña de Enmedio, Montaña Cerrillar) formed between 31 and 37 ky, which fed flows that extend down both flanks of the ridge into the Orotava Valley and towards the SE flank of the island (see Table 7.2 and Fig. 7.25). The intracaldera part of the rift is formed by younger eruptions. Basaltic vents (Montaña Mostaza and Montaña Arenas Negras) form the distal part of the intracaldera rift (Fig. 7.26), with ages constrained between the underlying Old Teide phonolites (33 ky), and the overlying lavas of Volcán del Portillo, dated by ^{14}C at 13,005 ± 132 and 13,889 ± 163 year BP (Fig. 7.25). A concordant $^{3}He/^{4}He$ exposure age of 15 ky for Montaña Mostaza is quoted in Ablay and Martí (2000).

A distinct bimodal distribution similar to that observed in the NWRZ occurs in the NERZ, with the felsic (phonolitic) eruptions located at the proximal end of the rift and the mafic eruptive vents along the distal part, close to the Caldera scarp.

7.6 Overview of the Volcanic Stratigraphy

A synthesis of the main volcano-stratigraphic units of the TVC is outlined in Fig. 7.27. Nested volcanism filling the 200 ky Icod collapse embayment is clearly separated from the pre-collapse, extra-Caldera formations.

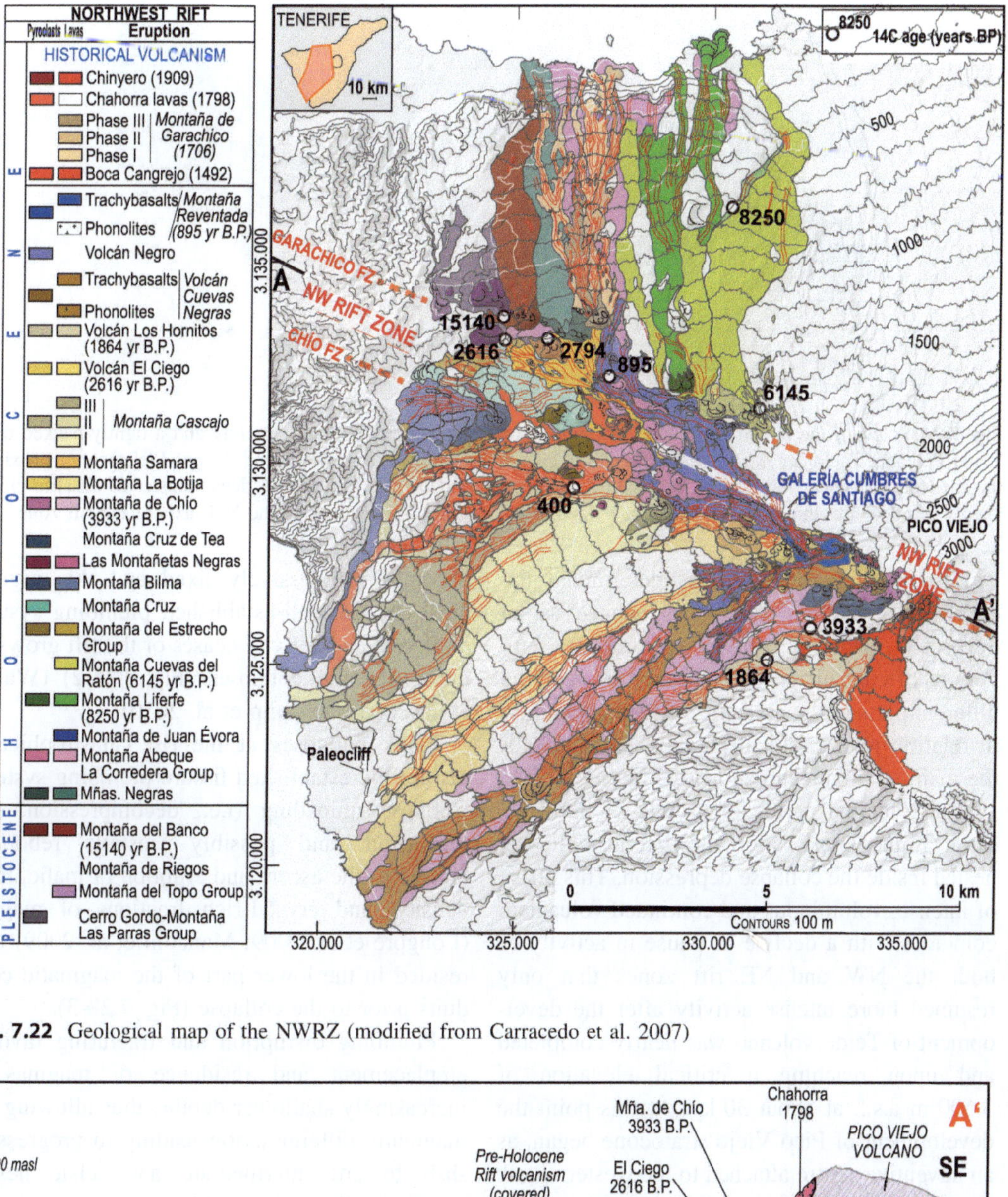

Fig. 7.22 Geological map of the NWRZ (modified from Carracedo et al. 2007)

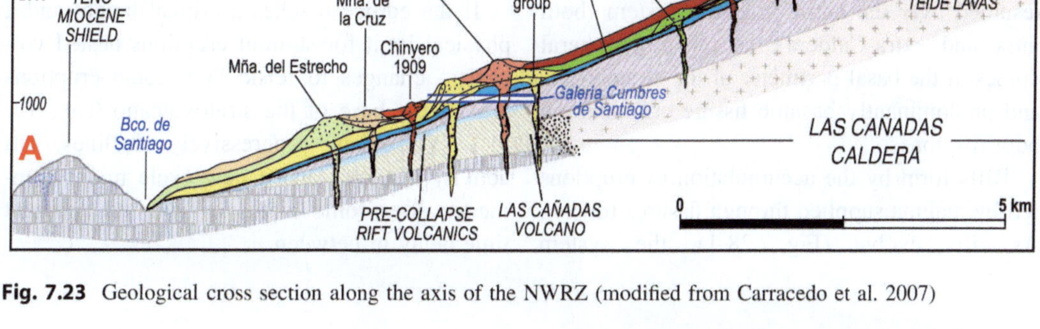

Fig. 7.23 Geological cross section along the axis of the NWRZ (modified from Carracedo et al. 2007)

Fig. 7.24 View of the NE rift zone from the summit of Teide Volcano. The older (Mid-Pleistocene) distal end of the rift (forested upper left part of the picture) has been deeply eroded by three successive giant landslides, showing its internal structure and a tightly-packed dyke swarm. The more recent (Upper Pleistocene) proximal part of the rift shows a dense cluster of basaltic cinder cones, oriented along the N–E axis of the rift zone

Based mainly on the ages determined the entire TVC and the rift zones behave as an integrated and interactive volcanic system. Periods of volcanic growth alternate between the different parts of the volcanic system, apparently in relation to the critical elevation attained for the main stratovolcano and ridge-like edifices.

The principal period of growth of the TVC starts immediately after the lateral collapse, nested inside the collapse depression. This phase of intense, voluminous and continued volcanism coincides with a decline or pause in activity of both the NW and NE rift zones that only resumed more intense activity after the development of Teide volcano was nearly completed and upon reaching a critical elevation of 3,500 m a.s.l. at about 30 ky. At this point the development of Pico Viejo stratocone began, as an adventive centre attached to the western flank of Teide, probably coinciding with renewal of eruptions at the NERZ. Subsequently, once Pico Viejo also attained a critical elevation, eruptions resumed over the entire volcanic system (both intra and extra-Caldera), as felsic peripheral domes at the basal perimeter of the stratocones, and predominantly basaltic fissure eruptions on both rift zones.

Rifts form by the accumulation of eruptions fed by magma supplied through fissures (dikes). As rifts evolve (Fig. 7.28-1), the system becomes progressively fixed and this regime persists as a well-established plumbing system until volcanic activity ceases or the rift grows to instability and collapses (Fig. 7.28-2) (Walter et al. 2005; Delcamp et al. 2010).

Flank collapses at the rift catastrophically disrupt the established fissure plumbing system. Collapse unloading (i.e., decompression, gas expansion, and possibly isostatic rebound) facilitates the ascent and eruption of mafic, high density, and crystal-rich fractions of magma (Longpré et al. 2009; Manconi et al. 2009) that resided in the lower part of the magmatic conduits prior to the collapse (Fig. 7.28-3).

Plumbing disruption and fracturing favours emplacement and residence of magmas at increasingly shallower depths, thus allowing for magmatic differentiation leading to progressive shift toward intermediate and felsic nested eruptions (Fig. 7.28-4). If volcanism continues, a central volcano may develop (e.g., the TVC) (Carracedo et al. 2011).

If the edifice reaches a critical height and a physical limit for summit eruptions nested volcanism changes to felsic lava dome eruptions around the base of the stratovolcano (e.g., Teide). Volcanism progressively declines, rift activity resumes, and a new cycle might commence after some hiatus or with no apparent time delay in between.

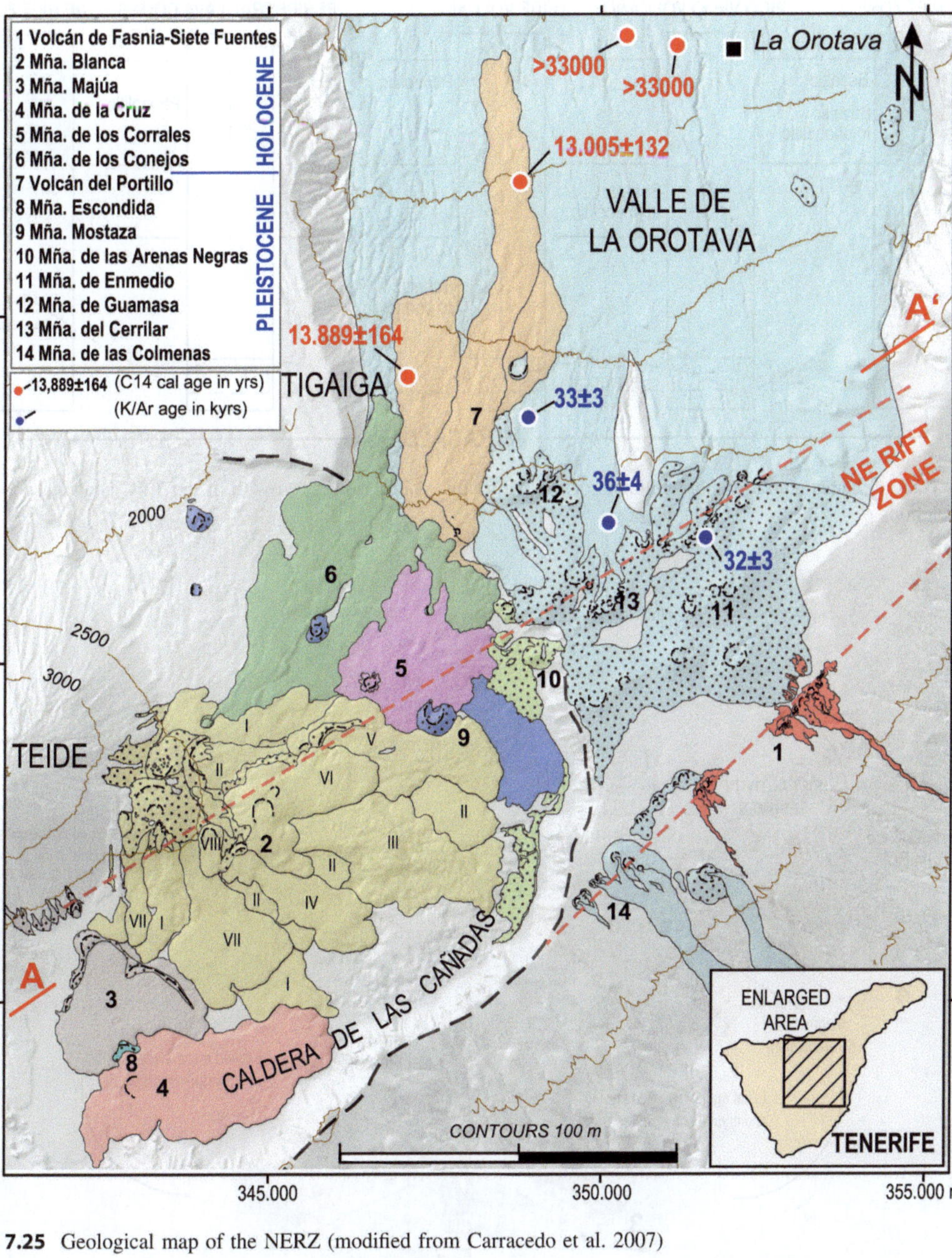

Fig. 7.25 Geological map of the NERZ (modified from Carracedo et al. 2007)

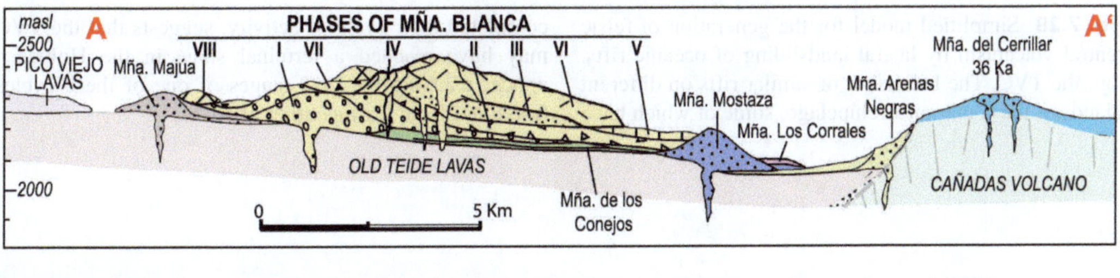

Fig. 7.26 Geological cross section along the axis of the NERZ (modified from Carracedo et al. 2007)

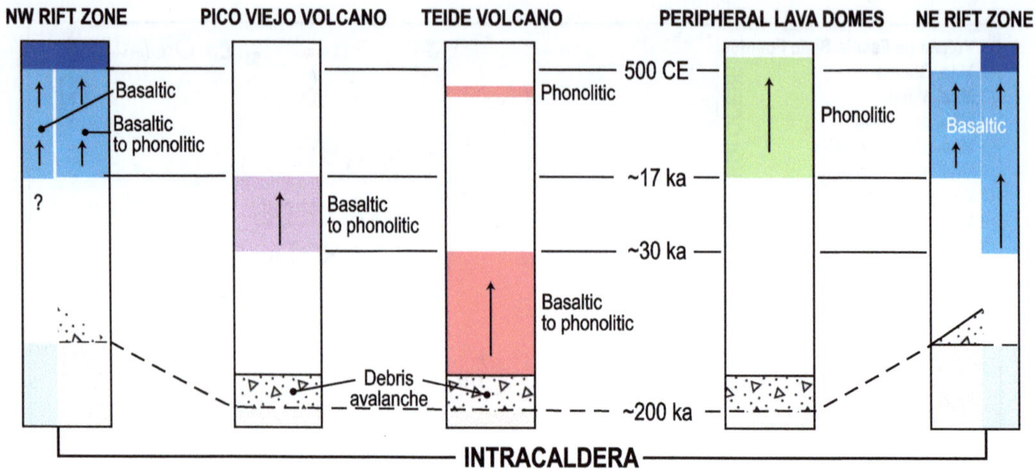

Fig. 7.27 Schematic stratigraphic columns of the main units and periods of growth of the TVC showing possible correlations

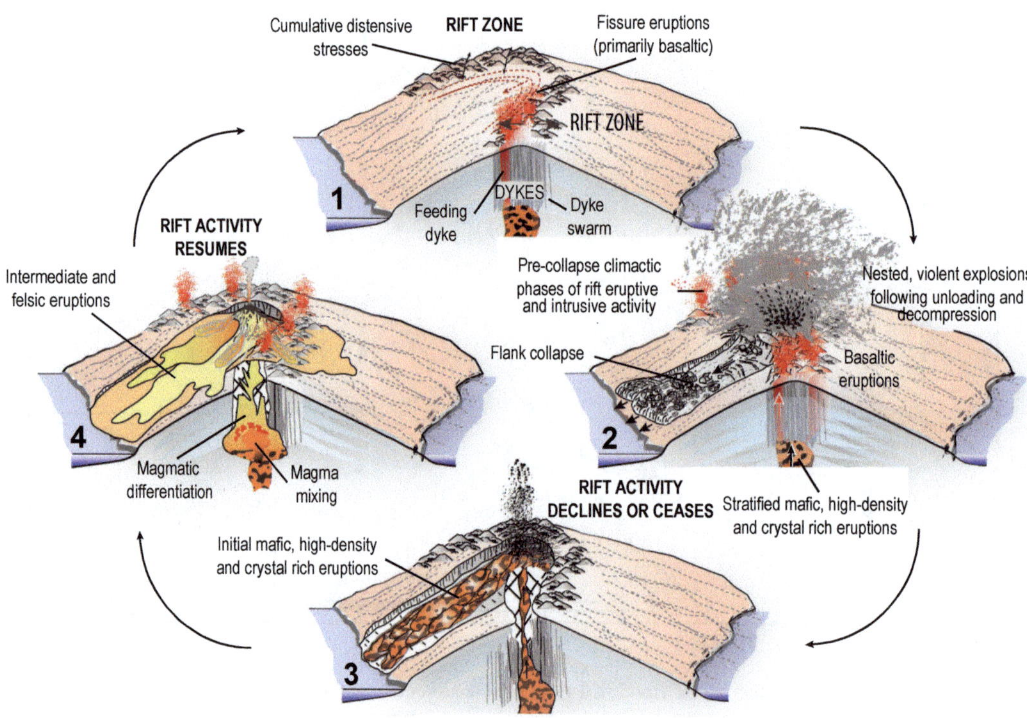

Fig. 7.28 Simplified model for the generation of felsic central volcanism by lateral landsliding of oceanic rifts, e.g., the TVC. The behaviour of similar rifts on different islands of the Canarian Archipelago, some of which have completed their cycle of activity, suggests that the TVC may have reached a terminal stage in the Holocene epoch, reaching the final stages of one of these cycles (modified from Carracedo et al. 2011)

References

Ablay GJ, Martí J (2000) Stratigraphy, structure and volcanic evolution of the Pico Teide-Pico Viejo formation, Tenerife, Canary Islands. J Volcanol Geotherm Res 103:175–208

Ablay GJ, Ernst GGJ, Marti J, Sparks RSJ (1995) The ~2 ka subplinian eruption of Montaña Blanca, Tenerife. Bull Volcanol 57:337–355

Ablay GJ, Carroll MR, Palmer MR, Martí J, Sparks RSJ (1998) Basanite-Phonolite lineages of the Teide-Pico Viejo volcanic complex, Tenerife, Canary Islands. J Petrol 39:905–936

Ancochea E, Huertas MJ, Cantagrel JM, Coello J, Fúster JM, Arnaud N, Ibarrola E (1999) Evolution of the Canadas edifice and its implications for the origin of the Canadas Caldera (Tenerife, Canary Islands). J Volcanol Geotherm Res 88:177–199

Araña V, Martí J, Aparicio A, García-Cacho L, García-García R (1994) Magma mixing in alkaline magmas: an example from Tenerife, Canary Islands. Lithos 32:119

Araña V, Felpeto A, Astiz M, García A, Ortiz R, Abella R (2000) Zonation of the main volcanic hazards (lava flows and ash fall) in Tenerife, Canary islands. A proposal for a surveillance network. J Volcanol Geotherm Res 103:377–391

Cantagrel JM, Arnaud NO, Ancochea E, Fúster JM, Huertas MJ (1999) Repeated debris avalanches on Tenerife and genesis of Las Canadas caldera wall (Canary Islands). Geology 27:739–742

Carracedo JC, Paterne M, Guillou H, Pérez Torrado FJ, Paris R, Rodríguez Badiola E, Hansen A (2003) Dataciones radiométricas (C^{14} y K-Ar) del Teide y el Rift NO, Tenerife, Islas Canarias. Estud Geol 59:15–29

Carracedo JC, Rodríguez Badiola E, Guillou H, Paterne M, Scaillet S, Pérez Torrado FJ, Paris R, Fra-Paleo U, Hansen A (2007) Eruptive and structural history of Teide Volcano and Rift Zones of Tenerife, Canary Islands. Geol Soc Am Bull 119:1027–1051

Carracedo JC, Guillou H, Nomade S, Rodriguez-Badiola E, Pérez-Torrado FJ, Rodriguez-Gonzalez A, Paris R, Troll VR, Wiesmaier S, Delcamp A, Fernandez-Turiel JL (2011) Evolution of ocean-island rifts: the northeast rift zone of Tenerife, Canary Islands. Geol Soc Am Bull 123:562–584

Davidson J, De Silva S (2000) Composite volcanoes. In: Sidgurdsson H (ed) Encyclopedia of volcanoes. Academic, San Diego, pp 663–681

Delcamp A, Petronis MS, Troll VR, Carracedo JC, de Vries BvW, Pérez-Torrado FJ (2010) Vertical axis rotation of the upper portions of the north-east rift of Tenerife Island inferred from paleomagnetic data. Tectonophysics 492:40–59

Farrujia I, Braojos JJ, Fernández J (2001) Ejecución de dos sondeos profundos en Las Cañadas del Teide. In: VII Simposio de Hidrogeología, Murcia, pp 661–671

Huertas MJ, Arnaud NO, Ancochea E, Cantagrel JM, Fúster JM (2002) Ar-40/Ar-39 stratigraphy of pyroclastic units from the Canadas Volcanic Edifice (Tenerife, Canary Islands) and their bearing on the structural evolution. J Volcanol Geotherm Res 115:351–365

Longpré M-A, Troll VR, Walter TR, Hansteen TH (2009) Volcanic and geochemical evolution of the Teno massif, Tenerife, Canary Islands: Some repercussions of giant landslides on ocean island magmatism. Geochem Geophys Geosyst 10:Q12017. doi: 10.1029/2009gc002892

Manconi A, Longpré M-A, Walter TR, Troll VR, Hansteen TH (2009) The effects of flank collapses on volcano plumbing systems. Geology 37: 1099–1102

Márquez A, López I, Herrera R, Martín-González F, Izquierdo T, Carreño F (2008) Spreading and potential instability of Teide volcano, Tenerife, Canary Islands. Geophys Res Lett 35:L05305. doi: 10.1029/2007GL032625

Martí J, Mitjavila J, Araña V (1994) Stratigraphy, structure and geochronology of the Las Cañadas Caldera (Tenerife, Canary Islands). Geol Mag 131:715–727

Mitjavila J (1990) Aplicación de técnicas de geoquímica isotópica y de geocronología al estudio vulcanológico del edificio de Diego Hernández y su relación con la Caldera de Las Cañadas (Tenerife). Ph. D. Thesis, Universidad de Barcelona, Barcelona

Mitjavila JM, Villa IM (1993) Temporal evolution of Diego Hernández formation (Las Cañadas, Tenerife) and confirmation of the age of the caldera using the $^{40}Ar/^{39}Ar$ method. Revista e la Sociedad Geológica de España 6(1–2):61–65

Navarro JM (1980) Plano geológico del complejo Teide-Pico Viejo (Islas Canarias, Excursión 121 A + C6, 26 Congreso Geológico Internacional, París). Bol Geol Minero 91:351–390

Navarro Latorre JM, Coello J (1989) Depressions originated by landslide processes in Tenerife. In: ESF Meeting on Canarian Volcanism, Lanzarote, 1989. European Science Foundation, Strasbourg, pp 150–52

Pérez Torrado FJ, Carracedo JC, Paris R, Hansen A (2004) Descubrimiento de depósitos freatomagmáticos en las laderas septentrionales del estratovolcán Teide (Tenerife, Islas Canarias): relaciones estratigráficas e implicaciones volcánicas. Geotemas 6:163–166

Pérez Torrado FJ, Carracedo JC, Paris R, Rodríguez Badiola E, Hansen A (2006) Erupciones freatomagmáticas del complejo volcánico Teide-Pico Viejo. In: Carracedo JC (ed) Los Volcanes del Parque Nacional del Teide. Serie Técnica. Organismo Autónomo Parques Nacionales, Ministerio de Medio Ambiente, Madrid, pp 345–357

Rodríguez Badiola E, Pérez Torrado FJ, Carracedo JC, Guillou H (2006) Petrografía y Geoquímica del edificio volcánico Teide-Pico Viejo y las dorsales noreste y noroeste de Tenerife. In: Carracedo JC (ed) Los volcanes del Parque Nacional del Teide/El Teide, Pico Viejo y las dorsales activas de Tenerife. Naturaleza y Parques Nacionales-Serie Técnica. Organismo Autónomo Parques Nacionales Ministerio de Medio Ambiente, Madrid, pp 129–186

Steiger RH, Jager E (1977) Subcommission on geochronology—convention on use of decay constants in geochronology and cosmochronology. Earth Planet Sci Lett 36:359–362

Valentin A, Albert Beltrán JF, Diez JL (1990) Geochemical and geothermal constraints on magma bodies associated with historic activity, Tenerife (Canary-Islands). J Volcanol Geotherm Res 44:251–264

Walter TR, Troll VR, Cailleau B, Belousov A, Schmincke H-U, Amelung F, Pvd Bogaard (2005) Rift zone reorganization through flank instability in ocean island volcanoes: an example from Tenerife. Canary Islands. Bull Volcanol 67(4):281–291

Watts AB, Masson DG (1995) A giant landslide on the northflank of Tenerife, Canary Islands. J Geophys Res 100:24487–24498

Wiesmaier S, Deegan F, Troll V, Carracedo J, Chadwick J, Chew D (2011) Magma mixing in the 1100 AD Montaña Reventada composite lava flow, Tenerife, Canary Islands: interaction between rift zone and central volcano plumbing systems. Contrib Mineral Petrol 162:651–669

The Last 2 ky of Eruptive Activity of the Teide Volcanic Complex: Features and Trends

8

Juan Carlos Carracedo

Abstract

Volcanic activity, involving both mafic and felsic eruptions, has been relatively frequent in the Teide Volcanic Complex over the last 2,000 years. Moderately explosive phonolitic events were endured by the aborigines, the most relevant being the Montaña Blanca subplinian event and the last summit eruption of Teide some 1,240 years ago. An eruption reported by Christopher Columbus, while passing the island in 1492, generally thought to relate to the latest eruption of Teide's summit, has been determined by radiometric dating to have occurred on the NW rift zone. This increases the number of Tenerife's eruptions during recorded historical time to five (1492, 1705, 1706, 1798, and 1909). Relationships and distribution of lava types show two main regimes at play; the basanite fed rift zones and the phonolite dominated central complex.

8.1 Introduction

The Puù Òò eruption is generally accepted as one of the longest-lived historical eruptions of Kilauea Volcano, in spite of the vent being almost 20 km distant from what can be considered the centre of the volcano, the Kilauea Caldera and the Halemaumau crater. However, it is less straightforward to associate the 1706 Garachico eruption on Tenerife with Teide Volcano, which is only 10 km distant (Fig. 8.1).

The most recent volcanism found in the volcanic stratigraphy of Tenerife is the "Serie Reciente", as derived by Fúster et al. (1968). Based mainly on lithological criteria, two coeval subseries were defined: the Recent Acid Series (*Serie Reciente Ácida*) and the Recent Basic Series (*Serie Reciente Básica*). On the geological map compiled by Fúster et al. (1968), the Recent Acid Series comprises all of the differentiated eruptions of Teide and Pico Viejo volcanoes, including the phonolitic peripheral domes and the latest eruption of Teide.

The *Serie Reciente Básica* includes the "well preserved" basaltic eruptions, comprising the historical (last 500 year) events along with Holocene and some pre-Holocene eruptions.

As outlined in Chap. 4, these recent (mainly Holocene) phonolitic and basaltic eruptions are end-members of a deep, primitive (rift) and a

J. C. Carracedo (✉)
Departamento de Física (GEOVOL), Universidad de Las Palmas de Gran Canaria, Las Palmas de Gran Canaria, Canary Islands, Spain
e-mail: jcarracedo@proyinves.ulpgc.es

J. C. Carracedo and V. R. Troll (eds.), *Teide Volcano*, Active Volcanoes of the World,
DOI: 10.1007/978-3-642-25893-0_8, © Springer-Verlag Berlin Heidelberg 2013

Fig. 8.1 The 1706 volcanic eruption of Garachico in an anonymous contemporary painting copied by Ubaldo Bordanova in 1898. The lava flows cascaded down the cliff and partially filled the harbour, the most important trading port with South America in the Canaries

shallow, differentiated (central volcano) magma series that form a bimodal compositional distribution, but with local mixing between the two. This intermix of the two sequences hinders an accurate chronological separation of these latest eruptive phases of the Teide Volcanic Complex.

Taking advantage of abundant new radiometric ages (Carracedo 2006; Carracedo et al. 2007b), geochronological criteria were used in Chap. 7 to define the recent –Holocene– volcanism of the TVC. However, we can apply a different, non-geological, approach and consider the last 2000 years, the period in which the island was already inhabited. Although the pre-Hispanic inhabitants (Guanches) did not possess a system of writing, some information about the eruptions they witnessed was retained through toponyms and oral traditions. The historical eruptions, described in eyewitness accounts, have been recorded since the Spanish colonization of the island in 1496.

The ten eruptions documented on Tenerife in the last 2 ky all belong to the TVC (Fig. 8.2). The historical events are exclusively fissural basaltic eruptions, generally strombolian, and of short duration (months); the pre-historical ones are predominantly phonolitic, longer-lasting (probably years), and considerably more explosive. These explosive eruptions took place during the Guanche period, probably explaining the word Echeide ("Hell" in the Guanche language) that gave rise to Mt. Teide.

8.2 The Pre-Historical Record: Mafic Eruptions of the TVC in the Last 2 ky

References to alleged eruptions of the TVC, mainly associated with Teide Volcano, are numerous but for the most part, likely erroneous. The frequent presence of orographic clouds (the well-known "*toca del Teide*" or Teide's headdress shown in Chap. 1, Fig. 1.5), phases of intense fumarolic activity within Teide's crater, or forest fires (either natural or caused by the pre-Hispanic Guanche inhabitants), may explain the recurrent references to possible eruptions of Teide. The majority of these are from fifteenth and sixteenth century sailors who passed by the island at a considerable distance from the coast.

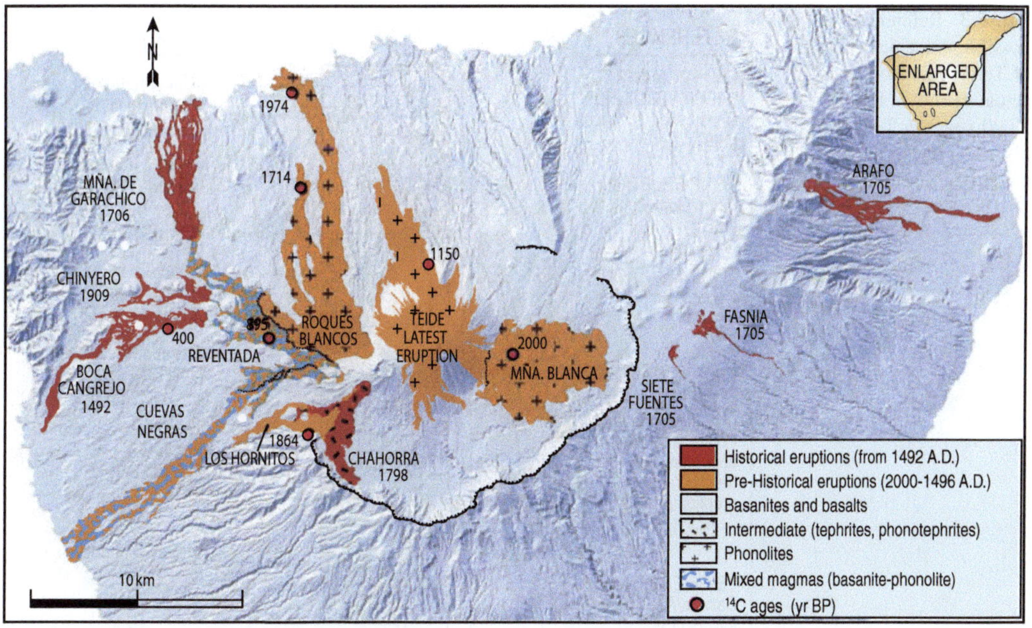

Fig. 8.2 Pre-historical (*orange*) and historical (*red*) eruptions of the TVC. Lava flows from most of these eruptions (9 out of 12) stopped on the flanks of the NW rift and did not reach the coast

Errors regarding historical eruptions are also due to the subsequent incorrect translation or interpretation of accounts written at the time of an event. A detailed study of the last stage of eruptive activity of Tenerife, particularly the Holocene, involving mapping and dating (using ^{14}C), now permits correlation of these possible eruptions of Teide to actual events (Fig. 8.3). The latest summit eruption of Teide is now established to have occurred in the Middle Ages at 1,150 ± 140 BP (Carracedo et al. 2007b).

8.3 Historical Eruptions of the TVC

An overestimation of the number of eruptions of the TVC from written accounts was apparent when the volcanic history was reconstructed in detail after mapping and dating of the successive events (Fig. 8.3).

The most controversial of these events was the latest eruption of Teide, generally related to a reference made by Christopher Columbus to "a fire on the sierra of Teide" when sailing past the south coast of Tenerife in August 1492. However, a ^{14}C age dated the last summit eruption to 1,240 ± 120 year BP (calibrated age 660 to 940 A.D.). This medieval age (Early Middle Ages) age excludes a Teide summit eruption in 1492.

Columbus probably intended a clear distinction when referring to the "fire" he reported to be located on the "sierra" (the ridge or mountain range) of Tenerife ("*scoglio*" in the original document, Colombo (1571), not on the Pico (the Peak), as the Teide Volcano was known in that epoch).

It is unlikely that an eruption, presumably of long duration and located at high altitude, would have passed unnoticed and unreported, since La Gomera and Gran Canaria, only 40 and 80 km distant, were colonized several years before.

8.4 The Christopher Columbus Eruption

Columbus sailed on August 24 from La Gomera to Gran Canaria to make the final preparations to his ships for his first voyage to America (Fig. 8.4). He

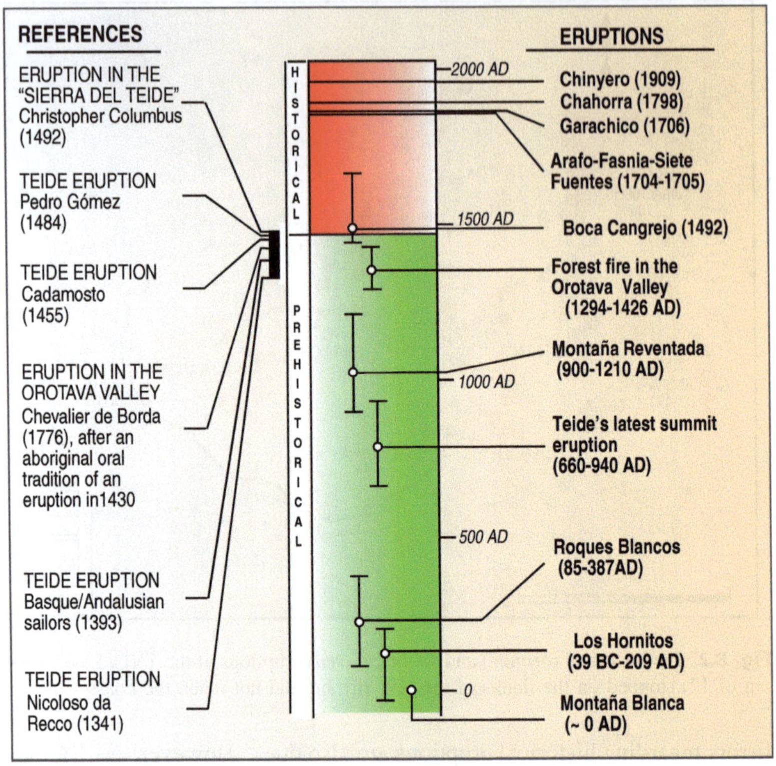

Fig. 8.3 References to possible volcanic eruptions of Teide volcano (*left*) and actual events in the last 2,000 years in Tenerife of historical age or C-14 dated (Carracedo et al. 2010)

reported in the ship's log that at night, passing close to Tenerife "...they saw a great fire on the sierra of Tenerife, very high..." ("*dal cui scoglio, che è altissimo, vediano uscir grossissime fiamme*"). To calm his crew's fears "...he explained to them the cause of that fire with the example of Mount Etna, in Sicily, and many others, where the same thing has been observed: "... *ad intendere il fondamento, e la causa dico tal fuoco, verificando il tutto con lo essempio del monte Etna in Sicilia, e di molti altri monti, dove si vedere il medesimo*" (Colombo 1571).

Once the successive eruptions of the TVC were dated and their stratigraphic position determined only one eruption complied with Columbus's description, the Boca Cangrejo Volcano, located 1.2 km to the south of Chinyero Volcano, that hosted Tenerife's latest eruption in 1909 (Fig. 8.5).

The 1492 eruptive centre is a multi-cratered breached vent. Tephrite lavas spread in numerous branches from the main vent along the NW rift zone, finally merging in a single flow that cascaded into the Santiago del Teide valley, and stopped about 600 m from the coast (Fig. 8.6).

Charcoal found underlying the lava, where abundant tree moulds were observed, gave a radiocarbon calibrated age of 400 ± 110 year BP, equivalent to the period 1,440 to 1,660 A. D (Carracedo et al. 2007a). This interval is overlapping with the report by Christopher Columbus in 1492. Therefore, the 1492 Boca Cangrejo eruption can be considered to be the first eruption in the historical record of Tenerife.

8.5 Eighteenth Century Eruptions in the TVC

Three of the five historical (<500 year) eruptions that took place in the TVC occurred in the eighteenth century and in different settings: in the NE and the NW rift zones in 1705 and 1706, respectively, and on the southern flank of Pico Viejo in 1798, although this is also likely to be related to the NW rift zone.

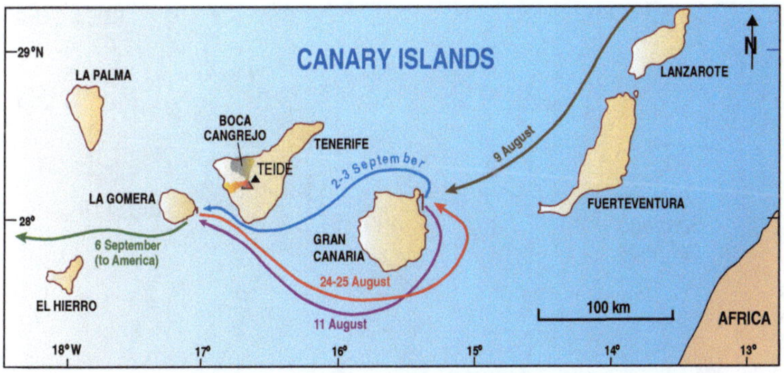

Fig. 8.4 Routes sailed by Columbus in August–September 1492 between La Gomera and Gran Canaria. On the night of August 24, they observed "a big fire on the sierra of Tenerife" (modified from Carracedo et al. 2007a)

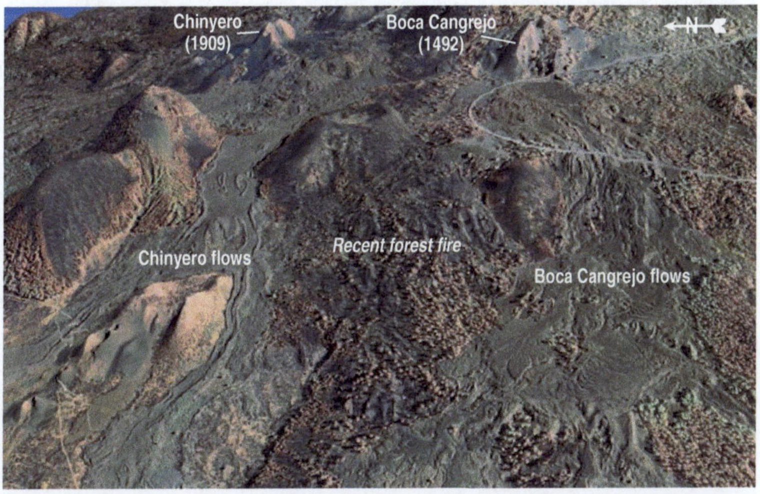

Fig. 8.5 Eruptive vent and lava flows of Boca Cangrejo Volcano, the eruption reported by Christopher Columbus in 1492. This volcano is very close (1 km to the north) to Chinyero Volcano, Tenerife's latest eruption in 1909 (Google Earth image)

8.5.1 The 1705 Fissure Eruption of Arafo-Fasnia-Siete Fuentes

Following a long period of eruptive repose (213 years after the 1492 Boca Cangrejo eruption), new vents opened after a week of strong seismicity that, according to eyewitness accounts (Viera y Clavijo 1776), caused more damage and concern in the population than the subsequent eruptive events themselves. Earthquake swarms of increasing intensity and number began on Christmas Eve 1704, and were particularly strong and frequent on 27 and 28 December. Seismicity affected the entire NE rift zone, where "… earthquakes shook and moved the earth making waves as in the sea, while churches and buildings collapsed in La Orotava, Los Realejos, Güímar and Candelaria". Apparently local intensities up to VII–VIII on the Mercali scale were recorded, similar to the Jedey earthquake (July 2) during the 1949 eruption of La Palma (Bonelli Rubio 1950). The absence of eruptive activity in this rift zone during the last 30 ky and the probable loss of "thermal memory" in the rift may explain the difficulties encountered by the basaltic feeder to progress, therefore causing the strong precursory seismic activity.

This volcanic event is a good example of basaltic fissure volcanism, with 3 vents aligned along a 10 km long volcanic trend parallel to the NE rift axis (Fig. 8.7a and b). The first vent to open, on Christmas Eve 1794, was Siete Fuentes volcano at the western end of the eruptive fissure (Fig. 8.7c), constructing a small cinder cone and erupting a short (1 km) lava flow.

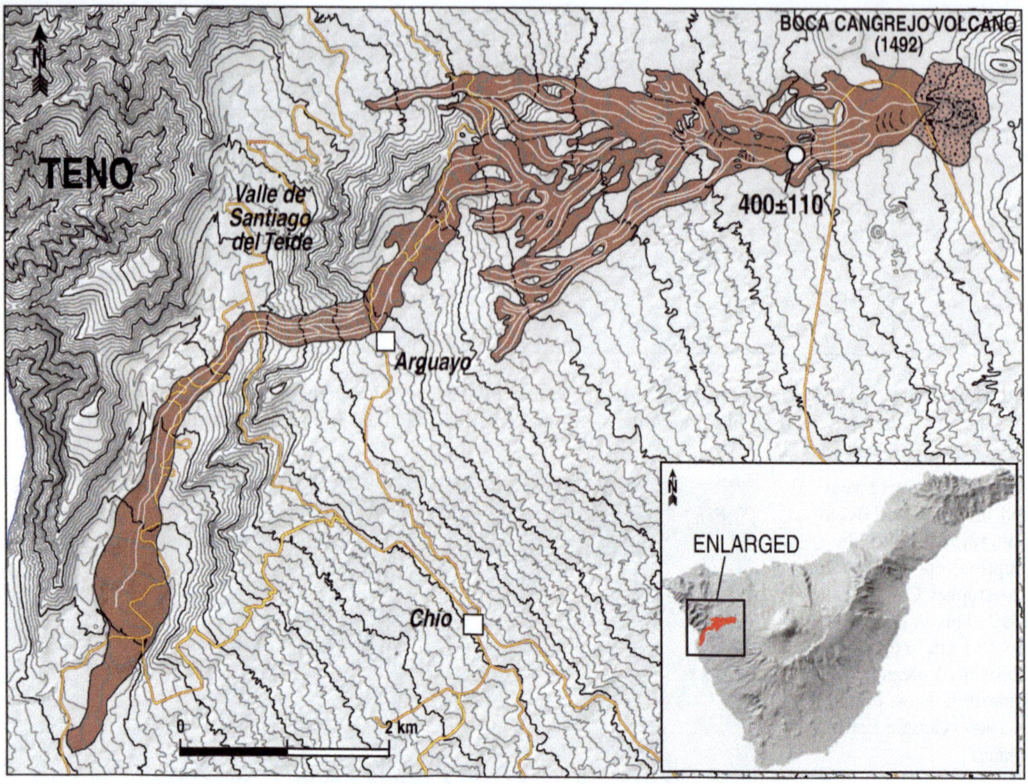

Fig. 8.6 Geological map of the Boca Cangrejo eruption, reported by Christopher Columbus during his first expedition to America in August 1492. This eruption is the first documented historical eruption of Tenerife (from Carracedo et al. 2007a)

On the 6 Jan 1705, the Volcán de Fasnia opened less than 2 km farther northeast along the eruptive fracture (Fig. 8.7c); some 30 initial small vents eventually concentrated to form 2 main cinder cones (Fig. 8.7d), erupting several short (<0.5 km) lava flows, and a longer one (~5 km), all channelled through the Barranco de Fasnia, also known as Barranco del Volcán (Fig. 8.8a). This episode ended on January 14.

Finally, on the 2nd February the last effusive vent opened on the southern flank of the NE rift zone, nested in the Güímar Valley, at an altitude of 1,500 m (Fig. 8.7e and f). A 100 m high cinder cone, Volcán de Arafo, formed and lava poured downslope forking in the valley at about 500 m.a.s.l. One flow stopped close to Güímar, while the second fork of the lava flow passed close to Arafo and stopped 1 km from the coast on 27 March, 1705 (Fig. 8.8b).

This 3 month-long eruption demonstrates the characteristic pattern of multi-vent fissure eruptions on the flank of rift zones, with the upper vents predominantly outgassing the volcanic system and generally accompanied by a limited eruptive volume, while the lower ones concentrate the emission of lavas (e.g., the 1712 and 1949 flank eruptions of Cumbre Vieja, La Palma).

8.5.2 The 1706 Eruption of Garachico Volcano

Although the 1705 eruption apparently ended in February (Cabrera Lagunilla and Hernández-Pacheco 1987) or March (Romero 1991),

Fig. 8.7 Different views of the eruptive centres of the 1705 eruption on Tenerife. **a** Aligned vents of the eruption viewed from the NE: *1* Siete Fuentes, *2* Fasnia, and *3* Arafo (Google Earth image). **b** Oblique aerial view from the E of the 1705 vents. **c** Another oblique view from the SW. **d** Cinder cones of Volcán de Fasnia. **e** Volcán de Arafo (or Montaña de Güímar), nested in the lateral collapse scar of the Güímar Valley. **f** Oblique view of Volcán de Arafo from the E

historical records report that seismic activity continued in Tenerife until a new vent, Montaña Negra, opened and the Garachico eruption started on May 5, 1706. Therefore, this new 1706 vent can be considered to be either the continuation of the 1705 eruption or a different eruption altogether. An unpublished contemporary document by Martínez de Fuentes reports: "Although finally that volcano (Montaña de Arafo) shut down, earthquakes persisted for more than a year… shaking buildings along with underground rattling, clearly indicating that the heated minerals were not yet completely evacuated, and were getting ready deep inside for an even more violent eruption".

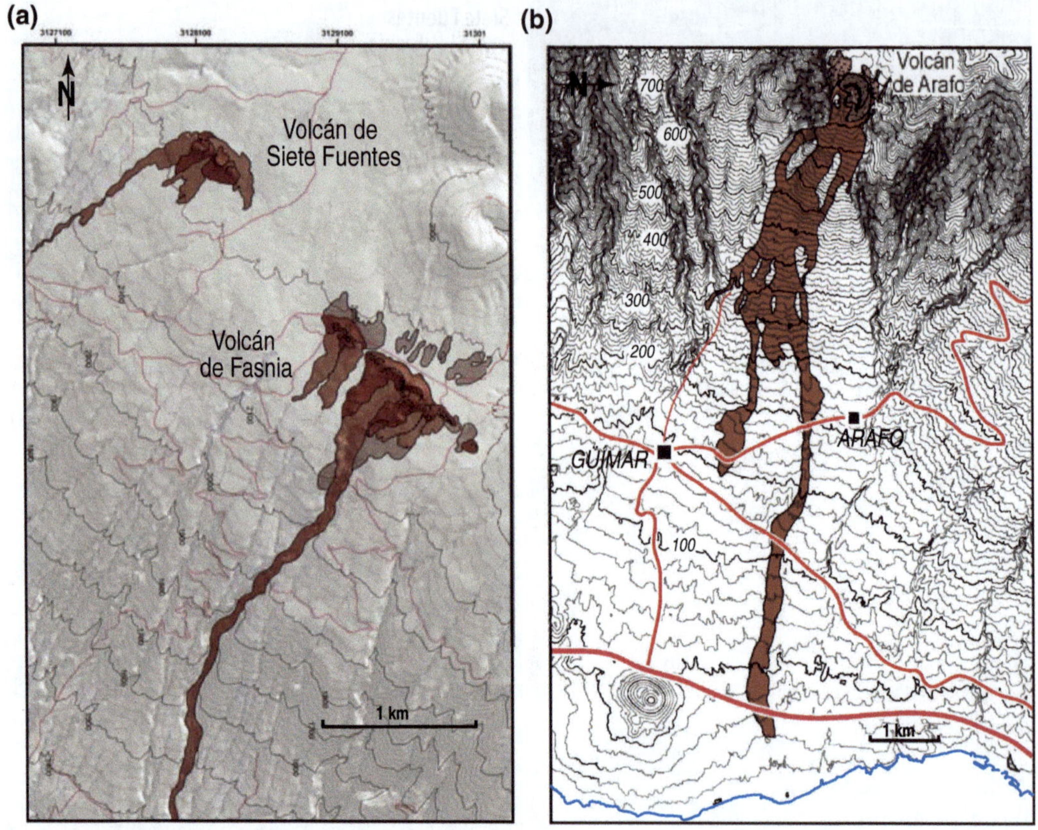

Fig. 8.8 **a** Geological map of Siete Fuentes and Fasnia volcanic vents from the 1705 eruption in the NERZ. **b** Geological map of Volcán de Arafo (or Montaña de Güímar), the easternmost vent of the 1705 eruption. Lava flows from these eruptions consistently ran out on the flank of the rift without reaching the coast

The eruption started on May 5, 1706 at 3:30 pm after a series of strong earthquakes the night before. The eruption began unexpectedly as earthquakes, frequent and persistent during the previous year, were no longer effective as a forewarning. In addition, the eruptive vent opened on a steep slope on the flank of the NW rift, only 6 km away from the town of Garachico (Fig. 8.9). In the afternoon lava flows were perched atop the cliff above Garachico, where they formed seven branches (see Fig. 8.1). The eastern ones partially filled the harbour, the most important port of the province of Tenerife in the sixteenth century (Fig. 8.10). Although this eruption caused moderate damage to the town and there were no casualties, it is considered to be the most disruptive of all historical eruptions of the Canary Islands, as the destruction of the port put a temporary stop to trade with the Americas and forced a complete reorganisation of the socioeconomic structure of the island. This setback was only fully resolved when the main port and capital were eventually established in Santa Cruz de Tenerife.

Eyewitness accounts primarily emphasise this damage to Garachico port. Viera y Clavijo (1776) described the devastation: "Vineyards, springs, birds, port, trade and neighbours disappeared and a lava branch filled the harbour and forced the sea to retreat, leaving a cove impracticable even for small boats".

In the early stages of the eruption a row of vents fed by a 500 m-long fissure emitted basaltic 'a'ā lavas for 8 days (Phase I in Fig. 8.9b). A new flow on May 12 progressed from the easternmost edge of the fissure towards

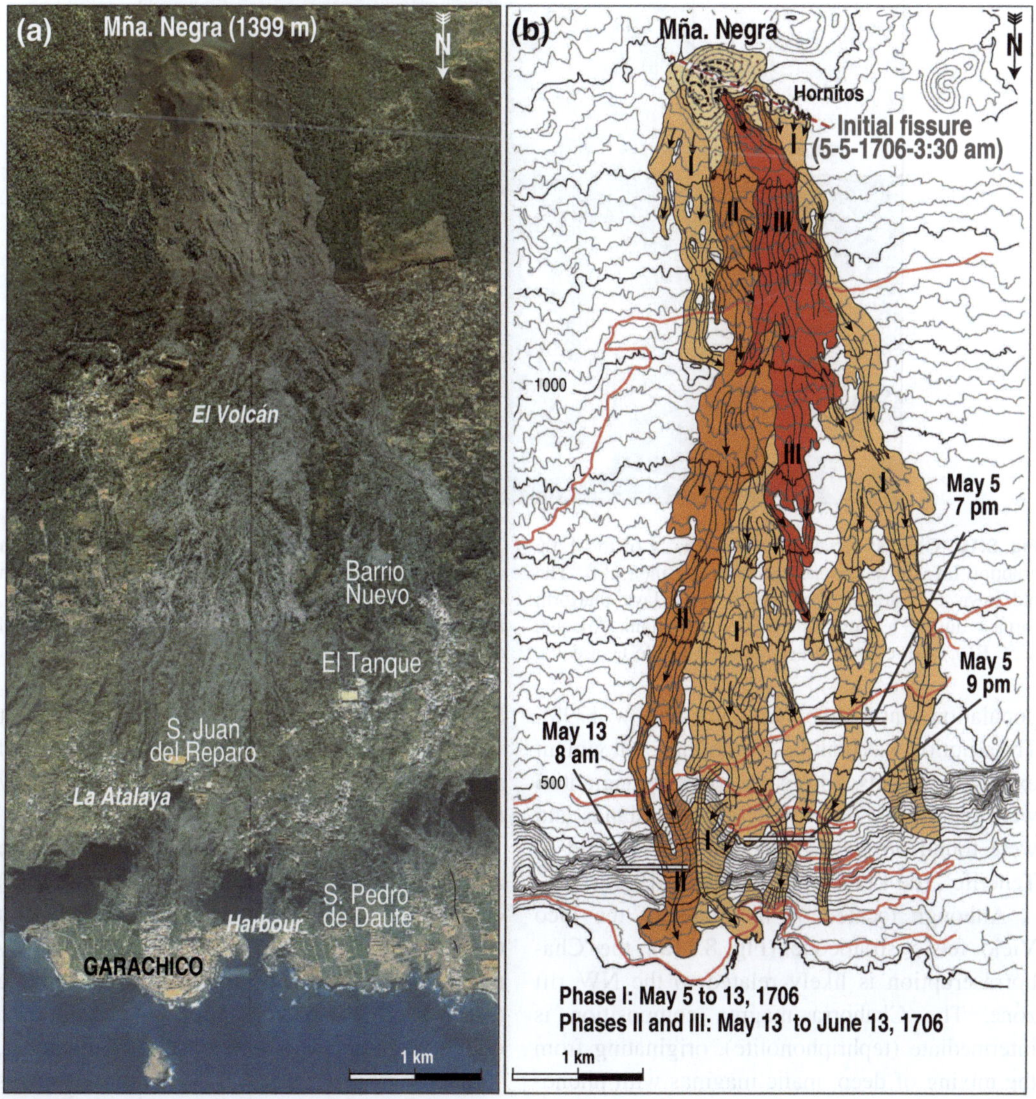

Fig. 8.9 **a** Google Earth image of the 1706 Garachico eruption. **b** Geological map showing the eruptive vent and the main phases of the eruption (from Carracedo 2006)

Garachico (Phase II in Fig. 8.9b), and destroyed the town centre. The later flows stopped close to the main vent without causing any further damage (Phase III in Fig. 8.9b). During the course of the eruption a 115 m-high volcanic cone (Montaña Negra) developed. The date of conclusion of the eruption is uncertain, although several contemporaneous accounts mention a duration of 40 days.

8.5.3 The 1798 Eruption of Chahorra Volcano

This late eighteenth century eruption (9 June 1798) has been described in detail in several contemporary accounts. The eruptive vents formed along a 1.2 km-long radial fracture on the southwest flank of Pico Viejo stratocone, following precursory seismicity felt by the

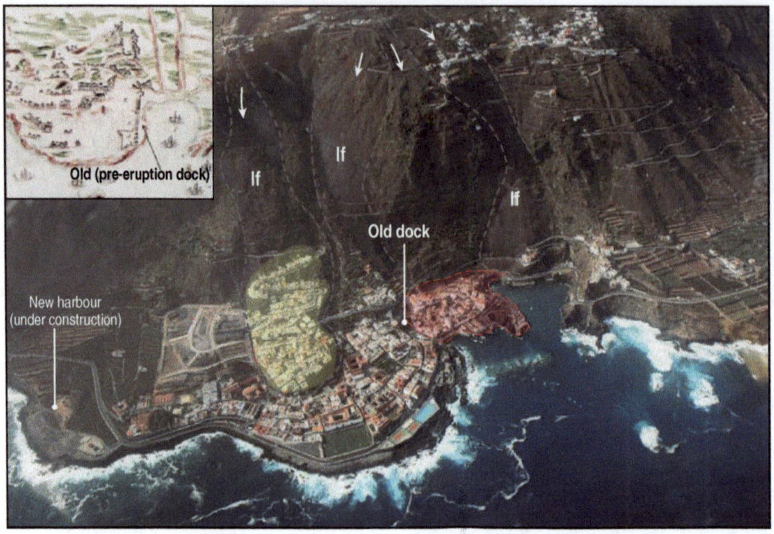

Fig. 8.10 Oblique aerial view from the N of the 1706 eruption lava flows cascading down the then cliff. The old town of Garachico was built on a lava platform formed during a previous eruption (Montaña Abeque, 8250 BP), with curved streets adapted to the lava delta. The figure shows the part of the town destroyed by lava and fires (in *yellow*), and the partially filled harbour (in *red*). Inset: An old picture showing the bay, pre-eruption, functioning as a natural harbour. The location of the old dock is also shown

population since 1795 (Cologan Fallow 1798). The eruption was mainly strombolian, but violent explosive events (probably phreatomagmatic) were reported: "Ashes fell in other islands and were found on leaves over the entire island of Tenerife" (de León and Cioranescu 1978).

Although located on the flank of the Pico Viejo felsic stratocone (Fig. 8.11a), the Chahorra eruption is likely related to the NW rift zone. The Chahorra magma composition is intermediate (tephriphonolite), originating from the mixing of deep, mafic magmas with phonolite magma from the shallow and differentiated magma reservoirs of the central felsic stratocones. This gives rise to a distinct bimodal series, with basanites and phonolites respectively as the two end-members, and intermediate eruptions found in the central part of the rift zones (See Fig. 4.15).

As occurred in La Palma in 1949 and 1712, the 1798 eruption on Tenerife is another example of a multi-vent fissure eruption located on a steep slope, where the upper vent degasses the system and is more explosive (strombolian), while the lower emits scoria and lava flows, and the lowest only lava flows. This is reported by Segundo de Franqui (1799) "the first and highest vent spewed dense spirals of fume, the second stones and fire, and the third and lowest one only fire".

Lava flows poured downslope and ponded to a considerable thickness (15–20 m in places) confined by the Las Cañadas Caldera rim (Figs. 8.11b, c and 8.12). The area covered by the lavas (4.9 km^2) and their estimated volume (30×106 m^3) are consistent with the duration of the eruption, which ended "mid September, 1798" (Cologan Fallow 1798) after 3 months and 6 days according to von Humboldt and Bonpland (1805). Thus, the 1798 Chahorra eruption is the longest among the historical eruptions of Tenerife and the second longest in the Canary Islands (after the Timanfaya eruption, Lanzarote, 1730–1736).

8.6 The Twentieth Century Eruption of the TVC

Only one eruption took place on Tenerife in the twentieth century, the Chinyero 1909 eruption, which was located at the western edge of the NW rift zone (see Figs. 8.2 and 8.13). Detailed

Fig. 8.11 The 1798 Chahorra eruption. **a** Eruptive vents viewed from the south (the main road). **b** Oblique view of the main, *upper* vent of the Chahorra eruption. **c** View from the summit of Pico Viejo of the vents and lava flows of the Chahorra eruption, nested within the Caldera de Las Cañadas

descriptions of this eruption were made and were widely distributed in both scientific articles and newspaper reports, providing the first ever photographs of an eruption in the Canary Islands (Fig. 8.14).

The eruption was preceded by more than one year of earthquake activity (since March, 1908), several individual events reaching intensities of VII on the Mercalli scale, were felt by the population mainly in the north of the island (Monge 1980). The eruption started in the afternoon (14:30) on November 18, 1909, in a remote and deserted, 1,490 m-high area of the NW rift zone (Ponte y Cologan 1911), as a W–E trending fissure with 5 different strombolian vents that merged into one main cinder cone, which eventually grew to an elevation of 80 m (Fig. 8.14a, b). Early in the eruption, a lava pond confined by adjacent cinder cones split into two branches and flowed towards the north and the west (Fig. 8.14c). The northern flow halted after about 1.7 km and solidified, while the westward one continued for another 3 km and then bifurcated, partly encircling Montaña Bilma (Figs. 8.14c, d and 8.15).

This basaltic eruption, studied and mapped by Fernandez Navarro (1911), lasted only 10 days (18 to 28 November, 1911) and covered an area of about 2.7 km^2, with a total volume of some 11×10^6 m^3. The eruption caused little damage (localised forest fires) and no casualties and was considered more as a spectacle and less as a hazard by the population.

Excursions were organised by boat from Santa Cruz de Tenerife to Garachico and from there to the eruption site, where many visitors admired the front of the advancing lava flows (Fig. 8.14d).

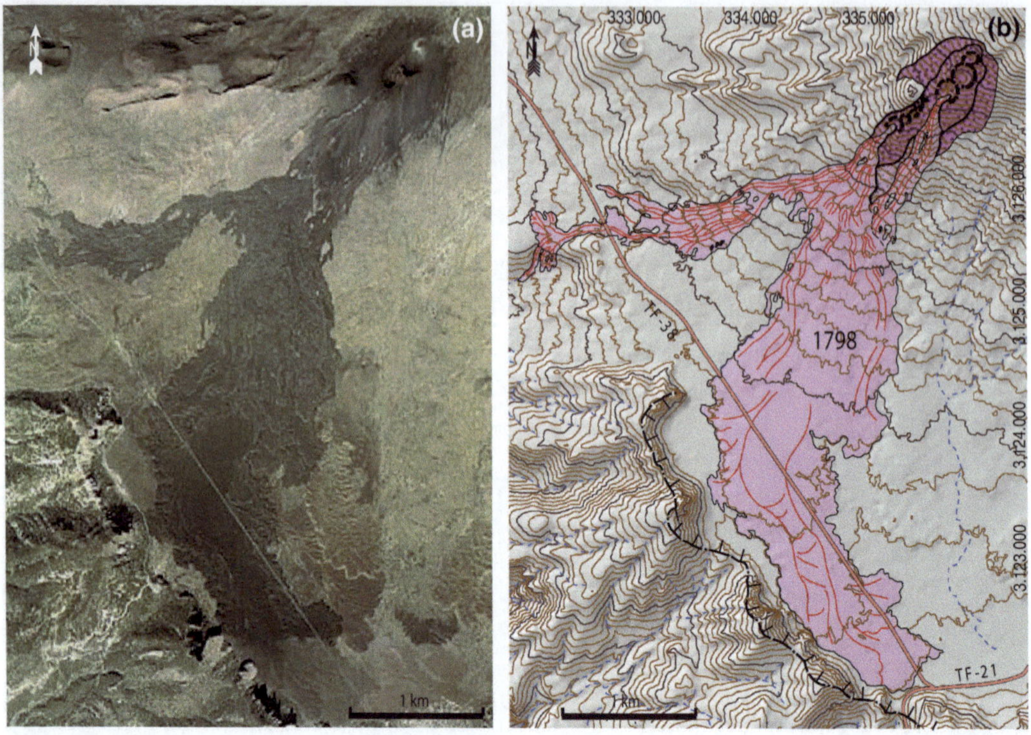

Fig. 8.12 **a** Google Earth view of the Chahorra eruption. **b** Geological map indicating the main vents and lava flows (from Carracedo 2006)

Fig. 8.13 Panoramic view of Chinyero vent and lava flows from Montaña de La Cruz (to the W), with Teide and Pico Viejo volcanoes in the background

8.7 Felsic Eruptions of the TVC

Felsic eruptions are an important aspect of TVC volcanism, constructing the later and more evolved parts of the Teide and Pico Viejo stratocones and the peripheral phonolitic lava domes around the two central volcanoes (see Chap. 7).

The four felsic eruptions that took place in the last 2 ky in the Teide Complex (Fig. 8.16) represent a range of styles, eruptive mechanisms

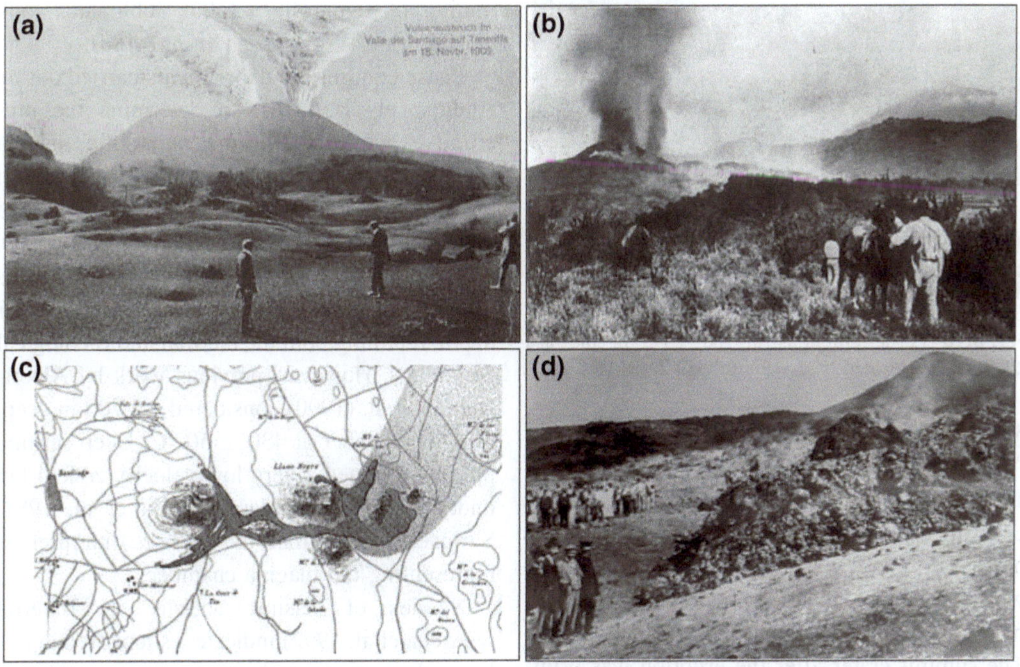

Fig. 8.14 **a**, **b**. Chinyero volcano (contemporary photographs, Centro de Fotografía Isla de Tenerife). **c** Map of the Chinyero eruption by Fernández Navarro (1911). **d** People closely watching a lava front of the 1909 eruption near Montaña Bilma (contemporary photograph, Centro de Fotografía Isla de Tenerife)

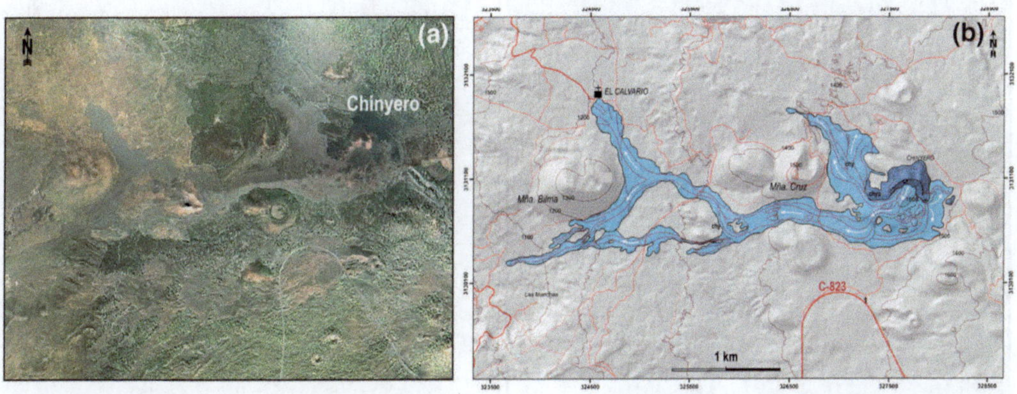

Fig. 8.15 The 1909 *Chinyero* eruption in Tenerife. **a** Google Earth view of the eruptive vent and lava flows. **b** Geological map showing the main eruptive vent and lava flows, encircling Montaña Bilma in their final course

and morphologies: (1) Summital eruption of the Teide stratocone (the latest Lavas Negras summit eruption), (2) Simple lava dome growth (Roques Blancos), (3) Composite lava dome growth (Montaña Blanca), and (4) Eruption with mixing of basaltic and phonolitic magmas (Montaña Reventada).

8.7.1 The Latest Eruption of Teide Volcano

The latest felsic volcanic activity of Teide occurred in 1150 BP (Carracedo et al. 2007a). This eruption built the 220 m summit cone of Teide, increasing its elevation from about

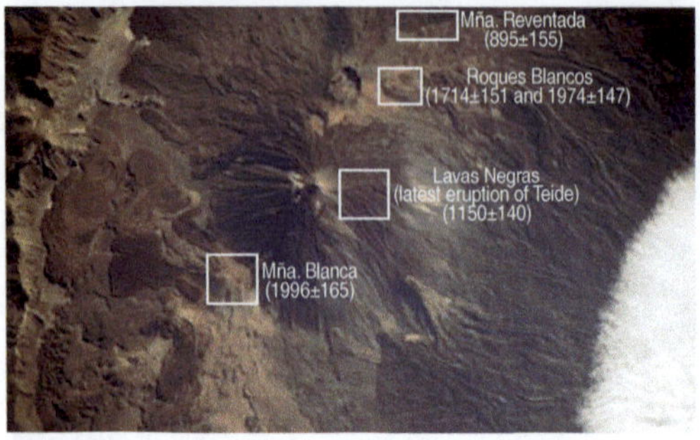

Fig. 8.16 Felsic eruptions of the TVC in the last 2000 years indicated on a satellite view by NASA

3,500 m to the present 3,718 m (Fig. 8.17). Approximately 0.66 km^3 (32.8 km^2) of lava spread radially from the summit crater, partially covering the former stratocone with thick, blocky flows of glassy (obsidianic) phonolites with an intense black colour (Lavas Negras, black lavas in Spanish) (Fig. 8.18a–f). The summit crater after the eruption was originally narrower and deeper before intensive mining of fumarolic sulphur sublimates took place during World War I (Fig. 8.18d).

Phase equilibrium experiments carried out by Andújar et al. (2010) to determine the pre-eruptive conditions of the phonolitic magma giving rise to Teide's latest eruption constrain the magma storage depth to about 5 ± 1 km below the current summit of Teide volcano. These authors suggest that magma from this eruption may still remain stored at a similar depth, retaining both the thermal and rheological conditions. However, previous work by Albert-Beltran et al. (1990) considered a 3,700 m deep, 30 km^3 chamber at 430 ± 50 °C. Deep seismic profiles detected a very low-velocity zone 4 km under the Teide caldera (Banda et al. 1981; Suriñach 1986), that the authors interpret as representing the magma chamber.

Studies of seismic activity in Tenerife (Mezcua et al. 1992) indicate a silent area in this low velocity zone and a concentration of hypocentres around the same, which is consistent

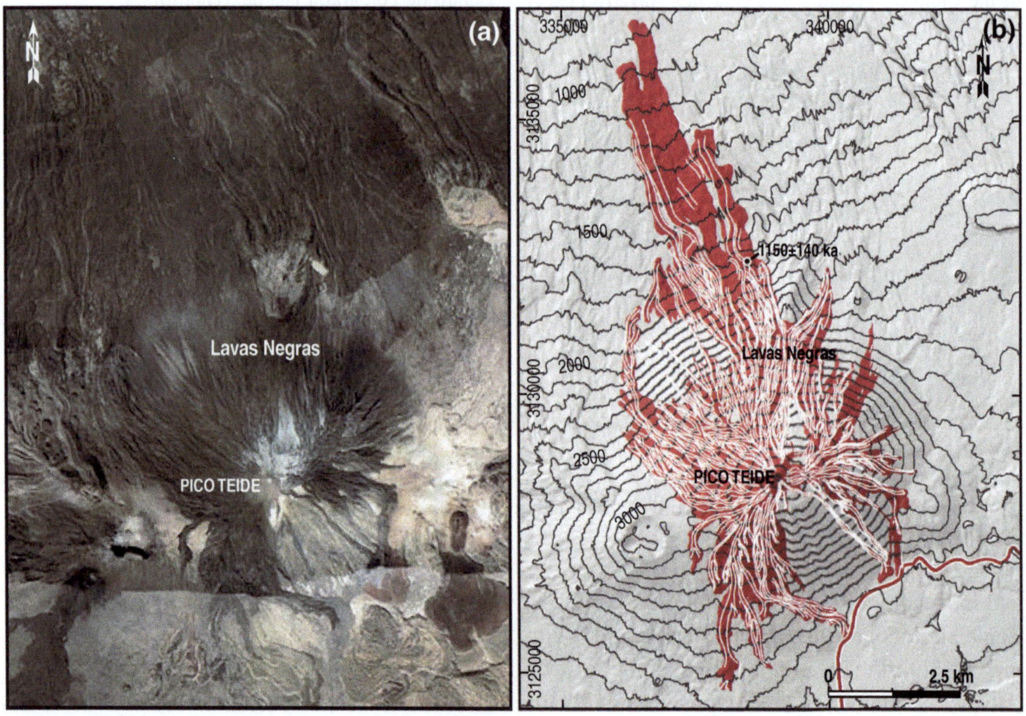

Fig. 8.17 The latest (medieval) eruption of Mt. Teide. **a** Google Earth view of the eruptive vent (the summit cone) and lava flows. **b** Geological map outlining the *Lavas Negras* phonolites forming a dense network of intertwined meandering flows with conspicuous levees

Fig. 8.18 Different views of Mt. Teide's latest summit eruption. **a** Southern face aerial view of the summit cone and the phonolitic flows encircling part of the "Old Teide" crater rim. **b** View of the stratocone from the W showing the channelled lava flows and their characteristic dendritic pattern. **c** Closer oblique aerial view of one of the southbound flows with a prominent lava channel confined between levees. **d** Aerial view from the SW of the interior of the Mt. Teide summit crater. The white colour is due to hydrothermal alteration. Patches of sulphur sublimate are seen in yellow. **e** Front of a lava flow (directed towards Pico Viejo) with its characteristic blocky morphology. **f** Obsidianic phonolite accretionary lava ball (locally Huevos del Teide or Teide's eggs) detached from the front of a lava flow descending the steep northern flank of Montaña Blanca

with the presence of this chamber (see Chap. 13 for further explanation).

8.7.2 Roques Blancos Lava Dome Eruption

The Roques Blancos lava dome is the youngest (^{14}C ages of 1714 ± 151 and 1974 ± 147 years BP) of the cluster of phonolitic lava domes encircling the central Teide and Pico Viejo stratocones. Located on the NW flank of Pico Viejo at an altitude of 2,700 m (see Figs. 8.16 and 8.19a), Roques Blancos is composed of a summit dome (Fig. 8.19b) and a complex system of ramified *coulées*, thick phonolitic flows with a blocky, steep front and conspicuous levees (Figs. 8.19c and 8.20c, d). A notable and characteristic feature of these flows is the abundance of obsidian, which commonly displays layered structures and folds due to differential laminar motion.

These viscous lavas rarely extend more than a few kilometres from their vents, often forming thick flows that move much slower than their basaltic and intermediate counterparts. The largest known Quaternary silicic lava body in the world is the 14 km-long coulée of Cerro Chao in north Chile (de Silva et al. 1994). However, the Roques Blancos phonolitic lava flow is longer, reaching the coast 14.7 km distant (Fig. 8.20).

The Roques Blancos flows and several similar lava domes (e.g., Pico Cabras, Abejera Alta) have similar lengths to the NW rift zone basanitic and transitional flows (see Fig. 14.10 and Table 14.1). Although the rheological behaviour allowing long flows (i.e. a low

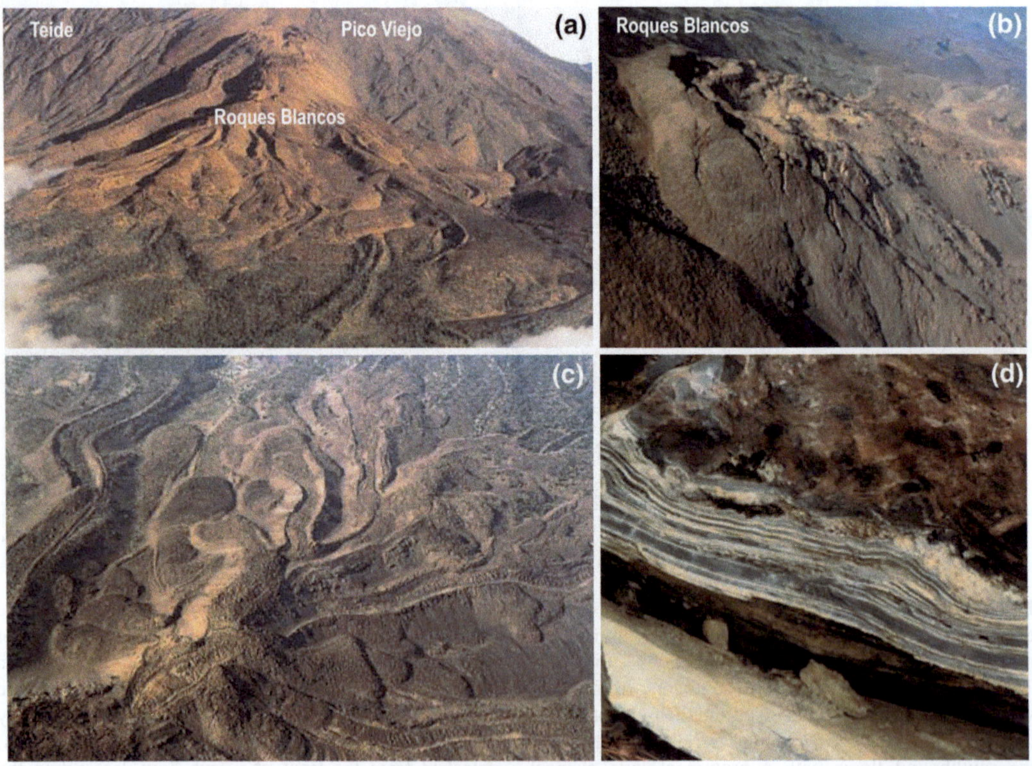

Fig. 8.19 The Roques Blancos eruption. **a** *Roques Blancos* dome and coulées viewed from the W. **b** Oblique aerial view from the north of the *Roques Blancos* dome. **c** Oblique view from the south of anastomosing lava flows from the *Roques Blancos* dome. **d** Characteristic banding in one of the *Roques Blancos* obsidianic phonolite flows. The alternating bands of different colours are caused by variations in microcrystallinity or microvesicularity. Often the layering can be distorted or folded

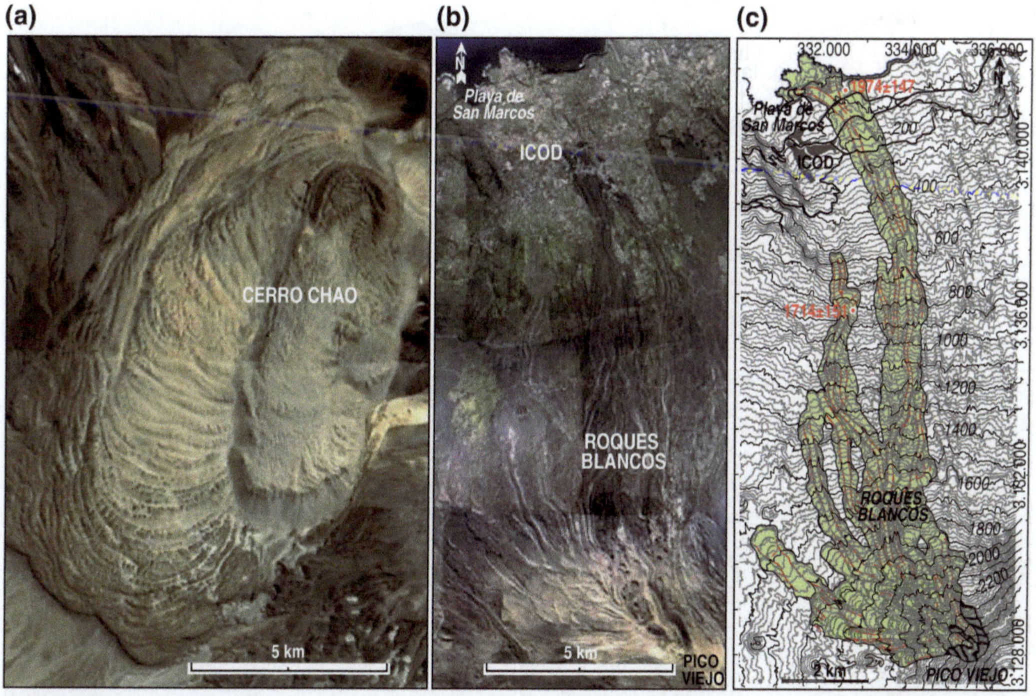

Fig. 8.20 **a** Google Earth view of the largest known Quaternary silicic (dacite) lava body in the world, the 14 km-long dacitic coulée of *Cerro Chao* in north Chile. **b** Google Earth view of *Roques Blancos*, a 14.7 km-long phonolitic coulée of the TVC. **c** Geological map of *Roques Blancos* showing the eruptive vent and branching and anastomosing lava flows. The ages of the northbound longer lava flows are also indicated

viscosity) is expected of basanitic and intermediate flows, it is unusual with these more silicic and more viscous phonolitic obsidian lavas.

The surface textures and internal structure of these phonolitic flows show characteristic features: thick flow margins that are tens of metres high and well-developed lava levees and flow ridges (ogives). The levees form as the central part of the flow keeps moving, whereas the margins of the flow subside as the lava cools down and solidifies (Fig. 8.21a). Ogives are transverse ridges on the central part of a phonolite flow that form as a result of compressional forces parallel to the course of the *coulée* (Fig. 8.21b).

These flows show a massive external lava carapace contrasting with a more brecciated interior (Fig. 8.21c): This outer shell likely produces an effective thermal isolation of the lava, allowing flows to remain hotter for longer and thus travel longer distances.

The duration of these silicic lava dome eruptions is probably considerably greater than that of basaltic and intermediate events, which generally last only a few months to, at the most, a few years (e.g., Timanfaya 1730–1736, Lanzarote). Indeed, the duration of the Chao eruption is thought to have been about 100 to 150 years (de Silva et al. 1994). The difference in the radiocarbon ages obtained for the two different Roques Blancos lava flows (see Fig. 8.20b) also suggests a long period of eruptive activity (Carracedo et al. 2007b). The anomalous size and duration of these eruptions may be related to the steep local slope onto which they were erupted ($\sim 10°$) and the available volume of magma.

8.7.3 Montaña Blanca Composite Lava Dome

Just as Roques Blancos may be considered a representative slope-lava dome eruption of the TVC, the Montaña Blanca eruption represents a lava dome located on the nearly flat floor of Las

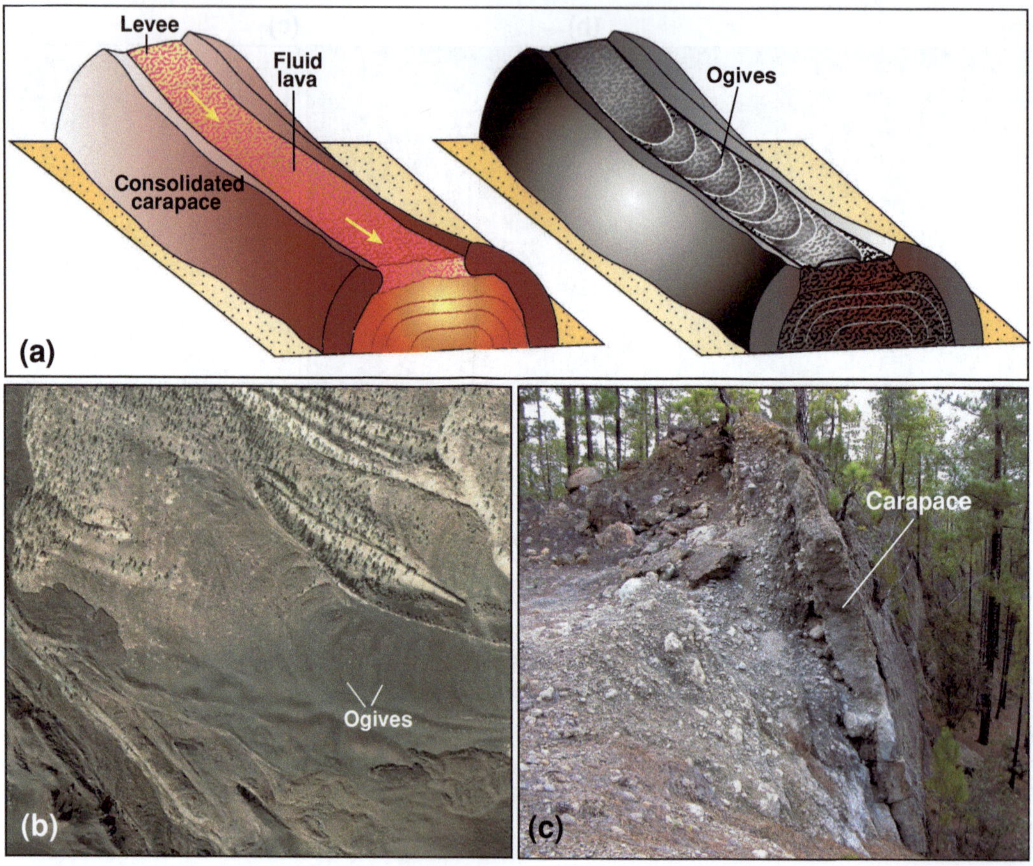

Fig. 8.21 **a** Schematic model of the formation of a channelled phonolitic *coulée*. The outer massive *carapace* thermally isolates the lava within, which can flow for distances up to 15 km. **b** Ogives formed in a thick *Roques Blancos* flow in response to compressional forces parallel to the course of the coulée. **c** Detail of the outer massive *carapace*

Fig. 8.22 The *Montaña Blanca* group of domes and coulées viewed from Teide summit

Cañadas Caldera. Here, flows tend to pile up over the vents building endogenous domes with a characteristic "rosette" structure (e.g., Montaña Rajada in Fig. 8.22), or alternatively run for short distances forming thick lava flows that are soon confined by the Las Cañadas Caldera rim (Figs. 8.22 and 8.23).

The Montaña Blanca eruption has been studied and mapped in detail. Ablay et al. (1995) defined three eruptive units, the uppermost one

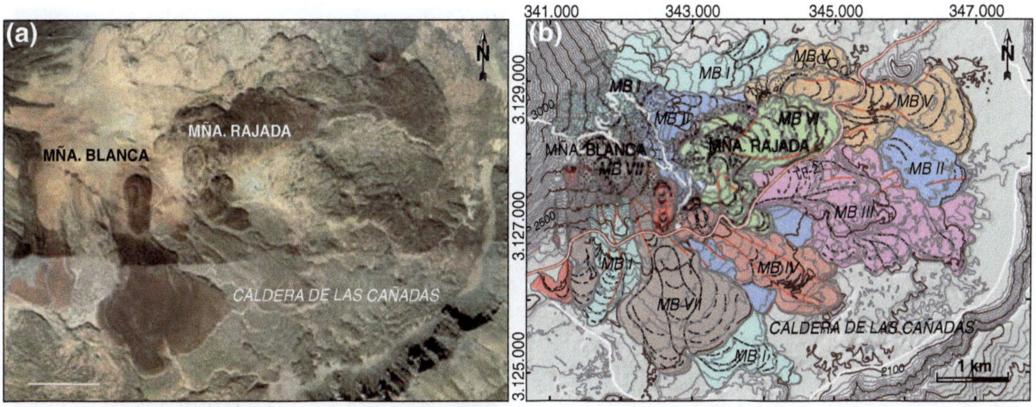

Fig. 8.23 The Montaña Blanca composite lava dome. **a** Google Earth view of the domes and lava flows. **b** Geological map of the *Mña. Blanca* composite cluster of endogenous domes. Located on the almost flat floor of Las Cañadas Caldera, lava flows run for short distances forming thick flows that pile close to the eruptive vents with a characteristic "rosette" structure

corresponding to the Montaña Blanca (MB) dome complex, subdivided by these authors into a Lower and Upper MB sub-members, the younger dated by ^{14}C at ~2 ky (Ablay et al. 1995). Carracedo et al. (2007b) defined eight different eruptive episodes during the Montaña Blanca events, which produced both lavas and pumice, although only the younger deposits are well exposed at the summit of Montaña Blanca. Charcoal under a Montaña Blanca-derived pumice deposit north of Teide has yielded a calibrated age of 1986 ± 165 year BP (Carracedo et al. 2007b) similar to that obtained by Ablay et al. (1995). The second youngest of the eight eruptive phases (VII in Fig. 8.23) is analogous to the unit dated at ~2 ky by Ablay et al. (1995). Lavas not covered by pumice air-fall are mainly short tongues extruded from a late flank fissure (Phase VIII).

Most of the Montaña Blanca pyroclasts comprise pumice lapilli and bombs of fallout origin. Isopach maps (see Fig. 14.18) indicate that the vents were located at the summit of Montaña Blanca (Ablay et al. 1995) and that the subplinian Montaña Blanca eruption is the only significant explosive event of the TVC.

8.7.4 The Montaña Reventada Magma Mixing Event

Dated using ^{14}C to 990 ± 70 year BP (900 to 1210 A.D. calibrated age), Montaña Reventada is the youngest pre-historical eruption of the TVC, and represented the final eruption before a ~600 year repose interval, which was eventually interrupted by the 1492 Boca Cangrejo event.

Montaña Reventada consists of a group of aligned eruptive vents located within the Northwest rift zone (Montaña Reventada and Las Lenguas) close to the central felsic Teide and Pico Viejo stratocones (see Fig. 8.16). Several lava flows from Montaña Reventada run along the flank and the saddle of the Northwest rift (Fig. 8.24), covering a surface area of about 22 km^2.

As described in detail in Chap. 11, the most interesting feature of this eruption is the presence of a composite flow or cooling unit with a lower felsic and an upper mafic member (Fig. 8.25), initially thought to reflect an eruption from a large, stratified magma chamber. However, the Montaña Reventada composite

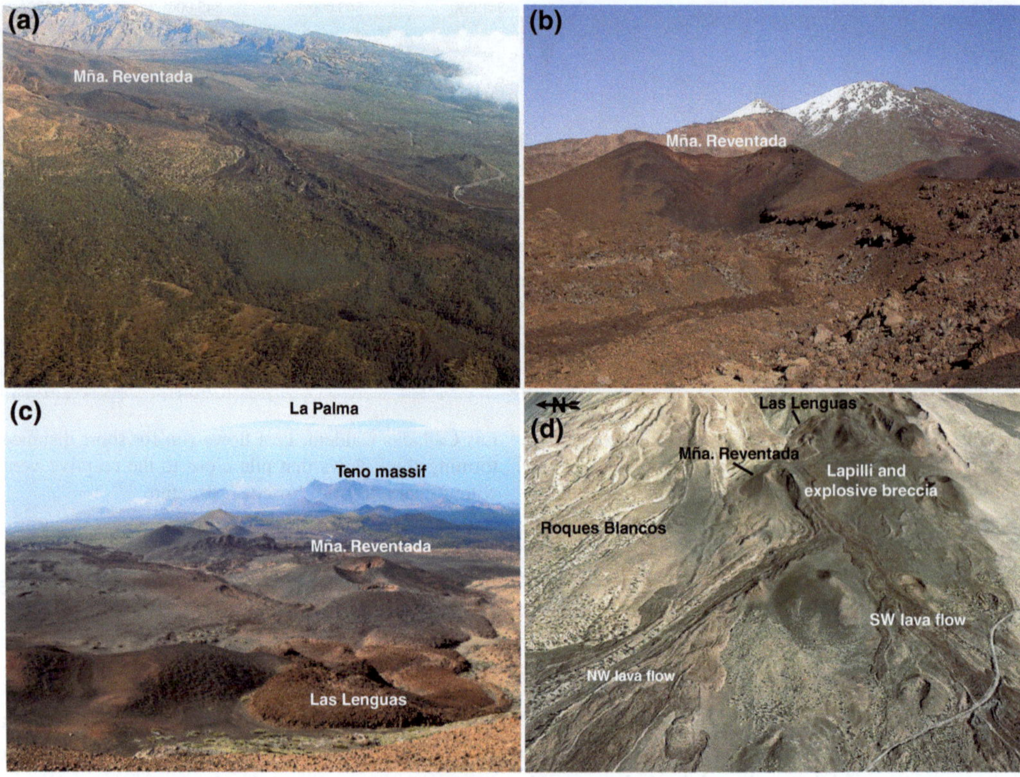

Fig. 8.24 The Montaña Reventada eruption. **a** Oblique aerial view from the NW of the eruptive vent and lava flows. **b** Close-up view from the west of Montaña Reventada, with Teide and Pico Viejo in the background. **c** View of Montaña Reventada from the summit of Pico Viejo. The three *coulées* of phonolite lava in the foreground belong to the Montaña Reventada eruption. **d** Oblique satellite view of the two-armed Montaña Reventada composite lava flow (image: Google Earth)

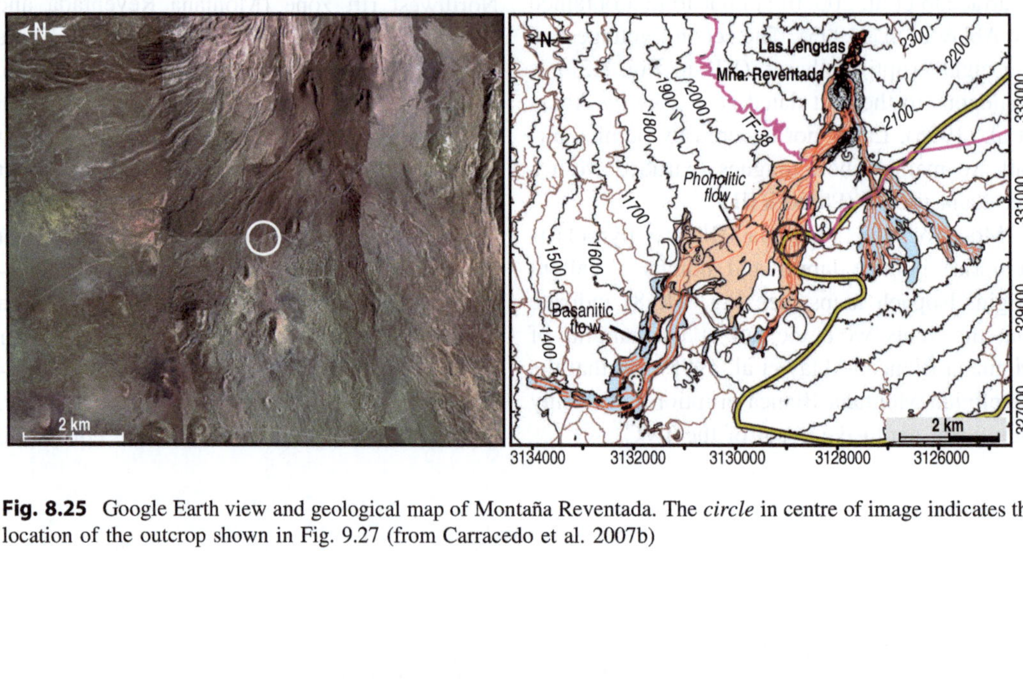

Fig. 8.25 Google Earth view and geological map of Montaña Reventada. The *circle* in centre of image indicates the location of the outcrop shown in Fig. 9.27 (from Carracedo et al. 2007b)

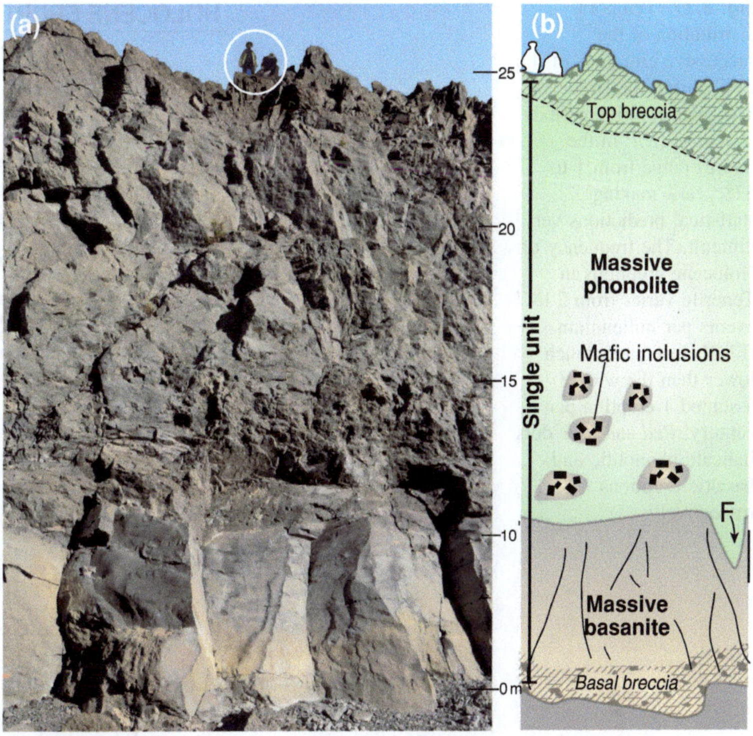

Fig. 8.26 **a** Photograph of the main outcrop of the Montaña Reventada composite flow with human figures on top for scale. **b** Simplified stratigraphic column of the outcrop. Note the opened fracture within the basanite (*F*) that has been filled with phonolite (after Wiesmaier et al. 2011)

flow is now thought to be composed of a basanite lava that erupted immediately prior to a phonolite lava, the latter with abundant dark inclusions (Araña et al. 1994; Wiesmaier et al. 2011). A sharp interface separates the basanite and phonolite domains, and chilled margins at this contact, as well as phonolite filled fractures within the basanite, indicate that the basanite was still hot upon emplacement of the phonolite. This means the two magmas erupted in very quick succession (Fig. 8.26).

Analysis of basanite, phonolite and inclusions for major, trace and Sr, Nd and Pb isotopes shows the inclusions in the phonolite to be derived from binary mixing of the basanite and phonolite endmembers (Wiesmaier et al. 2011). Although the basanite and phonolite magmas were in direct contact for at least a short time, contrasting $^{206}Pb/^{204}Pb$ ratios show that they are genetically distinct.

The process driving the formation of the Montaña Reventada composite flow is the physical mingling and progressive mixing accompanied by a decreasing temperature contrast between the basanite and phonolite endmembers. The Montaña Reventada basanite and phonolite first came into contact just prior to eruption and therefore had only limited interaction time. Montaña Reventada erupted within the transition zone between two plumbing systems, the phonolitic Teide-Pico Viejo complex and the basanitic Northwest rift zone (Carracedo et al. 2007b). A basanite dyke similar to those feeding eruptions of the NW and NE rift zones most likely intersected a previously emplaced phonolite magma chamber related to the central Teide-Pico Viejo volcanoes. This led to the eruption of geochemically and texturally unaffected basanite, with the inclusion-rich phonolite subsequently flowing into the established conduit (Wiesmaier et al. 2011).

The Montaña Reventada composite flow is not an isolated case in TVC volcanism. Other eruptions in the vicinity (e.g., Cuevas Negras volcano) show similar features. This marked compositional bimodality displaying a range from mafic lavas to highly differentiated felsic lavas with some transitional products can be

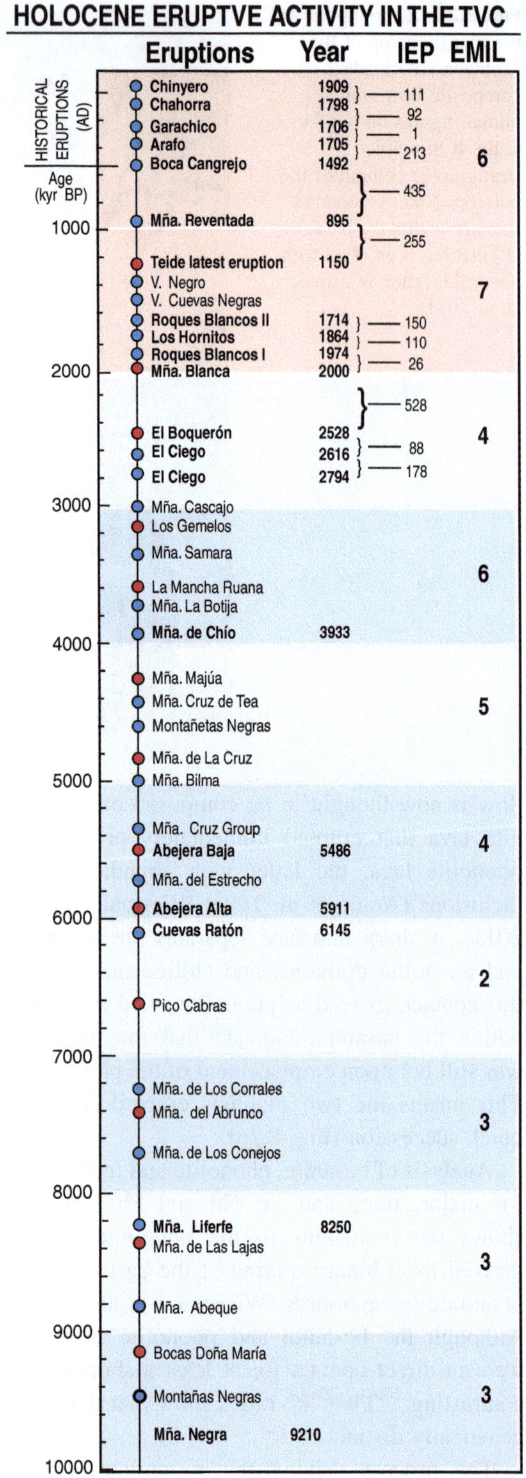

Fig. 8.27 Temporal distribution of the successive eruptions of the *TVC* during the Holocene. The inter-eruptive periods of repose (*IEP* in the figure) range from 1 to 528 years, making statistical predictions very difficult. The frequency of Holocene eruptions in Tenerife varies from 2 to 7 events per millennium (EMIL in figure), much lower than the widely assumed 1 eruption per century. *Red* and *blue* dots indicate phonolitic and basaltic eruptions respectively

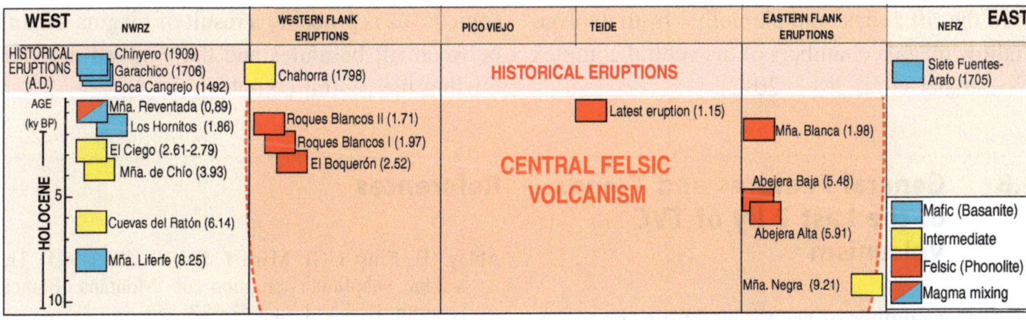

Fig. 8.28 Spatial distribution of Holocene eruptions of the TVC. Note the clustering of felsic lavas in the central Teide-Pico Viejo volcanoes. The mafic lavas define the rift zones, with the transitional lavas falling into the geographical and chemical transition zone between the rifts and the central felsic phonolitic lavas (data from Rodríguez-Badiola et al. 2006)

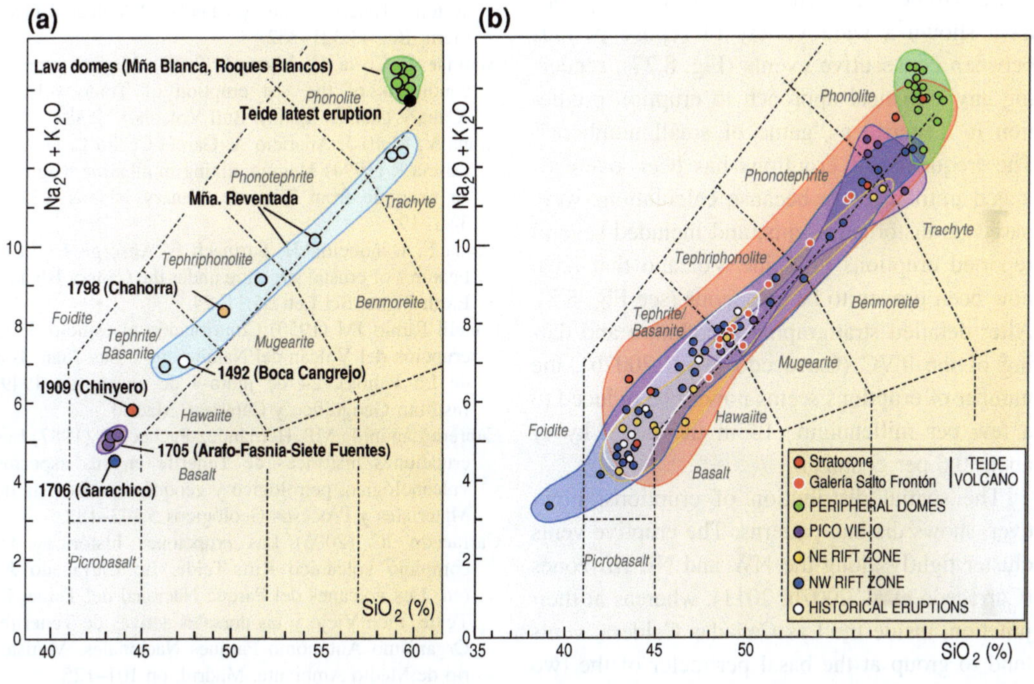

Fig. 8.29 **a** Samples of the <2 ky eruptions of the TVC plotted on a total alkali versus silica diagram (*TAS*) after Le Bas et al. (1986). Note the well-defined general bimodal trend and the specific case of the Montaña Reventada eruption. **b** Similar TAS diagram from lavas of the TVC throughout its development, closely mimicking the pattern of the latest volcanism (data from Rodríguez-Badiola et al. 2006)

clearly observed in this part of Tenerife. To highlight the temporal component, the compositional evolution of lavas at the TVC shows initially mafic lavas (200–30 ky) and late highly differentiated phonolite (30 ky–recent). This is discussed in greater detail in Chap. 7.

The effect of spatial distribution is illustrated by the compositional bimodality present in the NWRZ between the now-felsic Teide–Pico Viejo to the east, and the exclusively mafic lavas of the western edge of the rift. Intermediate composition magmas have erupted only in the middle part of the rift zone (Ablay et al. 1998; Carracedo et al. 2007b). Intermediate magmas on Tenerife may thus form through the interaction of two end-member-type magmas: basanite, from the rift zones, and phonolite, from evolved shallow magma chambers of the central complex (Wiesmaier et al. 2011, 2012).

8.8 General Features and Trends of the Last 2 ky of TVC Volcanism

The scant number of eruptions during the recent (last 2 ky) volcanism of the TVC prevents the possibility of obtaining significant statistics and trends in their temporal evolution. A more complete overview is gained by considering the entire Holocene, during which the eruptions have shown a wide variety of repose periods between consecutive events (Fig. 8.27), rendering any statistical approach to eruption prediction in Tenerife a "game of small numbers". The frequency of eruptions has been overestimated until recently because calculations were based on historical events, and included several reported eruptions of Teide Volcano that have now been proven to be erroneous (see Fig. 8.3). After detailed stratigraphic correlation and dating of the TVC (Carracedo et al. 2007b), the number of eruptions seems now to be reduced to a few per millennium (13 in the last 2 ky or about 0.6 per century).

The spatial distribution of eruptions, however, shows distinct patterns. The eruptive vents cluster tightly along the NW and NE rift zones (Carracedo et al. 2007b, 2011), whereas at their junction, inside the Las Cañadas Caldera, vents tend to group at the basal perimeter of the two main stratocones, Teide and Pico Viejo.

A distinct pattern is also observed in the petrological characteristics of the erupted lavas. The distal and central parts of both rift zones erupt only basanite lavas, with transitional eruptions located at the proximal edges of the rifts, close to the central volcanic complex, which itself erupts only differentiated phonolites (Fig. 8.28). This marked bimodality from mafic rift composition to highly differentiated central complex phonolite is clearly shown in the TAS diagram of lavas erupted from the TVC (Fig. 8.29). The transitional composition of some lavas is probably a result of magma mixing between rift basanites and the evolved magmas feeding the central complex.

References

Ablay GJ, Ernst GGJ, Marti J, Sparks RSJ (1995) The ~2 ka subplinian eruption of Montaña Blanca, Tenerife. Bull Volcanol 57:337–355

Ablay GJ, Carroll MR, Palmer MR, Martí J, Sparks RSJ (1998) Basanite-phonolite lineages of the Teide-Pico Viejo Volcanic Complex, Tenerife, Canary Islands. J Petrol 39:905–936

Albert-Beltran JF, Araña V, Diez JL, Valentin A (1990) Physical–chemical conditions of the Teide volcanic system (Tenerife, Canary-Islands). J Volcanol Geotherm Res 43:321–332

Andújar J, Costa F, Martí J (2010) Magma storage conditions of the last eruption of Teide volcano (Canary Islands, Spain). Bull Volcanol 72:381–395

Araña V, Martí J, Aparicio A, García-Cacho L, García-García R (1994) Magma mixing in alkaline magmas: an example from Tenerife, Canary Islands. Lithos 32:1–19

Banda E, Dañobeitia JJ, Surinach E, Ansorge J (1981) Features of crustal structure under the Canary Islands. Earth Planet Sci Lett 55:11–24

Bonelli Rubio JM (1950) Contribución al estudio de la erupción del Volcán del Nambroque o San Juan (Isla de La Palma), 24 de junio-4 de agosto de 1949. Instituto Geográfico y Catastral, Madrid

Cabrera Lagunilla MP, Hernández-Pacheco A (1987) Las erupciones históricas de Tenerife en sus aspectos vulcanológico, petrológico y geoquímico. Revista de Materiales y Procesos Geológicos 5:143–182

Carracedo JC (2006) Las erupciones históricas del complejo volcánico Rifts-Teide. In: Carracedo JC (ed) Los volcanes del Parque Nacional del Teide: El Teide, Pico Viejo y las dorsales activas de Tenerife. Organismo Autónomo Parques Nacionales, Ministerio de Medio Ambiente, Madrid, pp 101–125

Carracedo JC, Rodríguez Badiola E, Pérez Torrado FJ, Hansen A, Rodríguez González A, Scaillet S, Guillou H, Paterne M, Fra-Paleo U, Paris R (2007a) La erupción que Cristóbal Colón vio en la isla de Tenerife (Islas Canarias). Geogaceta 41:39–42

Carracedo JC, Rodríguez Badiola E, Guillou H, Paterne M, Scaillet S, Pérez Torrado FJ, Paris R, Fra-Paleo U, Hansen A (2007b) Eruptive and structural history of Teide volcano and rift zones of Tenerife, Canary Islands. Geol Soc Am Bull 119:1027–1051

Carracedo JC, Singer B, Jicha B, Pérez Torrado FJ, Guillou H, Badiola ER, Paris R (2010) Pre-Holocene age of Humboldt's 1430 eruption of the Orotava Valley, Tenerife, Canary Islands. Geol Today 26:101–104

Carracedo JC, Guillou H, Nomade S, Rodríguez-Badiola E, Pérez-Torrado FJ, Rodríguez-González A, Paris R,

Troll VR, Wiesmaier S, Delcamp A, Fernández-Turiel JL (2011) Evolution of ocean-island rifts: the northeast rift zone of Tenerife, Canary Islands. Geol Soc Am Bull 123:562–584

Cologan Fallow B (1798) La erupción del 9 de junio de 1798 de las Narices del Teide

Colombo F (1571) Historie del S. D. Fernando Colombo; Nelle quali s'ha particolare, et vera relatione della vita, et de'fatti dell'Ammiraglio D. Christoforo Colombo, suo padre: Et dello scoprimento, ch'egli fece dell'Indie Occidentali, dette Mondo Nuovo, hora possedute dal Sereniss. Re Catolico, nuovamente di lingua spagnuola tradotte nell'italiana dal S. Alfonso Ulloa. Apresso Francesco de' Franceschi Sanese, Venetia

de León FM, Cioranescu A (1978) Apuntes para la historia de las Islas Canarias, 1776–1868. Biblioteca Isleña. Aula de Cultura de Tenerife, 2nd edn. corr edition, p 404

de Silva SL, Self S, Francis PW, Drake RE, Ramirez C (1994) Effusive silicic volcanism in the Central Andes—the Chao dacite and other young lavas of the Altiplano-Puna Volcanic Complex. J Geophys Res Solid Earth 99:17805–17825

Fernandez Navarro L (1911) Erupción volcánica del Chinyero (Tenerife) en noviembre de 1909. Anales de la Junta para Ampliación de Estudios e Investigaciones Científicas Tomo V (Memoria 1ª):99

Fúster JM, Araña V, Brandle JL, Navarro JM, Alonso V, Aparicio A (1968) Geology and volcanology of the Canary Islands: Tenerife. Instituto Lucas Mallada, CSIC, Madrid

Le Bas MJ, Maitre RWL, Streckeisen A, Zanettin B, Rocks ISotSoI (1986) A chemical classification of volcanic rocks based on the total Alkali-Silica diagram. J Petrol 27(3):745–750

Mezcua J, Buforn E, Udías A, Rueda J (1992) Seismotectonics of the Canary Islands. Tectonophys 208:447–452

Monge F (1980) Sismicidad en el Archipiélago Canario. Tesis de Licenciatura. Univesrsidad Complutense, Madrid

Ponte y Cologan A (1911) Volcán del Chinyero. Memoria históricodescriptiva de esta erupción volcánica, acaecida en 18 de noviembre de 1909. Lit. A. J. Benítez, Tenerife

Rodríguez-Badiola E, Pérez-Torrado FJ, Carracedo JC, Guillou H (2006) Petrografía y Geoquímica del edificio volcánico Teide-Pico Viejo y las dorsales noreste y noroeste de Tenerife. In: Carracedo JC (ed) Los volcanes del Parque Nacional del Teide/El Teide, Pico Viejo y las dorsales activas de Tenerife. Naturaleza y Parques Nacionales-Serie Técnica. Organismo Autónomo Parques Nacionales Ministerio de Medio Ambiente, Madrid, pp 129–186

Romero C (1991) Las manifestaciones volcánicas históricas del Archipiélago Canario. Ph. D. Thesis, Universidad de La Laguna, San Cristóbal de La Laguna

Segundo de Franqui N (1799) Carta sobre la erupción del volcán de la montaña de Venge, cerca del Pico de Teide, en la isla de Tenerife, en 9 de junio de 1798. An Hist Nat 2:297–304

Suriñach E (1986) La estructura cortical del Archipiélago Canario. Resultados de la interpretación de perfiles sísmicos profundos. An Fís S B 82:62–77

Viera y Clavijo J (1776) Noticias de la Historia General de las Islas Canarias: Tenerife. Goya Ediciones, Sta. Cruz de Tenerife

von Humboldt A, Bonpland A (1805) Voyage aux régions équinoxiales du nouveau continent, fait en 1799–1804. Levrault, Paris

Wiesmaier S, Deegan FM, Troll VR, Carracedo JC, Chadwick JP, Chew DM (2011) Magma mixing in the 1100 AD Montaña Reventada composite lava flow, Tenerife, Canary Islands: interaction between rift zone and central volcano plumbing systems. Contrib Mineral Petrol 162:651–669

Wiesmaier S, Troll VR, Carracedo JC, Ellam RM, Bindeman I, Wolff JA (2012) Bimodality of Lavas in the Teide–Pico Viejo Succession in Tenerife—the Role of Crustal Melting in the Origin of Recent Phonolites. J Petrol. doi:10.1093/petrology/egs056

Timing, Distribution and Petrological Evolution of the Teide-Pico Viejo Volcanic Complex

9

Sebastian Wiesmaier, Valentin R. Troll, Eduardo Rodríguez-Badiola, and Juan Carlos Carracedo

Abstract

Several cycles of initially mafic to progressively felsic activity have given rise to large volume felsic deposits on Tenerife that serve as prime examples of pronounced magmatic differentiation in an ocean island setting. The Teide–Pico Viejo succession is the most recent of these cycles to show a systematic evolution from initially basanitic to phonolitic eruptions. Basanite lava flows bear olivine, pyroxene and occasionally plagioclase, while phonolites mainly display alkali feldspar with subordinate pyroxene, amphibole, biotite and oxides. Three groups of eruptives can be discerned based on their trace element composition: (1) Mafic lavas that show typical OIB signatures, (2) Transitional lavas, which are enriched in incompatible trace elements but may be depleted in Ba and Sr and (3) Phonolites, which are more enriched in incompatible trace elements, but show the strongest negative Ba and Sr anomalies. Linking the spatio-chronological distribution of eruptions with these compositional groups shows a progressive migration of mafic activity from the outskirts of the rift zones towards the central complex over the last 30 ka. The arrival of mafic activity at the central complex coincided with the onset of more evolved eruptions at Teide, thought to be triggered

S. Wiesmaier (✉)
Ludwig-Maximilians-Universität München,
Geo- and Environmental Sciences,
München, Germany
e-mail: sebastian.wiesmaier@min.uni-muenchen.de

V. R. Troll
Department of Earth Sciences, CEMPEG,
Uppsala University, 75236, Uppsala, Sweden

E. Rodríguez-Badiola
Museo Nacional de Ciencias Naturales del CSIC,
Madrid, Spain

J. C. Carracedo
Department of Physics (Geology), GEOVOL,
University of Las Palmas, Gran Canaria, Spain

J. C. Carracedo and V. R. Troll (eds.), *Teide Volcano*, Active Volcanoes of the World,
DOI: 10.1007/978-3-642-25893-0_9, © Springer-Verlag Berlin Heidelberg 2013

by mafic underplating. The distribution of mafic activity at the surface may thus be related to the volume of mafic underplating beneath the volcanic edifice at a given time.

9.1 Introduction

Magmatic differentiation is the modification of a magma's composition from its original primary make up. Primary magmas are melts that are assumed to have been generated by equilibrium partial melting of mantle rocks. As such, they can show certain variability in their composition, depending on mantle lithology and degree of partial melting involved. However, lava and intrusive rocks, which represent solidified magma, span a much wider compositional range. This implies that primary magmas are—more often than not—overprinted by processes that change their composition subsequent to their initial formation in the mantle.

The Teide–Pico Viejo succession represents a unique opportunity to investigate magmatic differentiation and the resulting felsic rock suites, in a young and very accessible ocean island setting. The Canary Islands form on top of slow-moving, thick oceanic lithosphere and show a relatively low magma-supply rate, only about one-tenth of that of the Hawaiian Islands (Sleep 1990). This increases the residence time of magma in the oceanic crust and the island edifices and results in cooling, crystallisation, magmatic differentiation and magma mixing. In contrast, highly productive hotspot settings, such as Hawaii, allow only for relatively short intervals of magma residence in the crust and thus reflect processes at greater depth, i.e. in the mantle. As a result, felsic material is nearly absent in the hotspot type-locality Hawaii, whereas large volumes of strongly differentiated, felsic rocks abound in the Canary Islands (e.g., Fúster et al. 1968; Schmincke 1969, 1976; Ridley 1970) and must represent the products of large-scale and voluminous magmatic differentiation.

The degree of differentiation that magma of a given volcano may reach directly influences that volcano's hazard potential. Differentiation generally increases the silica content of magma and so governs whether an eruption will be a comparatively passive outpouring of lava or, in contrast, a large cataclysmic explosion that may ravage proximal and distal (e.g., populated) territories. This is because silica increases the polymerisation in magma and can raise its viscosity by orders of magnitude (e.g., Hess and Dingwell 1996). Such high viscosity of e.g., rhyolitic or phonolitic magma is one of the fundamental drivers of explosive volcanism, because only high viscosity magma can be fragmented on a large scale during eruption (Webb and Dingwell 1990; Dingwell 1996), a phenomenon which is promoted at high silica concentration.

Two main processes of magmatic differentiation are known to change the composition of magma towards high silica values: fractional crystallisation and assimilation (e.g., DePaolo 1981). Fractional crystallisation, on one hand, is the commonly assumed way of magma evolution on cooling, whereby newly formed crystals are removed from the melt, for example by gravitational settling (e.g., Bowen 1928; Wager and Brown 1967). Because the most common magmatic crystals are combined of only a small number of components assembled in varying proportions, they deplete the remaining magmatic liquid in the elements they take up to grow, while the unused components become enriched in the magma. As a result, differentiated magma (often also called 'evolved') experiences a change in chemical composition, which usually manifests in an increased SiO_2 concentration associated with high concentrations of large ion lithophile elements such as Rb, Na and K. The magma becomes more felsic (Fig. 9.1).

Assimilation, on the other hand, describes the reaction of hot magma with the surrounding country rock that forms the boundary walls of a magma reservoir. Any magma that cleaves through the Earth's crust is likely to encounter a

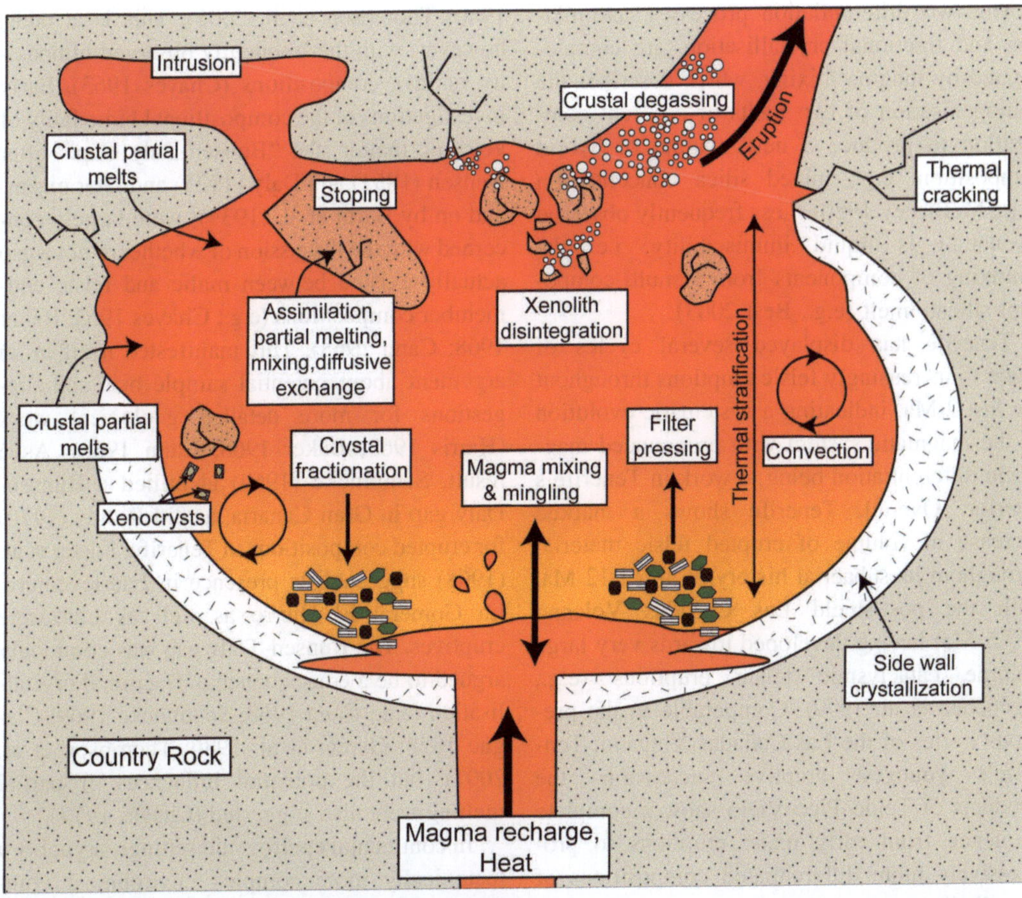

Fig. 9.1 Idealised sketch of differentiation processes that may occur in a crustal magma chamber (courtesy of F. Deegan)

variety of rock types that the crust is built of. At deep levels below Tenerife, the ultramafic rocks of the mantle grade into a sequence of rocks that are typical for the oceanic crust (e.g., gabbros, dykes and pillow lavas). Moreover, at shallow levels, the rocks from the volcanic edifice itself have to be passed by newly ascending magma. How intensely the ascending magma can be changed in composition by assimilation depends on three factors: the solidus of the wall rock, the compositional gradient between wall rock and magma, and the temperature of both magma and wall rock, i.e. the amount of energy released by the cooling magma that leads to melting of wall rock. The solidus of different wall rock lithologies varies strongly with their composition and the fusion enthalpies of their mineralogy, but most wall rocks will become partially unstable when exposed to temperatures of 1,200–1,300 °C (cf. Spera and Bohrson 2001). Wall rock compositions that are distinct from the magma may result in strong disequilibrium kinetic reactions that dissolve wall rock minerals prior to melting (e.g., Watson 1982; Deegan et al. 2010, 2012). Finally, the volume and temperature of magma versus that of the wall rock delimits the energy available for consumption of the wall rock. It follows that the type and the amount of rock that exists at depth beneath a volcano may have considerable influence on magma type and, hence, on its eruptive behaviour. This influence has to be determined for each volcano individually (Fig. 9.1).

The two differentiation processes, assimilation and fractional crystallisation, are complemented by magma mixing, which describes the homogenisation of two melts of different composition to produce a new magma of hybrid composition, e.g., mixed silica concentration (e.g., McBirney 1980). Less frequently observed in nature is liquid immiscibility, i.e. the unmixing of components from a multi-component silicate melt (e.g., Best 2003).

Tenerife has displayed several cycles of mafic to increasingly felsic eruptions throughout the last 2 Ma, indicating a systematic evolution of the magmatic system with pronounced magmatic differentiation being at work in Tenerife's interior. Overall, Tenerife shows a marked increase in volume of erupted felsic material throughout its subaerial history of about 12 Ma, with the post-shield Las Cañadas Volcano (<3.5 Ma) having developed towards very large volume cataclysmic felsic eruptions (e.g., Ancochea et al. 1999). Comparable to the previous cycles of the Las Cañadas Volcano, Tenerife's youngest eruptive succession, the products of Teide–Pico Viejo show a complete sequence from early mafic eruptions to progressively more differentiated compositions of the alkaline series up to phonolites. Because of their young age and a lack of overlying units to bury them, it has been possible to document the stratigraphy, age and compositional variation of this lava succession to superb detail (Rodríguez-Badiola et al. 2006; Carracedo et al. 2007, 2008). The age and stratigraphical constraints used here are presented in Chap. 7 and further information on the quantification of differentiation processes can be found in Chap. 10.

9.2 The Significance of Felsic Volcanism in Ocean Islands

In a setting devoid of large regional tectonic influences and thick continental crust, crystal fractionation had commonly been proposed as the principal differentiation process forming abundant felsic volcanic material (e.g., Cann 1968; Schmincke 1969; Clague 1978; Garcia et al. 1986; Thompson et al. 2001). This is at odds, however, with the frequently observed bimodality of lava compositions (Chayes 1963). Early investigations of the compositional bimodality of volcanic matter [the "Bunsen-Daly Gap", after Bunsen (1851) and Daly (1925) and later expanded on by Barth et al. (1939)] were largely concerned with the discussion of whether or not a gap actually existed between mafic and felsic end-member compositions (e.g., Chayes 1963; Baker 1968; Cann 1968). This manifested itself in an argument about potential sample bias and suggestions for more detailed geological work (Harris 1963; Baker 1968; Cann 1968). As a result, Schmincke (1969) identified a Bunsen-Daly gap in Gran Canaria, as did Ridley (1970) for erupted compositions in Tenerife. Fúster et al. (1968) suggested its presence in Fuerteventura, La Gomera, El Hierro as well as Lanzarote eruptives. The Bunsen–Daly gap was commonly argued to be a consequence of fractional crystallisation (e.g., Cann 1968; Schmincke 1969; Clague 1978; Garcia et al. 1986; Thompson et al. 2001), but the potential influences of partial melting were also suggested (Chayes 1977).

In continental epeirogenic settings, arguably a tectonically more complex environment compared to ocean islands, strong bimodalities also exist among the compositions of erupted material. For example, in the east African Gregory Rift Valley the eruption of Miocene flood basalts was rapidly followed by large volume plateau phonolites (Baker et al. 1971). Estimated ratios of felsic to mafic eruption volumes range between roughly 0.5 and 1.5 for between the Miocene to Holocene epochs, showing an overabundance of felsic material compared to what is expected from pure fractional crystallisation scenarios (Williams 1972). The lack of intermediate composition lavas was invoked to have originated either from partial melting of upper mantle by either a reduction in pressure by crustal upheaval or, alternatively, from an upper mantle, which is hotter and thus more susceptible to partial melting (Williams 1970). Hypotheses on the origin of Kenyan phonolites that invoked partial melting of upper mantle material were criticised by Lippard (1973). Due to the trace element patterns of these highly differentiated phonolites, it was argued that they cannot have formed by mere partial melting of mantle lithologies.

Explanations for this bimodality started to diversify considerably in the 1980s. Crystal fractionation models were modified to allow for large crystal loads that restrain convection and induce bimodality in erupted compositions (e.g., Marsh 1981; Brophy 1991 and references therein). Bailey (1987) suggested an upper mantle origin for trachytic melts, while Bonnefoi et al. (1995) invoked a model of critical cooling dynamics to be responsible for observed compositional relationships. More recently, assimilation of country rock has gained renewed momentum in differentiation models for magmas erupted in oceanic islands and is increasingly recognised (e.g., Thirlwall et al. 1997; Bohrson and Reid 1998; Garcia et al. 1998; O'Hara 1998; Harris et al. 2000; Troll and Schmincke 2002).

In Tenerife, too, fractional crystallisation was long considered the main differentiation process for the formation of felsic material (Wolff 1983; Wolff and Storey 1984). Relatively recently, recycling (i.e. partial or bulk melting) of rocks from within the island has been suggested to contribute to the evolution of felsic magmas in Tenerife (Wolff and Palacz 1989; Wolff et al. 2000), which would explain the large volumes of felsic material in some of the successions. However, the Teide–Pico Viejo stratovolcano was suggested to have mainly formed by fractional crystallisation with only minor assimilation of zeolite and/or hydrothermally altered material to explain deviations in trace element concentration from the fractionation trend (Ablay et al. 1998). Until then, systematic studies of the isotopic composition of Teide–Pico Viejo lavas to clarify the extent of assimilation at work had not been attempted.

9.3 Petrological History of Tenerife Island Prior to Teide Formation

Three overlapping shield volcanoes (Roque Del Conde, Teno and Anaga) formed between 11.9 and 4 Ma and constitute the subaerial foundation of Tenerife (Carracedo 1979; Ancochea et al. 1990; Thirlwall et al. 2000; Guillou et al. 2004; Paris et al. 2005; Walter et al. 2005). The first shield to emerge was the central shield, or Roque Del Conde, followed by the peripheral shields Teno and Anaga in the northwest and northeast, respectively.

As is typical for shield volcanoes, mainly mafic volcanism occurred, i.e. basanites, picrobasalts and alkali basalts. Picrobasalts are ankaramites with 40–60 % phenocrysts, mainly pyroxene and olivine, which are very likely of accumulative origin. All other volcanic rocks in Tenerife remain mostly below 20 % phenocrysts. Up-section in the stratigraphy of each of the three shield volcanoes, minor amounts of trachytic and phonolitic dykes, plugs and lavas and occasionally pyroclastic deposits are found (e.g., Walter et al. 2005; Longpré et al. 2009).

After the peripheral shields, Teno and Anaga, went extinct, the volcanic activity returned to the centre of the island at around 3.5 Ma, to build the Las Cañadas Volcano on top of the Roque Del Conde shield (Martí et al. 1994; Ancochea et al. 1999). Eruptive episodes in the Las Cañadas Volcano tended from scattered eruptive activity to distinct cycles of mafic to felsic eruptions, indicating the development of a central system of felsic magma chambers. The Las Cañadas Volcano consists of a complex Lower Group formation that comprises multiple and scattered eruptive centres and a three-cycle Upper Group, of which each cycle broadly evolved from subordinate, primitive effusion to large-volume felsic eruptions. The 2nd and 3rd cycle of the Upper Group culminated in the caldera-forming ignimbrite formations of Granadilla ($\sim$570 ka) and Abrigo ($\sim$180 ka), respectively (e.g., Wolff 1983; Martí et al. 1994, 1997; Bryan et al. 2000; Brown et al. 2003; Brown and Branney 2004; Edgar et al. 2007).

In the terminal stage of the Las Cañadas Volcano, the Diego Hernández Formation, a spectacular caldera-forming sequence, was erupted from various phonolite magma chambers, often, it seems, triggered by mafic recharge (e.g., Wolff 1983; Bryan et al. 1998; Wolff et al. 2000). For example, in the Poris member of the

Diego Hernández Formation, four types of magma have been distinguished; a high-Zr phonolite (most abundant), a low-Zr phonolite with higher MgO that is often intermingled with a tephriphonolite and, finally, a phonotephrite. The phonolite pumices contain alkali feldspar, biotite, sodic salite and Fe–Ti oxides, but in the high-Zr phonolites titanite also occurs. Intermediate composition pumices contain kaersutite, plagioclase, low-Na salite and Fe–Ti oxides, whereas the most mafic magma contains olivine, titanaugite and calcic plagioclase (Edgar et al. 2002). The large variability in lithic blocks contained in pyroclastic deposits of the Las Cañadas Volcano and the large volumes of the terminal members of individual Las Cañadas cycles (e.g., Granadilla, Abrigo ignimbrites) testify to the disruptive and catastrophic nature of these mixed eruptions, which most likely formed sub-circular caldera depressions (cf. Edgar et al. 2007). The Abrigo event, the last phase of cataclysmic eruptions of the Las Cañadas system, was probably closely associated with the ~200–180 ka Icod landslide (Watts and Masson 1995; Carracedo 1999; Carracedo et al. 2007; Márquez et al. 2008). The pre-existing vertical collapse structures probably contributed to the lateral instability of Las Cañadas volcano (e.g., Martí et al. 1994, 1997; Troll et al. 2002), but have probably been largely mass-wasted by the landslide. The Icod landslide relieved the magmatic plumbing system in the centre of the island of a large volume of overburden, unroofing the junction of the three-armed rift zone system and very likely, resetting the system to the eruption of mafic materials such as picrites and ankaramites (cf. Longpré et al. 2009; Manconi et al. 2009). Of the three rifts, the northwest and northeast arms are currently active and extend from underneath Teide–Pico Viejo volcano into the Teno and Anaga massifs (Carracedo et al. 2007, 2011; Chaps. 4 and 5). These rifts control the overall structure of Tenerife and show dominantly mafic volcanism in the form of basanites to rarer phonotephrites that erupt along elongated arrays of monogenetic cones (Chap. 4). The Icod landslide then triggered extensive amounts of mafic eruptions, as evidenced by the horizontal sections obtained from various *galerías* that penetrate the infill sequence (Chap. 7). The uncapping of the rift system marked the onset of a new cycle of eruptive activity in Tenerife by promoting rapid ascent of new and deep mafic magma into the collapse scar and thus initialising the construction of Teide–Pico Viejo central volcanic complex. Even more regular in its evolution from mafic to highly differentiated lavas than the previous cycles, the formation of Teide–Pico Viejo central complex concentrates Tenerife's non-rift activity into a single central and felsic edifice.

9.4 Petrological Description of the Teide–Pico Viejo Succession

The petrological traits of Teide–Pico Viejo have been the focus of several research publications (Fúster et al. 1968; Araña et al. 1989a, b; Ablay et al. 1998; Ablay and Marti 2000; Rodríguez-Badiola et al. 2006). Here, the main petrological features of the Teide–Pico Viejo succession are summarised, following the latest account of Rodríguez-Badiola et al. (2006).

9.4.1 Mafic Lavas

9.4.1.1 Basanites and Tephrites

Basanites are highly abundant in all volcanic sequences in Tenerife, both in the Teide–Pico Viejo succession as initial collapse fill and as the main constituent of the northeast and northwest rift zones. The initial mafic eruptions of Teide–Pico Viejo are petrographically indistinguishable from the synchronous rift zone eruptions. These early eruptions are the most primitive lavas found in the Teide–Pico Viejo succession and comprise basaltic rocks that bear olivine, clinopyroxene (titanaugite) and magnetite in variable proportions (locally reaching ankaramitic character) along with scarce plagioclase in a fine-grained groundmass. The matrix varies from micro- to hypocrystalline, is sometimes vesicular, and

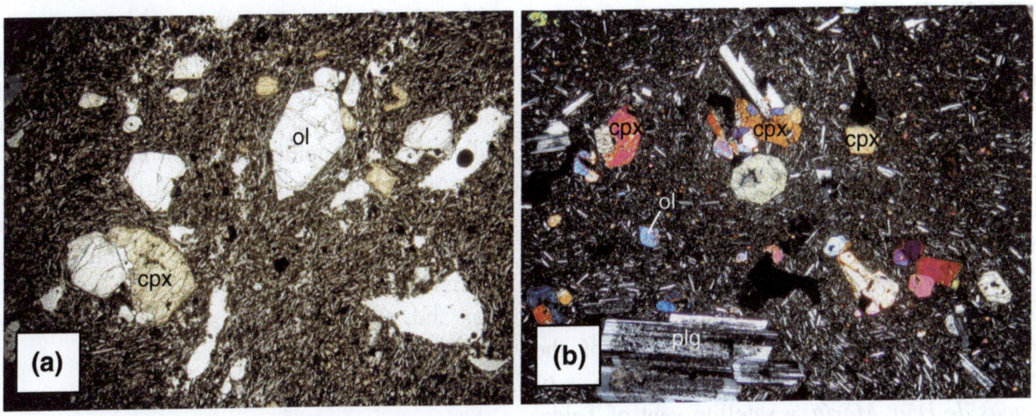

Fig. 9.2 **a** Olivine pyroxene basalt. Phenocrysts of titanaugite (*cpx*), olivine (*ol*) and opaque oxides in a hypocrystalline, mafic matrix (PP; 10×). **b** Olivine pyroxene basalt with plagioclase. Phenocrysts of augite (*cpx*), olivine (*ol*), sieve-textured plagioclase (plag) and opaque oxides in a hypocrystalline matrix with plagioclase microlites. (XP; 10×)

usually consists of laths of plagioclase and micrrolites of clinopyroxene, olivine and Fe–Ti oxides, but can also be more glassy locally (Fig. 9.2). This group also includes some olivine, clinopyroxene and plagioclase basalts with abundant plagioclase, indicating a slightly higher degree of differentiation within the basanite field, although still maintaining an overall mafic character.

9.4.1.2 Plagioclase Basalts

This group effectively combines all pāhoehoe flows of the Teide–Pico Viejo succession. These lavas are characterised by their long, tabular and thin plagioclase phenocrysts with frequent sieve textures, and they carry subordinate olivine and clinopyroxene, supported by a hypocrystalline and vesicular matrix. A variation of this type of lava is micro-plagioclase basalt, in which feldspar crystals of limited size are embedded in a hypocrystalline matrix of feldspathic composition (Fig. 9.3). On the surface, this lava type is only found within the initial flow sequence of Pico Viejo, but subterraneously this type abounds in the interior of *galerías* (e.g., Galería Salto del Frontón).

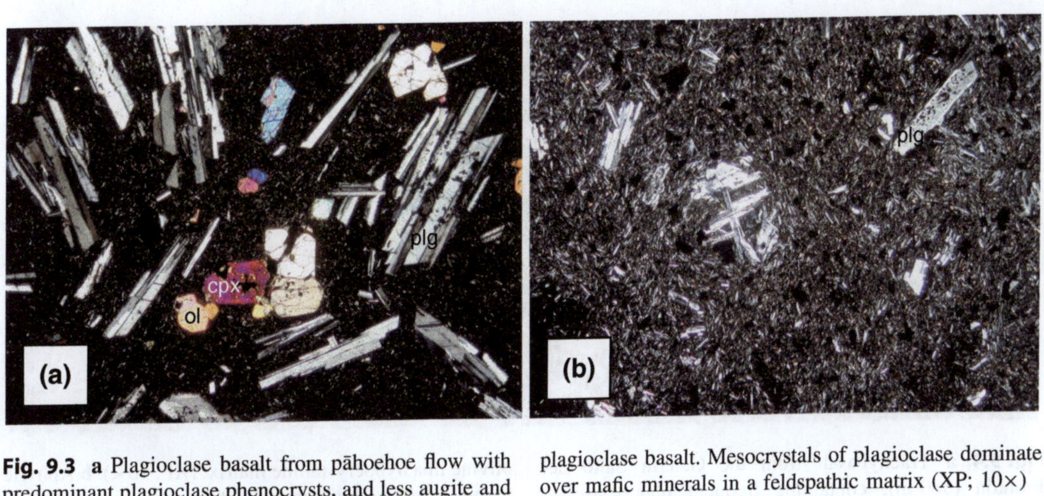

Fig. 9.3 **a** Plagioclase basalt from pāhoehoe flow with predominant plagioclase phenocrysts, and less augite and olivine in a hypocrystalline matrix (XP; 10×). **b** Micro-plagioclase basalt. Mesocrystals of plagioclase dominate over mafic minerals in a feldspathic matrix (XP; 10×)

9.4.1.3 Trachybasalts

This group includes hawaiites and potassic trachybasalts and was recognised in Tenerife by Fúster et al. (1968). Typical trachybasalts are characterised by abundant plagioclase phenocrysts of intermediate composition together with slightly sodic augite, oxides and subordinate olivine, suspended in a hypocrystalline matrix (Fig. 9.4). They are mainly found in *galerías*, for example in the Galería Salto del Frontón, where they occupy an extended part of the galería wall (500–1,000 m) and in the lavas of *Bocas de Dña. María*, a satellite vent of Teide. In some examples of this lava type, amphibole is present as stable phenocryst phase and these lavas may be intercalated with more evolved ones such as the intermediate Pico Viejo lavas that flowed south.

9.4.2 Transitional Lavas

Between 30 and 20 ka, the Teide–Pico Viejo edifice and its flank vents began to erupt more differentiated material (Carracedo et al. 2007). These lavas fall into the compositional fields of phonotephrite and tephriphonolite in the total alkali versus silica (TAS) diagram and show a large diversity in alkali concentration and petrography. At the mafic end of this group, alkaline feldspar (anorthoclase) occurs together with sodic augite, oxides and amphibole crystals with oxidation rims (Fig. 9.5). Often relict olivine is present. The more evolved lavas of this group show anorthoclase crystals together with euhedral amphibole, which indicates volatile saturation in the trachytic matrix at this stage of differentiation (Fig. 9.6). These lavas occur between the mafic and the felsic groupings in the stratigraphy of Teide–Pico Viejo central complex and, although rarely, are also found in the rifts.

9.4.3 Felsic Lavas

The highest degree of differentiation so far reached in the Teide alkaline series was the eruption of phonolite, the dominant lava composition of the last 20 kyr at Teide complex. These phonolites are occasionally intercalated in the stratigraphy with tephriphonolites and sometimes approach trachytic composition, as in the case of some domes of the Montaña Blanca sequence (Ablay et al. 1998). Typically, the phonolites contain alkaline feldspar (anorthoclase to sanidine), scarce aegirine-augite and alkaline amphibole, biotite and oxides embedded in a frequently glassy and often flow-banded matrix (Fig. 9.7).

During the progressive evolution of Teide–Pico Viejo from an initially mafic character to a

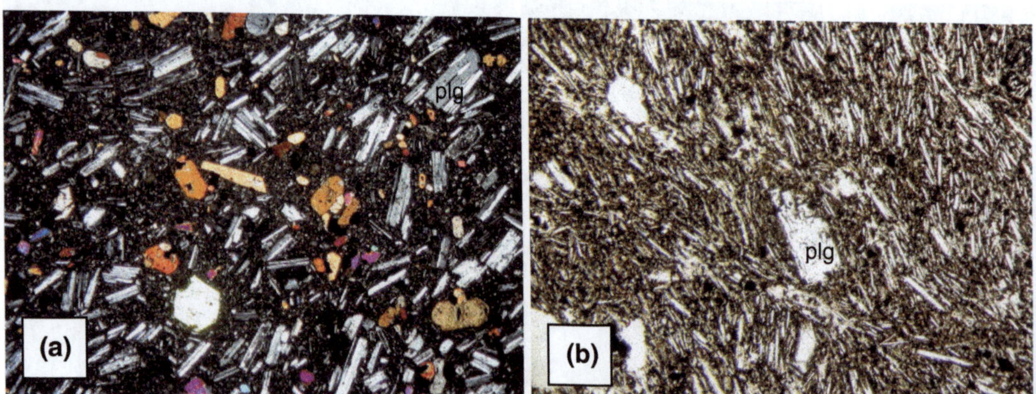

Fig. 9.4 **a** Trachybasalt from the Galería Salto del Frontón (560 m), containing frequent plagioclase of sodic composition together with augite and subordinate olivine in a hypocrystalline matrix. (XP; 10×). **b** Aphyric basanite bearing mesocrystals of plagioclase with scarce mafic microlites in a felsic matrix. (PP; 10×)

Fig. 9.5 Mafic tephriphonolite from the Galería Salto del Frontón (2,200 m), with plagioclase phenocrysts, titanaugite, Ti-amphibole with oxidation rims (kaersutite, kt), oxides and relict olivine in a hypocrystalline matrix with indications of immiscibility (**a** PP & **b** XP; 10×)

Fig. 9.6 Amphibole-bearing phonotephrite from the Galería Salto del Frontón (400 m). Anorthoclase phenocrysts (*afsp*) and ampibole in a microcrystalline matrix displaying flow lineations. (**a** PP & **b** XP; 10×)

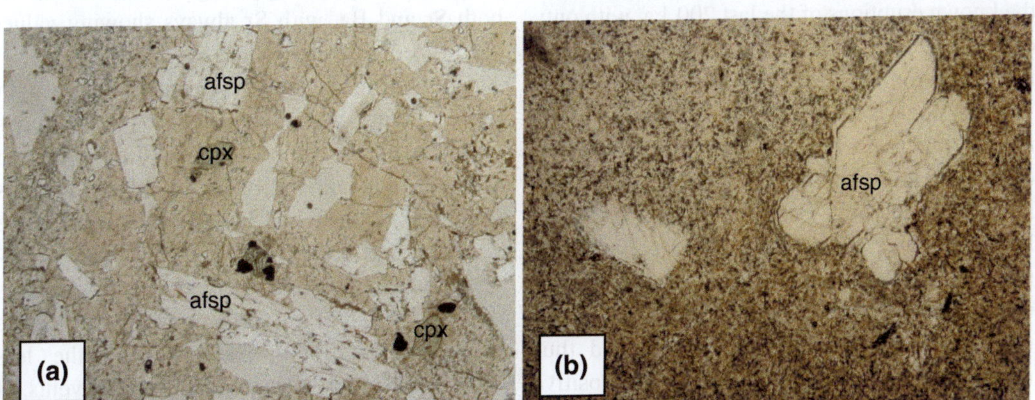

Fig. 9.7 **a** Lavas Negras, the product of the last eruption from the peak of Teide. Phenocrysts of anorthoclase and sanidine, aegirine-augite and scarce opaques in a hypocrystalline matrix. **b** Felsic phonolite from the Galería Salto del Frontón (360 m) showing anorthoclase and sanidine crystals in a trachytic matrix. (PP; 10×)

highly differentiated central volcano, the adjacent rift zones continued to produce mafic activity, with on average one eruption per century. In the boundary zone between the rift zones and the central complex, magma mixing of basanite (from the rift zone) and phonolite (from the central complex) frequently occurred (Araña et al. 1994; Wiesmaier et al. 2011, see Chap. 11), indicating that interaction between the two regimes is common. Ultimately, Teide–Pico Viejo has been constructed by initially mafic magma fed into the junction of the two rift zones (~200–40 ka) and has only subsequently developed highly differentiated compositions (~40–0 ka), while the rifts continued with mafic activity. Both regimes, rift zone and central volcano, are thus closely related and may be regarded as a single larger system.

9.5 Trace Element Characterisation of the Teide–Pico Viejo Succession

In order to compare the many lavas of one succession, it is useful to group those lavas sharing common geochemical traits. This is usually done in a TAS diagram (Le Bas et al. 1986). Rocks from the Teide–Pico Viejo succession plot as an alkaline trend from basanites to phonolites with some excursions into the trachytic series. The dataset comprises 97 % of the known eruptions of the last 200 ka, with only one sample per unit presented to avoid over-representation (Fig. 9.8).

At a single volcanic centre, additional information may also be gathered by grouping the eruptive deposits according to their trace element characteristics. Here, the Teide–Pico Viejo lavas are categorised by parallel trace element patterns, i.e. trends that do not cross-cut each other. Parallel patterns indicate combinations of trace element ratios that are related and thus represent similar geochemical histories. Positive or negative anomalies in the elements Pb, Ba and Sr further characterise the Teide–Pico Viejo lavas (Fig. 9.9). Three groups can be inferred from the trace element patterns, which we label the mafic, transitional and felsic lavas. Throughout this and the following chapter, the terms mafic, transitional and felsic are used in the context of the trace element classification devised below.

The mafic lavas (yellow squares) comprise all basanites and tephrites from the rift zones, one foidite, one basaltic trachyandesite and several intra-caldera and Teide–Pico Viejo phonotephrites and trachybasalts. They possess a negative Pb anomaly relative to the neighbouring elements in a multielement variation diagram. The mafic lavas are the most abundant (n = 40).

The transitional lavas (orange diamonds) range from trachyandesites to trachyte and from tephriphonolites to phonolites. These samples show positive Pb anomalies, negative Sr anomalies and a small to absent negative Ba anomaly. The transitional lavas (n = 10) erupted from the margins of Teide–Pico Viejo complex or vents inside the Las Cañadas Caldera and frequently define the compositional boundary between the rift zones and central complex.

The felsic lavas (red triangles) are exclusively comprised of phonolites that erupted from Teide–Pico Viejo or the flank vents of the volcano. The felsic lavas (n = 12) include, but are not limited to, Montaña Blanca, Roques Blancos and the last eruption from Teide‘s central vent (Lavas Negras, 1,150 ± 140 years BP). They are defined by strongly negative anomalies in both Sr and Ba, with Sr always showing values <10 (normalised to primitive mantle).

Combining the trace element grouping with the TAS diagram, mafic lavas show a continuous trend towards transitional compositions, but with a decrease in variability of the alkali elements (Fig. 9.9). Transitional lavas appear much more scattered and comprise loosely grouped tephriphonolites, phonolites and trachytes. Felsic lavas define a trend at a steep angle to the classic differentiation trend established by the less evolved samples. Geographically, the vents of the three compositional groups display a concentric arrangement around the central volcano Teide (Fig. 9.10).

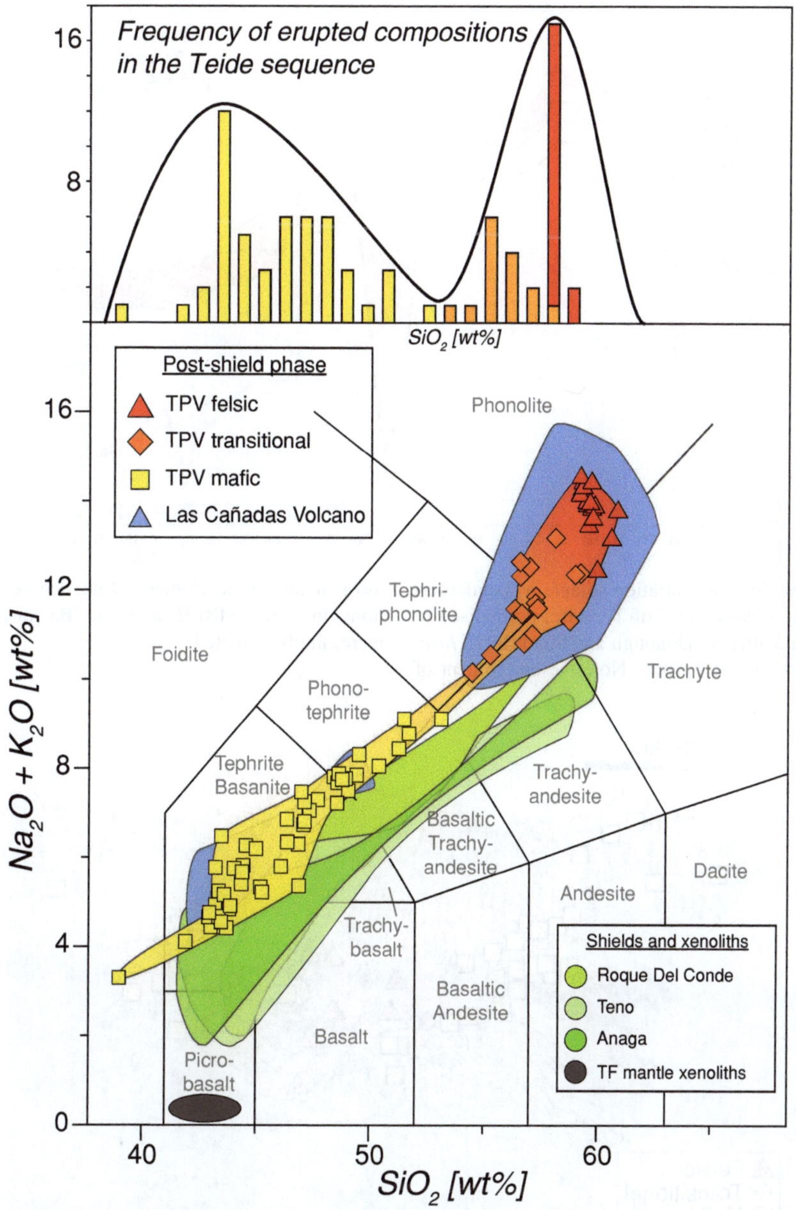

Fig. 9.8 Teide–Pico Viejo samples plotted on a total alkali versus silica diagram after Le Bas et al. (1986). Note the well-defined trend of the mafic lavas, the loose clustering of transitional lavas and the sub-vertical trend defined by the felsic group. Data from Rodríguez-Badiola et al. (2006), Thirlwall et al. (2000) and Neumann et al. (2002)

The strongly negative anomalies of Ba and Sr found in Teide felsic lavas were traditionally considered a hallmark of feldspar fractionation, the only mineral in these rocks to include these two cations in significant amounts. At Teide–Pico Viejo, this argument is seemingly supported by the fact that most of the differentiated lavas bear this mineral, so at first glance it appears that feldspar fractionation depleted Ba and Sr in Teide's phonolite magmas. Along these lines, Ablay et al. (1998) constructed a detailed model of fractional crystallisation,

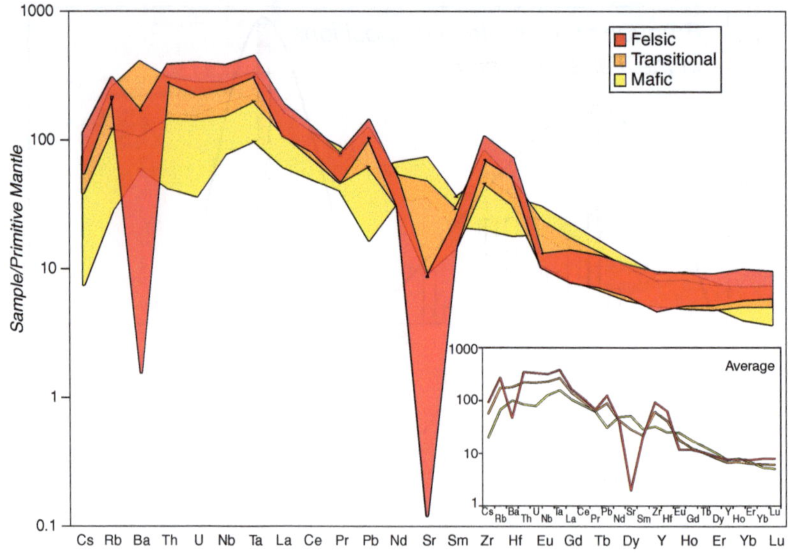

Fig. 9.9 Multi-element variation diagram (spiderdiagram) for post-collapse lavas of Tenerife, normalised to primitive mantle after McDonough and Sun (1995). *Inset* shows averages for each group. Note the enrichment of most incompatible elements with degree of differentiation. In turn, MREE and Sr, Ba and Eu become increasingly depleted

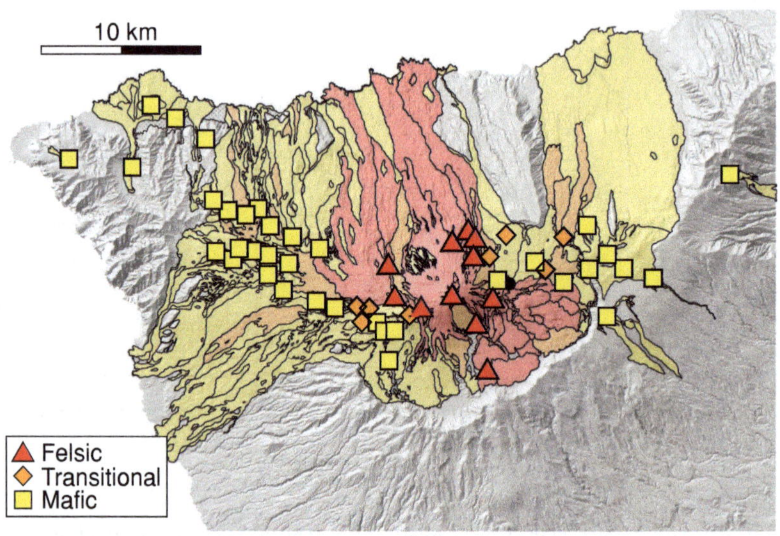

Fig. 9.10 Distribution of vents of the Teide–Pico Viejo succession in Tenerife. *Triangles* felsic lavas, *diamonds* transitional lavas, *squares* mafic lavas. The *black circle* indicates Montaña de los Conejos, an eruption for which whole-rock data is not available. Note the clustering of felsic lavas within the central complex. The mafic lavas define the rift zones, but also occur in the older products of the central complex. The transitional lavas fall into the geographical and chemical transition between the mafic and felsic regimes, i.e. the two rift zones and the central complex respectively

focussing on major and trace element data. As a result, they interpreted Teide phonolite lava to be a residual melt of around 12 % of the original mafic melt, the remaining 88 % having been removed from the magma by crystal separation. Disparities between natural and modelled trace element concentrations were explained by minor assimilation of zeolite or hydrothermally altered materials. End-member contaminant data were not presented nor were isotope data, which means that their interpretation can be tested by means of isotopic analyses (see Chap. 10).

9.6 Volumetric and Spatio-Chronological Characterisation of the Teide–Pico Viejo Succession

Tenerife is a mature volcanic island that underwent multiple mafic-to-felsic eruptive cycles in the post-shield Las Cañadas succession (e.g., Ancochea et al. 1999). The latest eruptive cycle of Teide–Pico Viejo thought to succeed Las Cañadas, has been volumetrically estimated at 160 km^3, divided into $\sim$62 km^3 of initial fill in the collapse embayment (dominantly mafic), $\sim$70 km^3 for the old Teide edifice (mafic to transitional), $\sim$15 km^3 for Pico Viejo volcano (mafic to transitional), $\sim$6.5 km^3 for Teide satellite vents (felsic) and $\sim$0.7 km^3 for the final construction of the Teide stratocone (felsic) (Carracedo et al. 2007). Rift zone eruptives are estimated during Teide's construction period to be around 32 and 9 km^3 for the NWRZ and NERZ, respectively. The distinction between mafic, transitional and felsic lavas is not always clearly defined in these estimates, due to partial or complete burial of units older than 15 ka. It was possible, however, to constrain the volume of individual, more recent eruptions at Teide–Pico Viejo (<15 ka) (Fig. 9.11; Carracedo et al. 2008).

Eruption frequency appears to decline with degree of differentiation. Teide–Pico Viejo phonolites erupted seldom ($\sim$every 1,000 years), but with much larger volumes compared to the small, mafic rift zone eruptions that occured up to 10 times as often. Eruptive frequency of transitional lavas also appears to have declined over the last 15 ka and fall below mafic lavas in terms of total volume. The felsic lavas, in turn, volumetrically dominated over the last 15 ka. Considering the large volumes of mafic lavas that initially filled the collapse scar, transitional lavas are probably subordinate to both felsic and mafic lavas. The compositional bimodality of the lava flows is thus evident in both, eruptive frequency and estimates of erupted volume.

The distribution of eruptive activity through time offers further insight. Figure 9.12 is a diagram of vent location (longitude) versus the age of an eruption. There, the evolution of Teide–Pico Viejo from eruption of mafic to felsic lavas results in a present day shadow zone of exclusively felsic eruptions (red), in which no mafic material is transported to the surface. The activity in the rift zones, however, has migrated through time (diagonal arrows). A systematic shift in activity has taken place in both rift zones from the distal parts of the rift to the more proximal parts, closer to the perimeter of Teide–Pico Viejo complex, beginning at around 35 ka in the NERZ and at 20 ka in the NWRZ. At approximately the same time mafic activity 'arrived' at the central complex, the first transitional lavas (PT-INE) erupted from the complex.

Crustal recycling very likely occurred during the petrogenesis of these transitional and felsic lavas (see Chap. 10). The close spatial association of mafic activity in the proximal parts of the rift and the initial transitional material erupted may indicate that mafic magma became increasingly arrested and that underplating became a significant phenomenon at that point. Underplating is the main heat source for crustal recycling and occurs when magma reaches a level in the crust of equal or lower density, with the resulting neutral buoyancy causing it to stall (cf. Walker 1993). As magma would continue to be fed into the system from below into the centre of the island, such a lens of underplated magma may have caused a diversion of mafic magma, initially from the centre of the island to the outskirts of the rift zones. The subsequent inward-shift of mafic activity towards the central phonolitic volcano implies that from then on mafic magma may have started to pool underneath this low density phonolite barrier and potentially reactivated solidifying and semi-solid felsic magma chambers that had formed previously. Such an interaction of rift zone and central volcano has been documented for

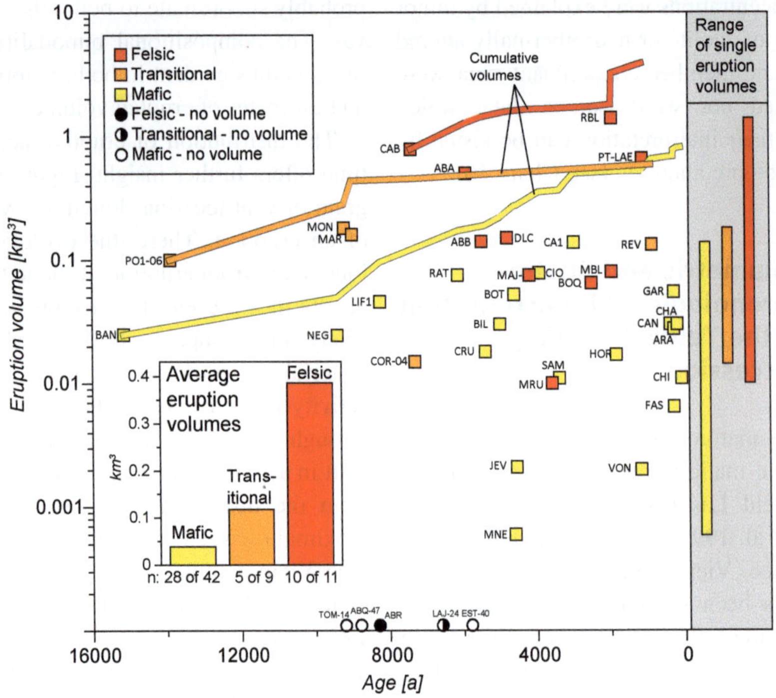

Fig. 9.11 Cumulative eruption volumes of the Teide–Pico Viejo succession of the last 15,000 years, grouped by composition. The inset shows the average eruption volumes of the three main groups, mafic, transitional and felsic lavas. Note how the degree of differentiation correlates with average and cumulative volumes. Mafic lavas have overtaken transitional compositions in cumulative volume due to their much higher eruptive frequency. Eruption volumes Carracedo et al. (2008), except for Montaña Blanca, which is from Ablay and Martí (2000)

example in the eruption of Montaña Reventada (Wiesmaier et al. 2011; see Chap. 11) and also manifests itself in the clustering of felsic and mixed eruptions along the projections of both the NERZ and NWRZ rift trends into the marginal areas of the central complex.

Based on the presented record, the future activity of Teide–Pico Viejo seems to be in slow decline. The record of explosive eruptions in the Teide–Pico Viejo succession alone is too ambiguous and with only a single explosive episode recorded, the statistics are inconclusive (see Chap. 13). However, the statistics could be interpreted in several ways, either as knell to more explosive activity, because one of the latest eruptions at Teide was sub-plinian (Ablay et al. 1995), or as the last gasp of a fading system. Teide–Pico Viejo may well thus indicate a consolidation of the Tenerife volcanic system towards lower magma production rates, but with a potential for development of small- to medium-scale explosive eruptions most likely from satellite vents around the central complex. Notably, the force and size of the Montaña Blanca eruption was considerably smaller than the large ignimbrite eruptions of the Pliocene Las Cañadas volcano. In contrast, the eruptive record of the last 1,000 years shows that mafic activity is scattered in comparison to the more localised earlier trends observed (Fig. 9.12). Apparently, the Teide phonolitic eruption has initiated a wider dispersal of volcanic activity (Fig. 9.12). For example, in the NERZ, mafic eruptions occurred again after a pause of nearly 10 ka and likewise in the NWRZ several historical eruptions occurred in the more distal parts of the rift, contrasting the previous trend that had converged towards the central volcano. This may be interpreted as a quasi-relaxation of the system, or as a new cycle of diversion of

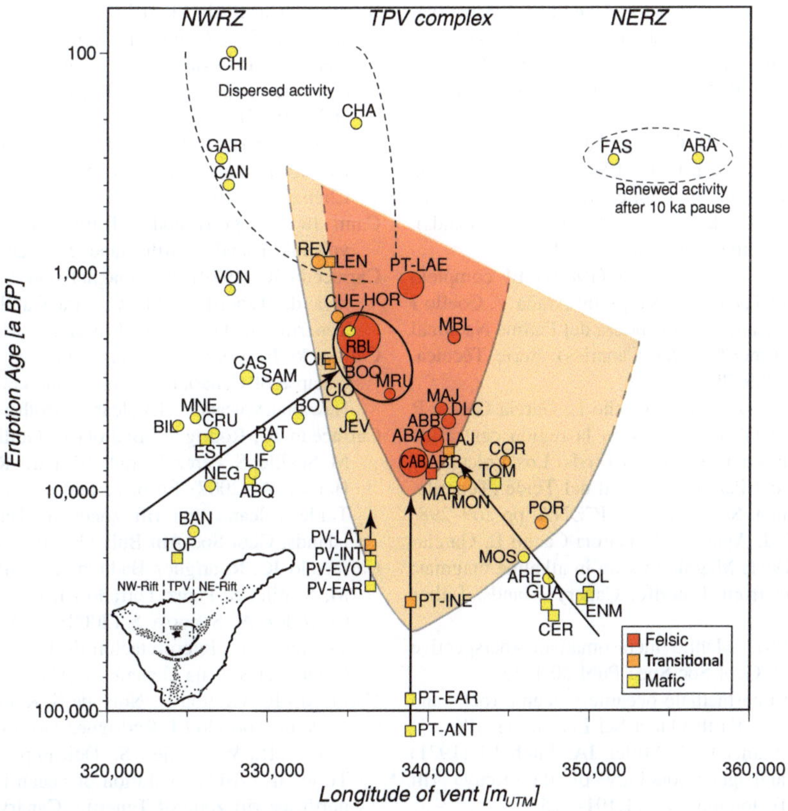

Fig. 9.12 Distribution of vent locations versus eruption age. *Round symbols* denote eruptions for which a volume estimate is available; in these cases, the size of the symbol correlates with the erupted volume. *Square* symbols have no estimate of eruption volume. Note how the activity of central edifice, NERZ and NWRZ migrate towards each other after 30 ka followed by a dispersal of rift zone activity after the central complex erupted significant volumes of felsic material from ~8 to 1 ka. Vent locations of rift zone eruptions prior to 30 ka are not preserved. Eruptions volumes Carracedo et al. (2008), except for Montaña Blanca, which is from Ablay and Martí (2000)

mafic magma to the sides of the central complex due to underplating below Teide's edifice. Since the volume of felsic magma generated by underplating is necessarily limited by the amount of mafic magma present for either fractionation or remelting, both models would imply a waning magmatic system at Teide–Pico Viejo complex. That said, Teide–Pico Viejo is well-monitored and any subsurface activity is certain to be detected weeks to months prior to a potential eruption. This has recently been evidenced in the prelude to the 2011–2012 submarine eruption at El Hierro, where seismic activity increased several months in advance of the eruption, despite the eruption itself being of relatively small volume and magnitude (Carracedo et al. 2012; Troll et al. 2012).

References

Ablay GJ, Martí J (2000) Stratigraphy, structure, and volcanic evolution of the Pico Teide-Pico Viejo formation, Tenerife, Canary Islands. J Volcanol Geotherm Res 103:175–208

Ablay GJ, Ernst GGJ, Marti J, Sparks RSJ (1995) The ~2ka subplinian eruption of Montaña Blanca, Tenerife. Bull Volcanol 57(5):337–355. doi: 10.1007/s004450050098

Ablay GJ, Carroll MR, Palmer MR, Martí J, Sparks RSJ (1998) Basanite-Phonolite Lineages of the Teide-Pico Viejo Volcanic Complex, Tenerife, Canary Islands. J Petrol 39:905–936

Ancochea E, Fúster J, Ibarrola E, Cendrero A, Coello J, Hernan F, Cantagrel JM, Jamond C (1990) Volcanic evolution of the island of Tenerife (Canary Islands) in the light of new K-Ar data. J Volcanol Geotherm Res 44:231–249

Ancochea E, Huertas MJ, Cantagrel JM, Coello J, Fúster JM, Arnaud N, Ibarrola E (1999) Evolution of the Canadas edifice and its implications for the origin of the Canadas Caldera (Tenerife, Canary Islands). J Volcanol Geotherm Res 88:177–199

Araña V, Barberi F, Ferrara G (1989a) El complejo volcánico del Teide-Pico Viejo. In: Araña V, Coello J (eds) Los Volcanes y La Caldera del Parque Nacional del Teide (Tenerife, Islas Canarias). Serie Técnica, ICONA, pp 269–298

Araña V, Aparicio A, Garcia Cacho L, Garcia Garcia R (1989b) Mezcla de magmas en la región central de Tenerife. In: Araña V, Coello J (eds) Los Volcanes y La Caldera del Parque Nacional del Teide (Tenerife, Islas Canarias). Serie Técnica, ICONA, pp 269–298

Araña V, Martí J, Aparicio A, García-Cacho L, García-García R (1994) Magma mixing in alkaline magmas: An example from Tenerife, Canary Islands. Lithos 32:1–19

Bailey DK (1987) Mantle metasomatism—perspective and prospect. Geol Soc Spec Publ 30:1–13

Baker I (1968) Intermediate oceanic volcanic rocks and the 'Daly gap'. Earth Planet Sci Lett 4:103–106

Baker BH, Williams LAJ, Miller JA, Fitch FJ (1971) Sequence and geochronology of the Kenya rift volcanics. Tectonophysics 11:191–215

Barth TFW, Correns CW, Eskola P (1939) Die Entstehung der Gesteine. Springer, Berlin

Best MG (2003) Igneous and metamorphic petrology. Blackwell, Oxford

Bohrson WA, Reid MR (1998) Genesis of evolved Ocean Island Magmas by deep- and shallow-level basement recycling, Socorro Island, Mexico: Constraints from Th and other Isotope signatures. J Petrol 39:995–1008

Bonnefoi CC, Provost A, Albarède F (1995) The 'Daly gap' as a magmatic catastrophe. Nature 378:270–272

Bowen NL (1928) The evolution of the igneous rocks. Princeton University Press, New Jersey

Brophy JG (1991) Composition gaps, critical crystallinity, and fractional crystallization in orogenic (calc-alkaline) magmatic systems. Contrib Mineral Petrol 109:173–182

Brown RJ, Branney MJ (2004) Event-stratigraphy of a caldera-forming ignimbrite eruption on Tenerife: the 273 ka Poris formation. Bull Volcanol 66:392–416

Brown RJ, Barry TL, Branney JJ, Pringle MS, Bryan SE (2003) The quaternary pyroclastic succession of Southeast Tenerife, Canary Islands; explosive eruptions, related caldera subsidence, and sector collapse. Geol Mag 140:265–288

Bryan SE, Martí J, Cas RAF (1998) Stratigraphy of the Bandas del Sur formation: an extracaldera record of quaternary phonolitic explosive eruptions from the Las Cañadas edifice, Tenerife (Canary Islands). Geol Mag 135:605–636

Bryan SE, Cas RAF, Martí J (2000) The 0.57 Ma plinian eruption of the Granadilla Member, Tenerife (Canary Islands): an example of complexity in eruption dynamics and evolution. J Volcanol Geotherm Res 103:209–238

Bunsen R (1851) Ueber die Processe der vulkanischen Gesteinsbildungen Islands. Annalen der Physik und Chemie 159:197–272

Cann JR (1968) Bimodal distribution of rocks from volcanic islands. Earth Planet Sci Lett 4:479–480

Carracedo JC (1979) Paleomagnetismo e historia volcánica de Tenerife. Aula Cultura Cabildo Insular de Tenerife, Santa Cruz de Tenerife, p 81

Carracedo JC (1999) Growth, structure, instability and collapse of Canarian volcanoes and comparisons with Hawaiian volcanoes. J Volcanol Geotherm Res 94:1–19

Carracedo JC, Rodríguez Badiola E, Guillou H, Paterne M, Scaillet S, Pérez Torrado FJ, Paris R, Fra-Paleo U, Hansen A (2007) Eruptive and structural history of Teide volcano and rift zones of Tenerife, Canary Islands. Geol Soc Am Bull 119:1027–1051

Carracedo JC, Rodríguez Badiola E, Guillou H, Paterne M, Scaillet S, Pérez Torrado FJ, Paris R, Rodríguez González A, Socorro S (2008) El Volcán Teide—Volcanología, Interpretación de Pasajes y Iterinarios Comentados. Caja Generál de Ahorros de Canarias

Carracedo JC, Guillou H, Nomade S, Rodríguez-Badiola E, Pérez-Torrado FJ, Rodríguez-González A, Paris R, Troll VR, Wiesmaier S, Delcamp A, Fernández-Turiel JL (2011) Evolution of ocean-island rifts: the northeast rift zone of Tenerife, Canary Islands. Geol Soc Am Bull 123:562–584

Carracedo JC, Pérez Torrado F, Rodríguez González A, Soler V, Fernández Turiel JL, Troll VR, Wiesmaier S (2012) The 2011 submarine volcanic eruption in El Hierro (Canary Islands). Geol Today 28:53–58

Chayes F (1963) Relative abundance of intermediate members of the oceanic basalt-trachyte association. J Geophys Res 68:1519

Chayes F (1977) The oceanic basalt-trachyte relation in general and in the Canary Islands. Am Miner 62:666–671

Clague DA (1978) The Oceanic Basalt-Trachyte association: an explanation of the Daly gap. J Geol 86:739–743

Daly RA (1925) The geology of Ascension Island. Proc Am Acad Arts Sci 60:3–80

Deegan FM, Troll VR, Freda C, Misiti V, Chadwick JP, McLeod CL, Davidson JP (2010) Magma-Carbonate interaction processes and associated CO_2 release at Merapi Volcano, Indonesia: insights from experimental petrology. J Petrol 51:1027–1051

Deegan FM, Troll VR, Barker AK, Harris C, Chadwick JP, Carracedo JC, Delcamp A (2012) Crustal versus source processes recorded in dykes from the Northeast volcanic rift zone of Tenerife, Canary Islands. Chem Geol 334(0):324–344

DePaolo DJ (1981) Trace element and isotopic effects of combined wallrock assimilation and fractional crystallization. Earth Planet Sci Lett 53:189–202

Dingwell DB (1996) Volcanic dilemma: flow or blow? Science 273:1054–1055

Edgar CJ, Wolff JA, Nichols HJ, Cas RAF, Martí J (2002) A complex quaternary ignimbrite-forming phonolitic eruption: the Poris Member of the Diego Hernández Formation (Tenerife, Canary Islands). J Volcanol Geotherm Res 118:99–130

Edgar CJ, Wolff JA, Olin PH, Nichols HJ, Pittari A, Cas RAF, Reiners PW, Spell TL, Martí J (2007) The late quaternary Diego Hernandez Formation, Tenerife: Volcanology of a complex cycle of voluminous explosive phonolitic eruptions. J Volcanol Geotherm Res 160:59–85

Fúster JM, Cendrero A, Gastesi P, Ibarrola E, López Ruiz J (1968) Geology and volcanology of the Canary Islands, Fuerteventura. Instituto Lucas Mallada, Madrid

Garcia MO, Frey FA, Grooms DG (1986) Petrology of volcanic rocks from Kaula Island, Hawaii. Contrib Mineral Petrol 94:461–471

Garcia MO, Ito E, Eiler JM, Pietruszka AJ (1998) Crustal contamination of Kilauea Volcano Magmas revealed by Oxygen Isotope analyses of glass and Olivine from Puu Oo Eruption Lavas. J Petrol 39:803–817

Guillou H, Carracedo JC, Paris R, Pérèz Torrado FJ (2004) Implications for the early shield-stage evolution of Tenerife from K/Ar ages and magnetic stratigraphy. Earth Planet Sci Lett 222:599–614

Harris PG (1963) Comments on a paper by F. Chayes, "Relative Abundance of Intermediate Members of the Oceanic bBasalt-Trachyte Association". J Geophys Res 68:5103–5107

Harris C, Smith HS, le Roex AP (2000) Oxygen isotope composition of phenocrysts from Tristan da Cunha and Gough Island lavas: variation with fractional crystallization and evidence for assimilation. Contrib Mineral Petrol 138:164–175

Hess KU, Dingwell DB (1996) Viscosities of hydrous leucogranitic melts: a non-Arrhenian model. Am Mineral 81:1297–1300

Le Bas MJ, Maitre RWL, Streckeisen A, Zanettin B, ISotSoI Rocks (1986) A chemical classification of volcanic rocks based on the total Alkali-Silica diagram. J Petrol 27:745–750

Lippard SJ (1973) The petrology of phonolites from the Kenya Rift. Lithos 6:217–234

Longpré M-A, Troll VR, Walter TR, Hansteen TH (2009) Volcanic and geochemical evolution of the Teno massif, Tenerife, Canary Islands: some repercussions of giant landslides on ocean island magmatism. Geochem Geophys Geosyst 10:Q12017. doi: 10.1029/2009gc002892

Manconi A, Longpré M-A, Walter TR, Troll VR, Hansteen TH (2009) The effects of flank collapses on volcano plumbing systems. Geology 37:1099–1102

Márquez A, López I, Herrera R, Martín-González F, Izquierdo T, Carreño F (2008) Spreading and potential instability of Teide volcano, Tenerife, Canary Islands. Geophys Res Lett 35:L05305, doi: 10.1029/2007GL032625

Marsh BD (1981) On the crystallinity, probability of occurrence, and rheology of lava and magma. Contrib Mineral Petrol 78:85–98

Martí J, Mitjavila J, Araña V (1994) Stratigraphy, structure and geochronology of the Las Cañadas caldera (Tenerife, Canary Islands). Geol Mag 131:715–727

Martí J, Hurlimann M, Ablay GJ, Gudmundsson A (1997) Vertical and lateral collapses on Tenerife (Canary Islands) and other volcanic ocean islands. Geology 25:879–882

McBirney AR (1980) Mixing and unmixing of magmas. J Volcanol Geotherm Res 7:357–371

McDonough WF, S-s Sun (1995) The composition of the Earth. Chem Geol 120:223–253

Neumann ER, Wulff-Pedersen E, Pearson NJ, Spencer EA (2002) Mantle Xenoliths from Tenerife (Canary Islands): Evidence for Reactions between Mantle Peridotites and Silicic Carbonatite Melts inducing Ca Metasomatism. J Petrol 43(5):825–857

O'Hara MJ (1998) Volcanic plumbing and the space problem-thermal and geochemical consequences of large-scale assimilation in Ocean Island development. J Petrol 39:1077–1089

Paris R, Guillou H, Carracedo JC, PerezTorrado FJ (2005) Volcanic and morphological evolution of La Gomera (Canary Islands), based on new K–Ar ages and magnetic stratigraphy: implications for oceanic island evolution. J Geol Soc (London, UK) 162:501–512

Ridley W (1970) The abundance of rock types on Tenerife, Canary Islands, and its petrogenetic significance. Bull Volcanol 34:196–204

Rodríguez-Badiola E, Pérez-Torrado FJ, Carracedo JC, Guillou H (2006) Petrografía y Geoquímica del edificio volcánico Teide-Pico Viejo y las dorsales noreste y noroeste de Tenerife. In: Carracedo JC (ed) Los volcanes del Parque Nacional del Teide/El Teide, Pico Viejo y las dorsales activas de Tenerife. Naturaleza Y Parques Nacionales Serie Técnica. Organismo Autónomo Parques Nacionales Ministerio De Medio Ambiente, Madrid, pp 129–186

Schmincke H-U (1969) Ignimbrite sequence on Gran Canaria. Bull Volcanol 33:1199–1219

Schmincke H-U (1976) Geology of the Canary Islands. In: Kunkel G (ed) Biogeography and ecology in the Canary Islands. W. Junk, The Hague, pp 67–184

Sleep NH (1990) Hotspots and Mantle Plumes: some phenomenology. J Geophys Res 95:6715–6736

Spera FJ, Bohrson WA (2001) Energy-constrained open-system magmatic processes I: general model and energy-constrained assimilation and fractional crystallization (EC-AFC) formulation. J Petrol 42:999–1018

Thirlwall MF, Jenkins C, Vroon PZ, Mattey DP (1997) Crustal interaction during construction of ocean islands: Pb—Sr—Nd—O isotope geochemistry of

the shield basalts of Gran Canaria, Canary Islands. Chem Geol 135:233–262
Thirlwall MF, Singer BS, Marriner GF (2000) 39Ar–40Ar ages and geochemistry of the basaltic shield stage of Tenerife, Canary Islands, Spain. J Volcanol Geotherm Res 103:247–297
Thompson G, Smith I, Malpas J (2001) Origin of oceanic phonolites by crystal fractionation and the problem of the Daly gap: an example from Rarotonga. Contrib Mineral Petrol 142:336–346
Troll VR, Schmincke H-U (2002) Magma Mixing and Crustal Recycling Recorded in Ternary Feldspar from Compositionally Zoned Peralkaline Ignimbrite 'A', Gran Canaria, Canary Islands. J Petrol 43:243–270
Troll VR, Walter TR, Schmincke H-U (2002) Cyclic caldera collapse: piston or piecemeal subsidence? Field and experimental evidence. Geology 30:135–138
Troll VR, Klügel A, Longpré M-A, Burchardt S, Deegan FM, Carracedo JC, Wiesmaier S, Kueppers U, Dahren B, Blythe LS, Hansteen TH, Freda C, Budd DA, Jolis EM, Jonsson E, Meade FC, Harris C, Berg SE, Mancini L, Polacci M, Pedroza K (2012) Floating stones off El Hierro, Canary Islands: xenoliths of pre-island sedimentary origin in the early products of the October 2011 eruption. Solid Earth 3:97–110
Wager LR, Brown GM (1967) Layered igneous rocks. Oliver and Boyd Ltd, Edinburgh
Walker GPL (1993) Basaltic-volcano systems. In: Prichard HM, Alabaster T, Harris NBW, Neary CR (eds) Magmatic processes and plate tectonics. Geol Soc Spec Publ 76:3–38
Walter TR, Troll VR, Cailleau B, Belousov A, Schmincke H-U, Amelung F, Pvd Bogaard (2005) Rift zone reorganization through flank instability in ocean island volcanoes: an example from Tenerife, Canary Islands. Bull Volcanol 67:281–291
Watson EB (1982) Basalt contamination by continental crust: some experiments and models. Contrib Mineral Petrol 80:73–87
Watts AB, Masson DG (1995) A giant landslide on the northflank of Tenerife, Canary Islands. J Geophys Res 100:24487–24498
Webb SL, Dingwell DB (1990) Non-newtonian rheology of igneous melts at high stresses and strain rates: experimental results for Rhyolite, Andesite, Basalt, and Nephelinite. J Geophys Res 95: 15695–15701
Wiesmaier S, Deegan F, Troll V, Carracedo J, Chadwick J, Chew D (2011) Magma mixing in the 1,100 AD Montaña Reventada composite lava flow, Tenerife, Canary Islands: interaction between rift zone and central volcano plumbing systems. Contrib Mineral Petrol 162:651–669
Williams L (1970) The Volcanics of the Gregory rift valley, East Africa. Bull Volcanol 34:439–465
Williams LAJ (1972) The Kenya Rift volcanics: a note on volumes and chemical composition. Tectonophysics 15:83–96
Wolff JA (1983) Petrology of Quaternary pyroclastic deposits from Tenerife, Canary Islands. University of London, London
Wolff JA, Palacz ZA (1989) Lead isotope and trace element variation in Tenerife pumices—Evidence for recycling within an ocean island volcano. Mineral Mag 53:519–525
Wolff JA, Storey M (1984) Zoning in highly alkaline magma bodies. Geol Mag 121:563–575
Wolff JA, Grandy JS, Larson PB (2000) Interaction of mantle-derived magma with island crust? Trace element and oxygen isotope data from the Diego Hernandez Formation, Las Cañadas, Tenerife. J Volcanol Geotherm Res 103:343–366

Magmatic Differentiation in the Teide–Pico Viejo Succession: Isotope Analysis as a Key to Deciphering the Origin of Phonolite Magma

10

Sebastian Wiesmaier, Valentin R. Troll, Juan Carlos Carracedo, Robert M. Ellam, Ilya Bindeman, John A. Wolff, and Frances M. Deegan

Abstract

In Tenerife, lavas of the recent Teide–Pico Viejo central complex show a marked bimodality in composition from initially mafic lava (200–30 ka) to highly differentiated phonolite (30–0 ka). Groundmass Sr–Nd–Pb–O and feldspar ^{18}O data demonstrate open system behaviour for the petrogenesis of Teide–Pico Viejo felsic lavas, but contamination by ocean sediment can be excluded due to the low $^{206}Pb/^{204}Pb$ ratios of North Atlantic sediment. Isotope mixing hyperbolae require an assimilant of predominantly felsic composition for the Teide–Pico Viejo succession. Unsystematic and heterogeneous variation of ^{18}O in fresh and unaltered feldspars across the Teide–Pico Viejo succession indicates magmatic

S. Wiesmaier (✉)
Geo- and Environmental Sciences, Ludwig-Maximilians-Universität München, München, Germany
e-mail: sebastian.wiesmaier@min.uni-muenchen.de

V. R. Troll
Department of Earth Sciences, CEMPEG, Uppsala University, 75236 Uppsala, Sweden

J. C. Carracedo
Department of Physics (Geology), GEOVOL, University of Las Palmas, Gran Canaria, Spain

R. M. Ellam
Scottish Universities Environmental Research Centre (SUERC), East Kilbride, Scotland, UK

I. Bindeman
Department of Geological Sciences, University of Oregon, Eugene, OR, USA

J. A. Wolff
Department of Geology, Washington State University, Pullman, WA, USA

F. M. Deegan
Laboratory for Isotope Geology, Swedish Museum of Natural History, Stockholm, Sweden

J. C. Carracedo and V. R. Troll (eds.), *Teide Volcano*, Active Volcanoes of the World,
DOI: 10.1007/978-3-642-25893-0_10, © Springer-Verlag Berlin Heidelberg 2013

addition of diverse ^{18}O assimilants, best matched by nepheline syenites that occur as fresh and altered lithic blocks in voluminous pre-Teide ignimbrite deposits. Rare earth element modelling indicates that nepheline syenite needs to be melted in bulk to form a suitable end-member composition. Energy-Constrained Assimilation Fractional Crystallisation (EC-AFC) modelling reproduces the bulk of the succession, which implies that the petrogenesis of Teide–Pico Viejo lavas is governed by the coupled assimilation of nepheline syenite during fractional crystallisation. The most differentiated (and most radiogenic) lava computes to >97.8 % assimilant, likely represented by a nepheline syenite bulk melt that formed by underplating with juvenile mafic material. These recent research developments therefore recognise a wider variability of magmatic differentiation processes at Teide–Pico Viejo than previously considered.

10.1 Introduction

In oceanic islands, the origin of large volumes of felsic volcanic rock has long been a matter of debate (Chap. 9). Due to the absence of large regional tectonic influences and, for most ocean islands, large sedimentary sequences, it had been proposed that crystal fractionation must be the dominant mechanism of differentiation (e.g., Cann 1968; Clague 1978; Garcia et al. 1986; Thompson et al. 2001). At Teide–Pico Viejo, too, the accepted model for magmatic differentiation was for a long time one of pure fractional crystallisation, essentially because detection of assimilation was not possible by traditional petrological means (e.g., Ablay et al. 1998). Major and trace element variations of the Teide–Pico Viejo complex combined with geothermobarometry and estimates of pre-eruptive volatile contents have allowed the establishment of a fractionation sequence that invoked 88 % removal of crystals for the generation of phonolite. Moreover, the presence of amphibole in some Pico Viejo lavas indicated the presence of two distinct petrological lineages for Pico Viejo and Pico Teide and, hence, two plumbing systems were postulated to exist with different depths for their main magma chambers. This interpretation of Teide–Pico Viejo magmatic differentiation processes involved fractional crystallisation exclusively, and hence the model described Teide as a closed system, i.e., a system in which no external material enters the magma and only energy is exchanged with the surrounding wall rock (e.g., cooling). Ablay et al. (1998), however, noted that certain trace element variations were not explained by the pure crystal fractionation model. The authors invoked small additions of zeolite and hydrothermally altered material to explain these variations, but until recently no in-depth petrogenetic investigation using isotopic methods was available to test this model. The focus of this chapter is the quantification of crustal influences using Sr, Nd, Pb and O isotope geochemistry.

10.2 The Application of Radiogenic Isotopes in Igneous Petrology

A reliable way to resolve whether a magmatic system is open or closed to the influx of foreign material is the study of radiogenic isotopes ratios in minerals and rocks. If the mantle-derived magmatic source is assumed to have a relatively constant isotope composition at a certain point in time, as is the case in Tenerife (Simonsen et al. 2000), isotope ratios in magma may change only by two principal means, radioactive decay and/or mixing with material of differing isotopic composition.

Radioactive decay requires millions of years to change the isotope ratios of Sr, Nd and Pb at a detectable level. For example, the half-life of

^{87}Rb, which decays to ^{87}Sr and thus slowly increases the ratio of $^{87}Sr/^{86}Sr$, is thought to be 48.8 Ga (e.g., Dickin 2005). However, over the geologically brief episode in which the Teide–Pico Viejo edifice has been constructed (<200 ka), radioactive decay is too minute to significantly affect the $^{87}Sr/^{86}Sr$ ratios of the Teide–Pico Viejo rocks and, thus, the ratios would effectively remain constant in a closed system situation.

In turn, if the magmatic system is open, assimilation of external material may change the isotopic composition of magma when material of a distinct isotopic composition is admixed. Consider a magma with an Sr isotope ratio of 0.703 that assimilates material, e.g., country rock with an Sr isotope ratio of 0.705, in this case more ^{87}Sr than ^{86}Sr is fluxed into the magma, consequently raising the ratio of $^{87}Sr/^{86}Sr$ in the now contaminated magma mixture (e.g., Duffield and Ruiz 1998). The final Sr isotope ratio of this mixture will fall somewhere between 0.703 and 0.705, depending on the respective Sr concentrations in magma and country rock and the relative volumes of magma and country rock involved.

As isotope ratios are virtually unaffected by radiogenic decay in a young system such as Teide-Pico Viejo, they become an extremely powerful tool for the detection of assimilation. Measured isotope ratios that are different from the mantle source must have formed by modification of the magma by the addition of external material with a different isotopic composition. Isotope ratios can thus be used as fingerprints to identify rock types that may have been assimilated during magma evolution and furthermore allow calculation of the amount of material that must have been assimilated (e.g., Troll et al. 2005; Meade et al. 2009). Assimilation can strongly alter and accelerate magmatic differentiation and promotes felsic magma production, making this an important aspect of magma development to understand.

As the Teide–Pico Viejo system is geologically speaking very young (<200 ka), it provides an ideal testing ground for the application of radiogenic isotopes in constraining magma chamber processes and potential crustal influences on ocean island magmatic suites. Here, a thorough update of the magmatic differentiation processes that are recognised at Teide–Pico Viejo using Sr–Nd–Pb and O isotopes is summarised from the work of Wiesmaier (2010) and Wiesmaier et al. (2012).

10.3 Previous Work and Research Techniques

In a recent study, the radiogenic isotope composition of 58 of the 64 known eruptions of the Teide–Pico Viejo succession have been analysed (Wiesmaier et al. 2012). Some of these eruptions comprise multiple phases (Carracedo et al. 2007). A total of 61 groundmass samples were analysed from the three groups of Teide–Pico Viejo lavas (mafic, transitional and felsic), for their Pb, Sr and Nd isotopic composition. Fresh groundmass represents the very last melt composition before complete solidification and is, when properly separated, completely unaffected by any masking effects from the accumulation of pheno-/xenocrysts and thus allows melt processes to be constrained reliably (e.g., Marsh 2004; Kinman et al. 2009). Moreover, five sedimentary rocks of exhumed pre-island seafloor from Fuerteventura were analysed to serve as potential end-members for the isotopic composition of non-magmatic rocks that may lie underneath or within the islands (e.g., Stillman et al. 1975; Hobson et al. 1998; Hansteen and Troll 2003).

Feldspar separates from 15 lava flows were also analysed for their $\delta^{18}O$ composition at the University of Oregon, USA. Plagioclase phenocrysts were analysed by laser fluorination to determine $\delta^{18}O$ values, the maximum analytical uncertainties on $\delta^{18}O$ measurements are estimated at 0.15 ‰. For details of the analytical methods for both radiogenic and stable isotopes the reader is referred to Wiesmaier (2010) and Wiesmaier et al. (2012). A summary of results for all three sets of radiogenic isotope systems and oxygen isotopes are reported in Table 10.1.

Table 10.1 Overview of the isotope composition of Teide–Pico Viejo deposits and Fuerteventura sediments.

Isotope systems	Mafic	Transitional	Felsic	Sediment
$^{87}Sr/^{86}Sr$	0.703040(21)–0.703229(18)	0.703094(20)–0.703332(17) *comparable to mafic*	0.703091(2)–0.704900(16) *higher than mafic/transitional*	0.703473(21)–0.707684(21)
$^{143}Nd/^{144}Nd$	0.512901–0.512991 (± 6–45)	0.512916–0.512956 *narrower range than mafic*	0.512924–0.512950 *yet narrower range*	
$^{206}Pb/^{204}Pb$	19.5050(22)–19.8142(22)	19.7493(26)–19.7743(26) *narrow range compared to mafic*	19.7541(28)–19.7816(24) *narrow range compared to mafic*	18.5307(13)–19.7447(01)
$^{207}Pb/^{204}Pb$	15.5919(22)–15.6456(24)	15.6102(24)–15.6337(28) *narrower range than mafic*	15.5288(22)–15.6174(26) *much lower than mafic/trans.*	15.6263(26)–15.6767(14)
$^{208}Pb/^{204}Pb$	39.4490(68)–39.6371(68)	39.5536(70)–39.6067(80) *narrower range than mafic*	39.5316(72)–39.6001(68) *narrower range than mafic*	38.2678(70)–39.6810(42)
$\delta^{18}O$ [‰]	5.43–5.84	5.46–5.88 *comparable to mafic*	5.88–5.99 *higher values*	

Results from groundmass isotope analyses. Errors for Sr, Nd and Pb isotope ratios are 2 × Standard Deviation. Felsic lavas show an elevation in Sr isotope ratios compared to the other lavas. Compared to mafic lavas, in turn, transitional and felsic lavas show a strong confinement in $^{143}Nd/^{144}Nd$ and $^{206}Pb/^{204}Pb$ ratios. Oxygen isotope values of feldspar phenocrysts show rather heterogeneous values in mafic and transitional lavas, and a small increase in felsic lavas. *Data* Wiesmaier (2010) and Wiesmaier et al. (2012)

10.4 Sr-Nd-Pb-O Systematics at Teide-Pico Viejo

Mafic and transitional lavas overlap considerably in $^{87}Sr/^{86}Sr$ ratios, whereas the felsic lavas yield $^{87}Sr/^{86}Sr$ ratios of 0.703091(2)–0.704900(16), significantly higher than the mafic and the transitional compositions. The sediment samples are higher than both mafic and transitional samples, but partly overlap with the felsic lavas.

Mafic lavas yield variable $^{143}Nd/^{144}Nd$ ratios of 0.512901–0.512991. Transitional samples also show some variation, but are confined to a smaller interval of $^{143}Nd/^{144}Nd$. Felsic samples plot in an even tighter $^{143}Nd/^{144}Nd$ range from 0.512924 to 0.512950.

The $^{206}Pb/^{204}Pb$ ratios of transitional and felsic lavas show a peculiar behaviour compared to the mafic lavas. Mafic lavas show a spread in values that is consistent with previously published data from other mafic deposits in Tenerife. The transitional and felsic lavas, in turn, are confined to a very narrow interval of $^{206}Pb/^{204}Pb$ values, with felsic lavas yielding much lower values for $^{207}Pb/^{204}Pb$ than what is found in the mafic lavas. In $^{208}Pb/^{204}Pb$, mafic lavas completely encompass the felsic and transitional lava data range in their variability.

Oxygen isotope ratios in feldspars from mafic lavas (presented relative to the international standard SMOW) range from +5.43 ± 0.11 ‰ to +5.84 ± 0.02 ‰. Feldspars from transitional lavas display values from +5.82 ± 0.29 ‰ to +5.88 ± 0.11 ‰, with the exception of one outlier at +5.46 ± 0.11 ‰. The felsic lavas give feldspar oxygen isotope values that range from +5.88 ± 0.11 ‰ to +5.99 ± 0.11 ‰, i.e., only marginally higher than those from mafic samples.

Fig. 10.1 The twin Pico Viejo and Teide volcanoes are a spectacular example of felsic strato-cones in oceanic islands (www.fotosaereasdecanarias.com)

10.5 Discussion

The data permit several simple qualitative deductions that will set the scene for a more in-depth interpretation. Most importantly, the Sr, Nd, Pb and O isotope data of Teide–Pico Viejo lavas show a strong variability in isotope ratios, which exceeds the variability of the uncontaminated mantle signal in Tenerife. For example, the $^{87}Sr/^{86}Sr$ ratio of the mantle signal is thought to range between 0.7031 and 0.7033 (Simonsen et al. 2000). Although this is consistent with the bulk of Teide lavas that show a common $^{87}Sr/^{86}Sr$ range of 0.7031–0.7033, two felsic lavas (phonolites) exceed this Sr isotope baseline considerably and reach up to 0.7049 (Table 10.1). This implies that at some stage during the evolution of Teide magma, material of a different isotopic composition was introduced. The magmatic plumbing system beneath Teide–Pico Viejo must therefore be open in order to undergo chemical exchange with fluids and/or surrounding wall rock (Fig. 10.1).

The Pb isotope data of felsic Teide–Pico Viejo lavas have a very uniform $^{206}Pb/^{204}Pb$ ratio. This is, once more, distinct from the mafic lavas, which are variable in $^{206}Pb/^{204}Pb$ and so delineate a field of 'baseline' Pb isotope ratios comparable with other ocean islands (Fig. 10.2). This alone would not be unusual, but in addition, the felsic lavas possess much more variable $^{207}Pb/^{204}Pb$ ratios than the mafic lavas. Although the $^{206}Pb/^{204}Pb$ and $^{207}Pb/^{204}Pb$ ratios in felsic lavas are consistent with the range of Pb isotope data so far found in Tenerife rocks, the combination of uniform $^{206}Pb/^{204}Pb$ and very low $^{207}Pb/^{204}Pb$ composition in the Teide felsic lavas is uncommon. The Sr and Pb isotope ratios therefore indicate that the Teide–Pico Viejo succession has formed by open-system processes, assimilating a component of distinct isotope composition during the differentiation of the most evolved (i.e., most felsic) magmas at the Teide–Pico Viejo central complex.

10.5.1 Sediment Contamination?

Judging from the Sr isotope data alone, it appears possible that sediment may have been taken up by magma consequently altering its $^{87}Sr/^{86}Sr$ composition. The Jurassic oceanic crust below Tenerife is overlain by thick sedimentary sequences with elevated $^{87}Sr/^{86}Sr$ that magma has to traverse before reaching the surface (Sun 1980; Hoernle et al. 1991; Hoernle 1998; Hansteen and Troll 2003; Aparicio et al. 2010; Troll et al. 2012). Atlantic sediment may also be uplifted into the island's core and is thus potentially present during magmatic ascent in the Canaries (Stillman et al. 1975; Robertson and Stillman 1979; Hansteen and Troll 2003).

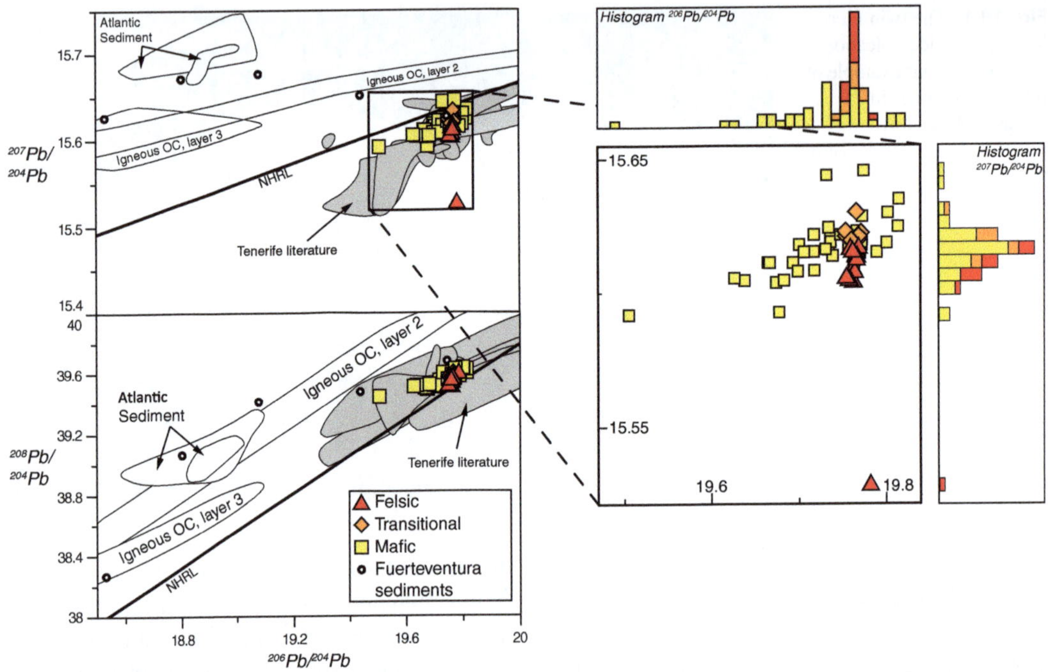

Fig. 10.2 Results from Pb isotope analyses combined with histograms of $^{206}Pb/^{204}Pb$ and $^{207}Pb/^{204}Pb$ ratios of the Teide–Pico Viejo succession. All errors are 2SD and are included in the symbols. Note the lack of a trend in the Teide–Pico Viejo succession towards Atlantic sediment compositions. Transitional and felsic lavas show a peak in $^{206}Pb/^{204}Pb$ ratio distribution at around 19.76, similar to mafic lavas. In contrast, felsic lavas are off this mafic peak in $^{207}Pb/^{204}Pb$ ratio, indicating an open system. Note the confinement of transitional and felsic lavas to a narrow range of $^{206}Pb/^{204}Pb$ ratios. Data for Tenerife are from Sun (1980), Simonsen et al. (2000), Abratis et al. (2002), Gurenko et al. (2006), and Wolff et al. (2000); data for oceanic crust and Atlantic sediment were taken from Sun (1980), Hoernle et al. (1991) and Hoernle (1998)

However, since sediment also contains significant amounts of Pb, sediment contamination would have to be detected in both the Sr and the Pb isotope systems, but this is not the case. Atlantic sediments always have lower $^{206}Pb/^{204}Pb$ and much higher $^{207}Pb/^{204}Pb$ compositions (Fig. 10.2). In the Teide–Pico Viejo succession, a Pb isotopic trend toward sediment composition is not discernible. Despite the comparable $^{87}Sr/^{86}Sr$ ratios in sediment and phonolite, Atlantic sediments are therefore unlikely to have entered the magmatic plumbing system of Teide–Pico Viejo in appreciable quantities. This implies, in turn, that the high Sr isotope ratios found in some of Teide's most evolved phonolites must stem from material other than sediment.

10.5.2 Constraints on the Assimilant

So far we have established that the open system of Teide–Pico Viejo volcano assimilated a non-sedimentary component. This yet to be defined assimilated material (or *assimilant*), can be further constrained by Sr–Nd isotope modelling. Isotope modelling allows us to calculate the isotopic composition of a mixture of two different end-members. The resulting mixing curves, or mixing hyperbolae, trace the possible mixtures of the two end-members, from 100:0 to 0:100 (end-member A:B). Of course, magmatic processes are not bound to simple mixing, but this type of model helps to detect first order genetic relationships in a sequence of rock samples and helps to further constrain the composition of the assimilant.

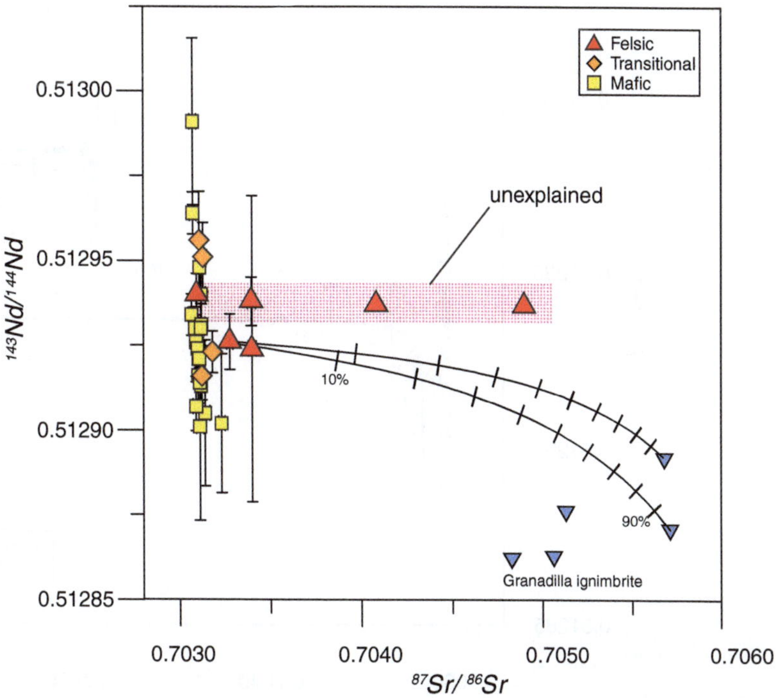

Fig. 10.3 $^{87}Sr/^{86}Sr$ versus $^{143}Nd/^{144}Nd$ of Teide–Pico Viejo lavas compared to data from the Granadilla ignimbrite (Palacz and Wolff 1989). Granadilla ignimbrite data are the only available high $^{87}Sr/^{86}Sr$ data from Tenerife, but are insufficient to explain the Teide–Pico Viejo isotopic variations in full

In the case of Teide–Pico Viejo, a high $^{87}Sr/^{86}Sr$ assimilant is required. As sediment has been ruled out, we need to look for an end-member that may be part of the island itself. The ~570 ka Granadilla ignimbrite is the only known deposit in Tenerife with $^{87}Sr/^{86}Sr$ ratios sufficiently high to serve as a potential assimilant (Palacz and Wolff 1989). The Granadilla ignimbrite is the product of a caldera-forming eruption within the Upper Group of the Las Cañadas Volcano deposits (Wolff and Palacz 1989; Martí et al. 1994, 1997; Bryan et al. 2002) and bears abundant nepheline syenite blocks of co-genetic composition. This material very likely abounds at depth below Teide volcano, and is thus theoretically available as an assimilant.

However, the Granadilla ignimbrite data also fail to reproduce the full pattern defined by the Teide–Pico Viejo lavas. The two Granadilla–Teide mixing curves in Fig. 10.3 demonstrate this. Furthermore, Granadilla ignimbrite samples also possess lower $^{206}Pb/^{204}Pb$ ratios than the felsic lavas (19.621–19.734 versus 19.754–19.782). Thus, Teide–Pico Viejo differentiated magma has not been strongly affected by material of the Granadilla eruption.

However, this exercise does provide additional information on the assimilant. We know that the assimilant must have had a higher $^{87}Sr/^{86}Sr$ ratio than sample MB7 ($^{87}Sr/^{86}Sr = 0.7049$). From 153 measurements found in the literature and the data presented here, we also know that in Tenerife, high $^{87}Sr/^{86}Sr$ ratios are only found in rocks of low Sr concentration. This implies a highly differentiated rock, comparable to the Granadilla ignimbrite. Moreover, $^{206}Pb/^{204}Pb$ values of the contaminant are required to be similar to the ratios found in felsic lavas but need to be lower in $^{207}Pb/^{204}Pb$. Few rocks have these characteristics, but the Diego Hernández Formation basalts for example, provide suitable Pb isotope ratios. Finally, mafic Las Cañadas rocks possess appropriate $^{143}Nd/^{144}Nd$ values, which are higher than the relatively low Granadilla values (see Simonsen et al. 2000).

Mixing modelling can also be performed with a hypothetical assimilant. The aim is to ascertain what type of assimilant would best reproduce the compositional features of the Teide–Pico Viejo succession. This hypothetical assimilant is based on the Granadilla ignimbrite composition, but

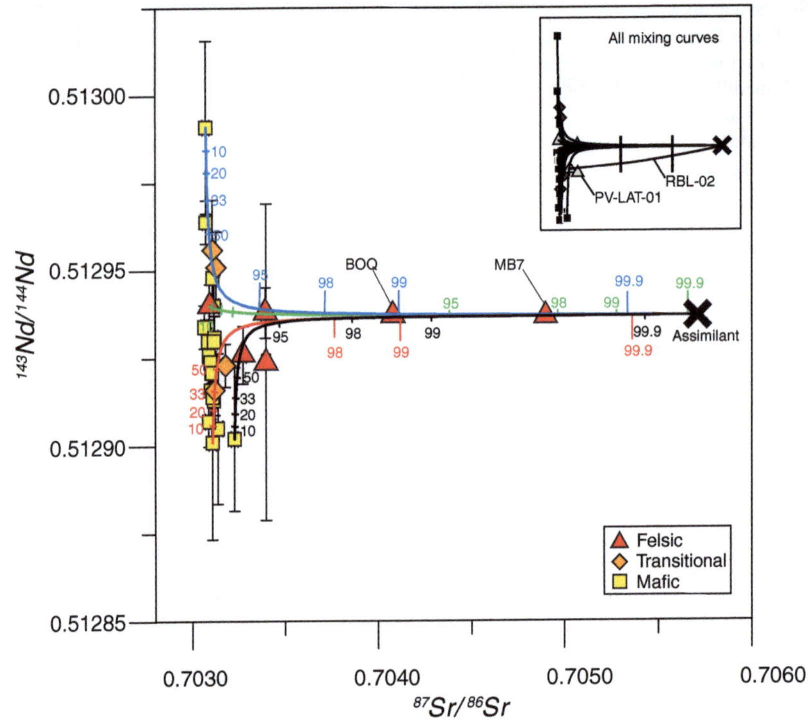

Fig. 10.4 Plot of $^{87}Sr/^{86}Sr$ versus $^{143}Nd/^{144}Nd$ of Teide–Pico Viejo lavas, modelled by using a hypothetical assimilant (*black cross*). Four representative mixing hyperbolae are presented between the hypothetical assimilant and three mafic lavas (*curves*: *red*, *black* and *blue*) and one felsic lava (*green curve*). *Numbers* indicate percentage of assimilant involved in the mixture. Hyperbolae of all remaining calculations are shown in the *inset*. The true assimilant likely was of felsic composition as low Sr/Nd ratios are required by the shape of the mixing hyperbolae. Note that to successfully model the sample MB7, between 97.8 and 99.5 % assimilant is required

with higher $^{143}Nd/^{144}Nd$ and suitable $^{206}Pb/^{204}Pb$ values. The mean Nd ratio of the Teide-Pico Viejo felsic lavas was assumed as the $^{143}Nd/^{144}Nd$ value for the assimilant (0.512937, n = 6), and the uniform $^{206}Pb/^{204}Pb$ ratios of the same felsic lavas were used as the Pb isotope ratio of the assimilant. Using comparable Sr and Nd concentrations and, consequently, a low Sr/Nd ratio in the hypothetical assimilant ensures that the assimilant bears the Sr, Nd and Pb signatures of an idealised highly differentiated Tenerife rock.

The resulting mixing hyperbolae not only encompass the felsic lavas within a single mixing relationship, but also include all transitional and mafic lavas, that is, the entire differentiation sequence from mafic to felsic lavas (Fig. 10.4). Being able to model an entire volcanic succession by invoking a single, isotopically feasible, end-member suggests a convincing genetic relationship for the Teide–Pico Viejo lavas and indicates that similar material was assimilated throughout the 200 ka of Teide–Pico Viejo evolution. In addition, the curvatures of the mixing hyperbolae provide strong evidence for the composition of the assimilant. To reproduce the Teide–Pico Viejo data, the mixing hyperbolae require a low Sr/Nd ratio in the assimilant, if this ratio is changed, the curves no longer reproduce the Teide data. A low Sr/Nd most likely represents an assimilant of felsic composition.

Furthermore, mafic lavas seem to include rather little assimilant, but the more felsic the lavas become, the higher the percentage of assimilant they appear to have incorporated. The Sr–Nd fingerprints of felsic lavas contain between 77 and 99.78 % of this hypothetical assimilant, regardless of whether the composition of a mafic, a transitional or a low-$^{87}Sr/^{86}Sr$ felsic lava is used as magmatic end-member. The felsic lava with the highest $^{87}Sr/^{86}Sr$ ratio, sample MB7, requires an end-member percentage of >97.8 % of the hypothetical felsic composition. The isotopic composition of Teide–Pico Viejo felsic lavas therefore displays a strong influence from a highly differentiated assimilant that increases dramatically with degree of differentiation.

These indications for a felsic assimilant during Teide–Pico Viejo magmatic evolution are consistent with phase equilibria considerations, melt inclusion analysis and geothermobarometry experiments, which indicate a shallow depth of origin for felsic magmas in Tenerife (Wellman 1970; Andújar et al. 2008, 2010). From these experiments, it is possible to infer the pressure at which a certain phenocryst or a crystal assemblage may have formed. The experimental results suggest that nepheline syenite (the deep plutonic equivalent of the pre-Teide Las Cañadas ignimbrites) and Teide phonolite both crystallised at 4–6 km depth, i.e., when Teide phonolite magma crystallised inside the volcano it was likely surrounded by these pre-Teide nepheline syenite rocks.

10.5.3 Heterogeneous Oxygen Isotope Composition of the Assimilant

The $\delta^{18}O$ ratios in feldspar crystals and groundmasses are consistent with open system differentiation. Closed-system fractional crystallisation is thought to continually raise $\delta^{18}O$ values at a small, but defined rate with differentiation (grey fields in Fig. 10.5), whereas uptake of non-magmatic material usually generates a deviation from this pattern. At Teide–Pico Viejo both low and high $\delta^{18}O$ components are discernible, albeit only in single eruptions and hence not systematically. Most data points plot within error of the closed system field for the alkali-rich series, with three samples yielding values significantly lower than that (Fig. 10.5). At very high degrees of differentiation, two groundmass samples show high $\delta^{18}O$ values, which are not reproduced by the feldspar crystals hosted in these samples, indicative of the late-stage addition of a high-$\delta^{18}O$ component to the melt.

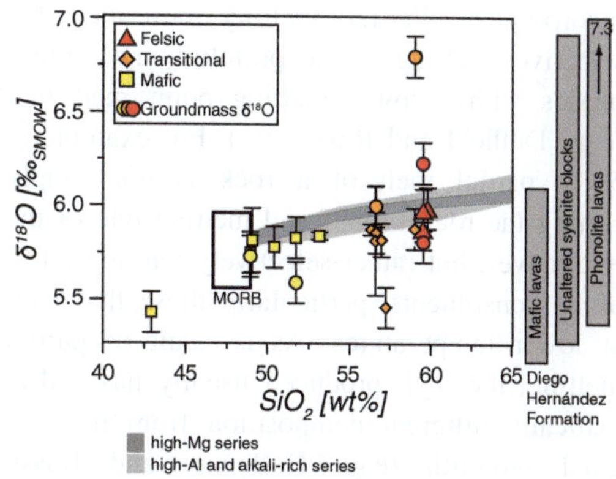

Fig. 10.5 Oxygen isotope ratios of feldspar and groundmass separates versus whole-rock SiO_2. *Triangles*, *squares* and *diamonds* represent feldspars, *circles* are groundmass. *Grey shaded arrays* denote closed-system oxygen isotope fractionation (from Bindeman 2008). Teide–Pico Viejo data deviate from these arrays and thus demonstrate an open system. *To the right*: *grey bars* are data from the Diego Hernández Formation from Wolff et al. (2000). Teide data overlap with the range of values found in Diego Hernández ignimbrites and nepheline syenites, consistent with Diego Hernández-type rocks being a potential assimilant

The $\delta^{18}O$ data of the highly differentiated members of the Diego Hernández Formation and the enclosed nepheline syenite blocks (Wolff et al. 2000) show an even larger variability in $\delta^{18}O$, which moreover, overlaps the values found for the Teide–Pico Viejo succession. Such a heterogeneous assimilant is likely to produce variability in $\delta^{18}O$ ratios across a series of eruptions separate in time. The $\delta^{18}O$ signatures found in Teide–Pico Viejo lavas are therefore best explained by the uptake of highly differentiated material of a heterogeneous $\delta^{18}O$ composition.

10.5.4 Bulk Melting of Country Rock

An important question for the assimilation of foreign material into magma is whether entire fragments of country rock (i.e., xenoliths) or a partial melt of this country rock is incorporated. Partial melting describes the selective melting of a protolith, i.e., those phases with a lower melting point melt first (e.g., Duffield and Ruiz 1998). For example, a 50 % partial melt of a rock is not simply cutting the rock in half and melting one of the two halves, but rather selectively melting 50 % of its constituents, particularly those that melt at lower temperatures. As a result of partial melting, the melt produced usually has a dramatically different composition from the original protolith (e.g., Holloway and Bussy 2008). This is best reflected in the trace element concentrations of a partial melt. Incompatible elements are trace elements that do not fit well into crystal lattices of common magmatic crystals because of their large ionic radius. Instead, these elements prefer the liquid phase and will therefore be enriched in the partial melt and depleted in the solid residue. Thus incompatible elements are particularly useful as tracers of partial melting.

Batch partial melting calculations of the Diego Hernández nepheline syenite allow us to estimate both the degree of partial melting and the composition of the partial melt produced. The Rare Earth Elements (REE) are especially well-suited for use in partial melting calculations, because of their systematic variation of ionic radius at a near-constant valent state, the so-called *lanthanide contraction*. This means that the relative behaviour of two neighbouring elements is very predictable, as these cations systematically partition into crystals or melt. In a plot of Gd/Yb versus La/Sm, for example, lower degrees of partial melting cause higher Gd/Yb and La/Sm ratios (and vice versa).

Using partial melts of the Diego Hernández Formation nepheline syenite as an assimilant, the mixing model only reproduces Teide–Pico Viejo phonolites above a melt fraction of ca. F = 0.95. Using bulk melts of nepheline syenite however, (F = 1) reproduces the felsic Teide lavas well. Note that 'altered' nepheline syenite provides the best fit in these calculations, and is thus more likely than assimilating a 'fresh' nepheline syenite (Fig. 10.6).

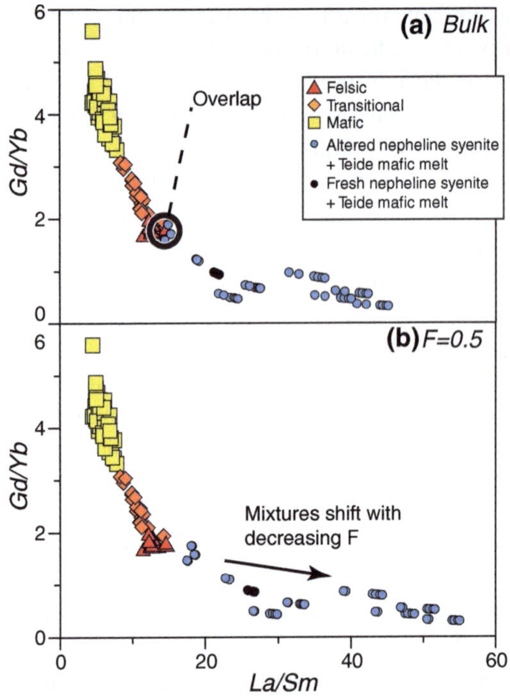

Fig. 10.6 Results of rare earth element modelling, using the mixing percentages derived from isotope modelling. To constrain the degree of partial melting of country rock, Teide–Pico Viejo lavas (*squares*, *diamonds*, *triangles*) are compared with mixtures of Diego Hernández batch melts and Teide–Pico Viejo lavas (*crosses*). The degree of partial melting of Diego Hernández rocks is given in the *upper right-hand corner*. Teide felsic lavas are best reproduced by mixing Teide mafic lava compositions with bulk melts of Diego Hernández samples. Lower degrees of partial melting of Diego Hernández rocks (< 0.95) shifts the resulting mixtures towards higher La/Sm ratios that increasingly differ from the observed Teide lava compositions

10.5.5 Quantification of Differentiation Processes at Teide–Pico Viejo

In the preceding discussion, radiogenic and stable isotope data from Teide–Pico Viejo lavas were shown to demonstrate the influence of a highly differentiated igneous assimilant. Moreover, the underlying calculations support the bulk incorporation of that assimilant. While these mixing calculations are a perfectly valid first order approximation, magmatic differentiation is commonly attributed to assimilation and fractional crystallisation (DePaolo 1981). To quantify the interplay of fractionation and assimilation at Teide–Pico Viejo, the EC-AFC model after Spera and Bohrson (2001) was employed. This complex but user-friendly code embeds assimilation and fractional crystallisation into a geochemical, isotopic and thermodynamic framework and thus permits more realistic assumptions on the intensity and interplay of the various differentiation processes (Spera and Bohrson 2001, 2002, 2004).

Here, an EC-AFC model is shown for one batch of hot, mafic magma that thermally equilibrates with a surrounding, cooler and highly differentiated assimilant of nepheline syenite composition. All information that has been gathered so far is included in this model: isotope and geochemical data on both Teide–Pico Viejo and the hypothetical assimilant, crystallisation depths and a steep geothermal gradient of 100 °C/km, observed at Teide (http://www.petratherm.com.au/_webapp_117699/Canary_Islands). Specific heat capacities and crystallisation/fusion enthalpies were calculated using the average compositions of Tenerife Granadilla ignimbrite.

The felsic assimilant and the chosen thermal parameters reproduce the geochemical and isotopic variations in Teide–Pico Viejo lavas well (Fig. 10.7). The EC-AFC curves match the entire differentiation sequence in Pb, Zr and Hf composition. Sr, Ba and Nd curves reproduce all compositions except for the samples of intermediate concentrations. This, however, is an artefact of the EC-AFC code, because abrupt changes in bulk compatibility of trace elements, a likely phenomenon in nature, are not accounted for.

Isotope ratios are well reproduced, too. As Pb and Nd ratios are very variable in mafic lavas, several mafic end-member compositions were tested (see model curves in Fig. 10.7). All of the mafic end-members trend towards the same felsic end-member isotope composition during thermal equilibration and replicate the naturally occurring pattern of isotope compositions. Equally, the $^{87}Sr/^{86}Sr$ ratios of the Teide succession are well explained by these mixing hyperbolae, except for one outlier. By and large, the EC-AFC model demonstrates that the formation of felsic Teide–Pico Viejo magma is consistent with assimilation of felsic rock with simultaneous crystal fractionation from a mafic parent.

The model also confirms details revealed by the REE batch melting calculations and Sr–Nd isotope modelling results. Firstly, the EC-AFC model requires felsic country rock to be melted wholesale, and secondly, assimilation appears more dominant with higher differentiation of magma.

10.5.6 Mechanisms for Crustal Melting

The only sample composition that is not being replicated by EC-AFC is one sample from Montaña Blanca (MB7). This is the lava with the highest degree of differentiation found in the entire Teide–Pico Viejo suite. Perhaps processes additional to combined assimilation and fractional crystallisation have influenced this particular magma.

In a plot of $^{87}Sr/^{86}Sr$ versus $1/^{86}Sr$, the felsic lavas either show high $^{87}Sr/^{86}Sr$ ratios, low ^{86}Sr concentrations (high $1/^{86}Sr$) or a combination of both. The felsic samples RBL-02 and BOQ plot to the right hand side at high $1/^{86}Sr$ values (Fig. 10.8). Using the felsic assimilant deduced from isotope and trace element constraints, the EC-AFC curves reproduce the samples BOQ and RBL-02, but fall short of explaining the Sr isotope composition of sample MB7. Instead, four of the felsic lavas (including Montaña Blanca) define a linear array, which indicates a mixing relationship with the felsic assimilant. We previously mentioned that fractional crystallisation and assimilation would be the main drivers of magmatic differentiation. For this *individual* lava, however, the isotope data point to more than 98 % of assimilant involved in the formation of sample MB7 (Montaña Blanca). This supports the view that phase 7 of the Montaña Blanca volcano was formed by large scale crustal melting with only minor amounts ($\sim$2 %) juvenile magma. Essentially, such large amounts of assimilant imply that the MB7 magma was a crustal melt and was 'contaminated' to a small percentage by mafic magma.

It is therefore maybe not surprising that the EC-AFC model failed for this sample. The EC-

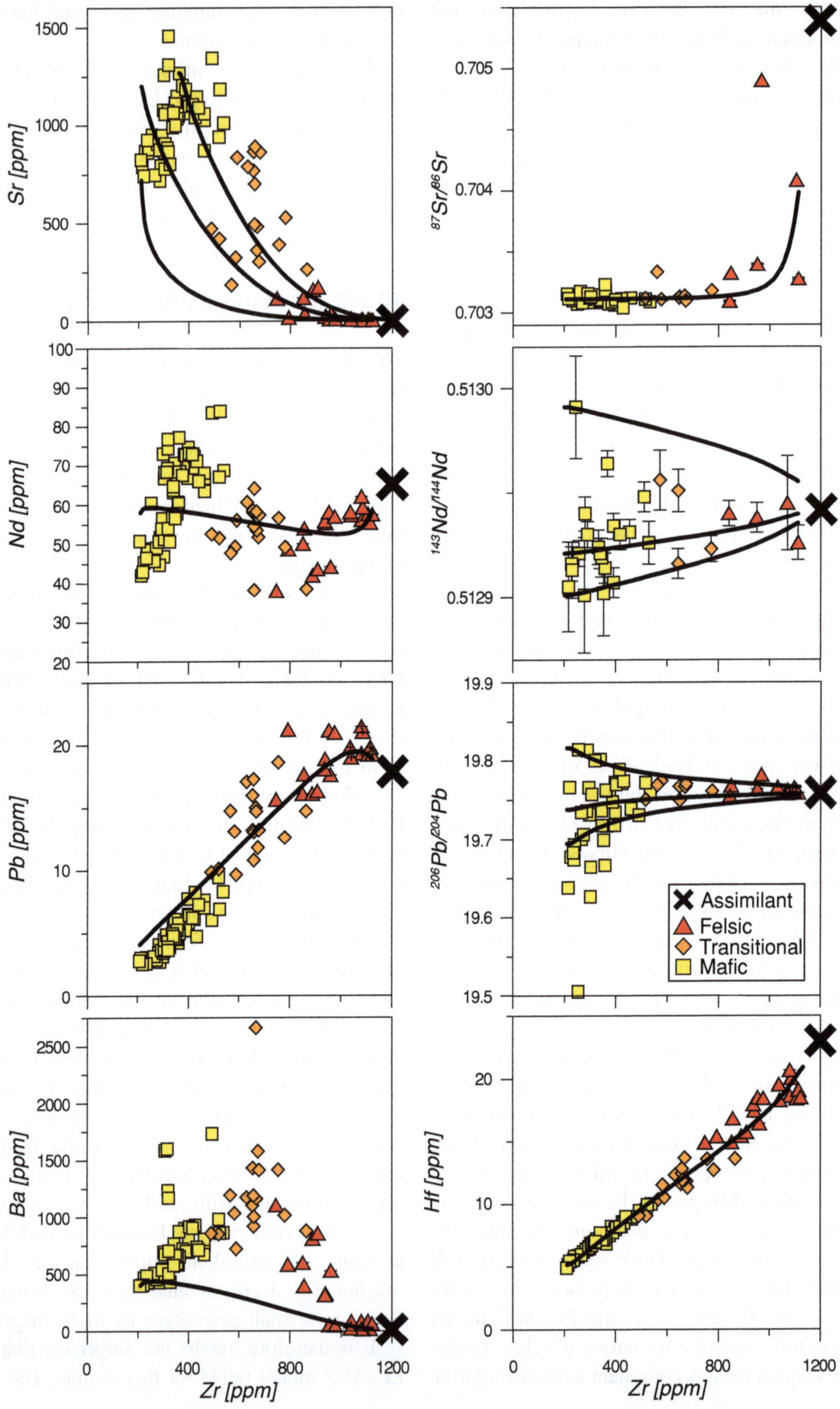
Sr [ppm]
87Sr/86Sr
Nd [ppm]
143Nd/144Nd
Pb [ppm]
206Pb/204Pb
Ba [ppm]
Hf [ppm]
Zr [ppm]
Zr [ppm]
Assimilant
Felsic
Transitional
Mafic

◀ **Fig. 10.7** Results from EC-AFC modelling. Teide–Pico Viejo succession in *yellow-orange-red symbols*, "x" marks the assimilant composition. Zirconium was used as an index of differentiation on the x-axis. Trace elements Sr, Nd, Pb, Zr, Hf and Ba and isotope ratios $^{87}Sr/^{86}Sr$, $^{143}Nd/^{144}Nd$ and $^{206}Pb/^{204}Pb$ have been modelled; *black* curves are model results. All trace elements and isotope ratios are well reproduced, indicating a strong agreement between our model and the inferred differentiation processes at work. One exception however, is, when bulk partition coefficients change from incompatible to compatible, at this point intermediate compositions are not well reproduced by the code employed. Note that the sample with the highest $^{87}Sr/^{86}Sr$ ratio (MB7) is also not always reproduced by the model (see text for details)

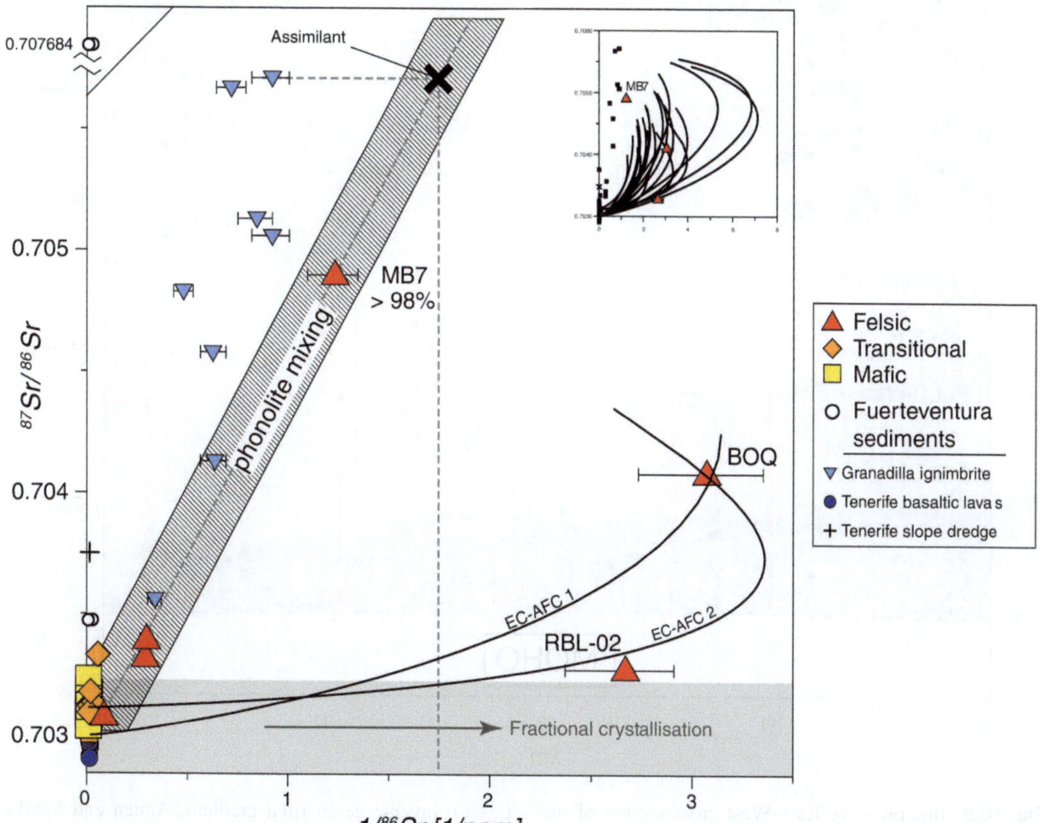

Fig. 10.8 Groundmass $^{87}Sr/^{86}Sr$ versus $1/^{86}Sr$ of Teide–Pico Viejo lavas. The Sr concentration of the hypothetical end-member used for EC-AFC modelling was computed to be 5.89 ppm, using Granadilla ignimbrite Sr isotope ratios. Errors (2SD) are contained within symbols when invisible. Strontium concentrations of Fuerteventura sediments are from Deegan (unpublished data). Samples BOQ and RBL-02 plot at low ^{86}Sr concentrations and are successfully modelled by EC-AFC. Sample MB7 cannot be modelled by EC-AFC using the thermal constraints inherent to the Tenerife setting. Instead, this sample opens a straight mixing array with other felsic lavas and the assimilant, indicating a crustal melt origin. End-member data are from Palacz and Wolff (1989), Simonsen et al. (2000) and Abratis et al. (2002)

AFC model describes the combined effects of wall-rock assimilation and fractional crystallisation within a single magma chamber. However, when pockets of felsic melt (e.g., molten country rock) form around mafic magma chambers without touching them or if they are physically unable to mix (e.g., Marshall and Sparks 1984), it is mainly heat that is transferred from the magma for assimilation and no significant chemical exchange takes place (Petford and Gallagher 2001; Holloway et al. 2008). In this case, AFC models are not

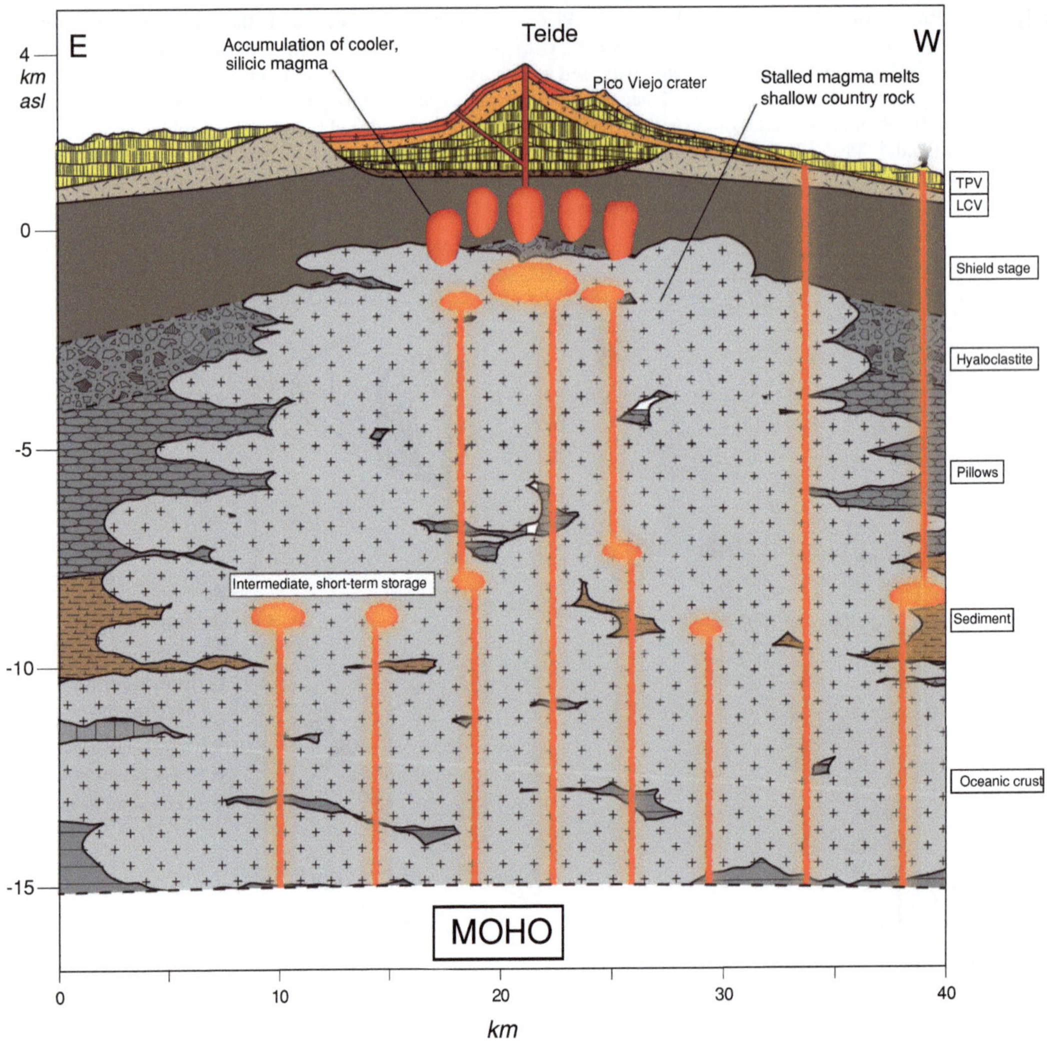

Fig. 10.9 Interpretative East–West cross-section of the current plumbing system of Tenerife (1.8 × vertical exaggeration). Teide–Pico Viejo deposits are colour-coded after compositional group. Note the greater degree of differentiation up-section. Deep island core and ocean crust lithology are represented in this sketch for orientation, but may in fact have been largely reworked by Tenerife's igneous activity since Miocene times. Included is information from the following publications: crustal structure, Krastel and Schmincke (2002); seamount sequence, Staudigel and Schmincke (1984); Teide–Pico Viejo and underlying units, Carracedo et al. (2007); height of landslide breccia, Márquez et al. (2008); inverse geothermal gradient, Annen and Sparks (2002). Lower crustal/upper mantle as the main crystallisation level has been invoked by multiple workers (Ablay et al. 1998; Hill et al. 2002; Spera and Bohrson 2004). In the Canary Islands specifically, intermediate levels of short-term magma residence were inferred from the re-equilibration of CO_2-inclusions in mineral phases (Klügel et al. 2005b; Galipp et al. 2006; Longpré et al. 2008). Underplating of hot, mafic material underneath the central Teide–Pico Viejo complex may cause formation of crustal melt pockets that are sometimes decoupled from the juvenile material providing the heat (see text for details)

applicable, since the lack of a common diffusive interface between magma and assimilant will preclude the uptake of crustal melts into resident magma. In other words, while fractional crystallisation is necessarily associated with the liquid magma, crustal melting may occur in the absence of direct contact with the heat-providing magma.

Because AFC processes and assimilant variability are both insufficient to explain the composition of phase 7 of the Montaña Blanca eruption in full, Wiesmaier et al. (2012) suggested that this lava is the product of a >98 % melt of pre-Teide nepheline syenite with very little juvenile material directly involved. All other phonolites and less differentiated lavas can be derived from juvenile mafic magma by various degrees of assimilation and fractional crystallisation. Direct mixing between mantle and crustal melts is also a possibility and may have modified some samples considerably (see also Chap. 11).

10.6 Petrogenesis at Teide–Pico Viejo

In the terminal stage of the pre-Teide Las Cañadas Volcano, the caldera-forming Diego Hernández phase was accompanied by the Icod landslide, which unroofed the junction of the two active rift zones and probably caused accelerated ascent of mafic magma from depth (cf. Longpré et al. 2008, 2009). This led to renewed and abundant mafic activity (Carracedo et al. 2007), and resulted in the "old" Teide eruptions (~200–100 ka), which were basanitic in composition. At this stage, older felsic intrusive material was not subjected to large-scale remelting. This may be because mafic eruptions in the Canaries are known to ascend swiftly via dyke systems (rift zones) with little to no residence time at shallow depth (cf. Klügel et al. 2005a; Stroncik et al. 2009; Troll et al. 2012) and because the magma supply may have been strong directly after the landslide, 'flushing' the system with mafic melt (e.g., Longpré et al. 2009; Manconi et al. 2009).

The increasing portion of crustal melts involved in the formation of higher differentiated lavas, all younger than 15–30 ka, indicates that at some point during the Teide–Pico Viejo history, felsic country rock began to be recycled to a significant degree. This was probably triggered by an increase in residence time of magma at shallow crustal levels. A potential explanation for this could be density filtering due to the increasing load exerted by the growing Teide–Pico Viejo edifice (cf. Pinel and Jaupart 2000) or, alternatively, the formation of 'density barriers' of partially molten and thus less dense felsic country rock that causes denser mafic magma to become neutrally buoyant (Huppert and Sparks 1988). The resulting underplating of mafic melts would have steepened the geothermal gradient and thus helped to eventually provide the energy for the wholesale-melting of the country rock (Fig. 10.9), cf. Annen and Sparks (2002).

The onset of crustal assimilation must have been progressive with the initial and relatively little contaminated transitional lavas erupted at around 30 ka, i.e., before the first felsic lavas appeared. The culmination of felsic activity at Teide took place with the eruption of high $^{87}Sr/^{86}Sr$ deposits around 2,000 years ago (Montaña Blanca and El Boquerón). According to EC-AFC modelling, the amount of pre-Teide felsic assimilant in these rather recent lavas is quite large. This implies that juvenile mafic magmas act as a heat source at depth to melt country rock, but are not always directly in contact with this newly forming crustal melt, or are unable to mix with it (Fig. 10.9). New, largely autonomous pockets of felsic magma are thus thought to be involved in the eruption of recent Teide phonolite.

Compositional bimodality has been demonstrated to exist in the case of Teide–Pico Viejo. The most primitive compositions evolve towards intermediate levels mainly by means of fractional crystallisation and variable, but small degrees of assimilation. In turn, highly differentiated magma broadly formed by AFC processes or by wholesale melting of felsic country rock and incomplete mixing. The observed contrast in differentiation processes between mafic and felsic lavas explains the bimodality and the temporal sequence of erupted lavas in the Teide–Pico Viejo succession and thus allows to establish a temporally, compositionally and geographically consistent model for the nature of phonolite volcanism on recent Tenerife.

References

Ablay GJ, Carroll MR, Palmer MR, Martí J, Sparks RSJ (1998) Basanite-phonolite lineages of the Teide-Pico Viejo volcanic complex, Tenerife, Canary Islands. J Petrol 39:905–936

Abratis M, Schmincke HU, Hansteen T (2002) Composition and evolution of submarine volcanic rocks from the central and western Canary Islands. Int J Earth Sci 91:562–582

Andújar J, Costa F, Martí J, Wolff JA, Carroll MR (2008) Experimental constraints on pre-eruptive conditions of phonolitic magma from the caldera-forming El Abrigo eruption, Tenerife (Canary Islands). Chem Geol 257:173–191

Andújar J, Costa F, Martí J (2010) Magma storage conditions of the last eruption of Teide volcano (Canary Islands, Spain). Bull Volcanol 72:381–395

Annen C, Sparks RSJ (2002) Effects of repetitive emplacement of basaltic intrusions on thermal evolution and melt generation in the crust. Earth Planet Sci Lett 203:937–955

Aparicio A, Tassinari CCG, García R, Araña V (2010) Sr and Nd isotope composition of the metamorphic, sedimentary and ultramafic xenoliths of Lanzarote (Canary Islands): implications for magma sources. J Volcanol Geotherm Res 189:143–150

Bindeman IN (2008) Oxygen isotopes in mantle and crustal magmas as revealed by single crystal analysis. Rev Mineral Geochem 69:445–478

Bryan SE, Martí J, Leosson M (2002) Petrology and geochemistry of the Bandas del Sur Formation, Las Cañadas Edifice, Tenerife (Canary Islands). J Petrol 43:1815–1856

Cann JR (1968) Bimodal distribution of rocks from volcanic islands. Earth Planet Sci Lett 4:479–480

Carracedo JC, Rodríguez Badiola E, Guillou H, Paterne M, Scaillet S, Pérez Torrado FJ, Paris R, Fra-Paleo U, Hansen A (2007) Eruptive and structural history of Teide volcano and rift zones of Tenerife, Canary Islands. Geol Soc Am Bull 119:1027–1051

Clague DA (1978) The oceanic basalt-trachyte association: an explanation of the daly gap. J Geol 86: 739–743

DePaolo DJ (1981) Trace element and isotopic effects of combined wall rock assimilation and fractional crystallization. Earth Planet Sci Lett 53:189–202

Dickin AP (2005) Radiogenic isotope geology. Cambridge University Press, Cambridge

Duffield WA, Ruiz J (1998) A model that helps explain Sr-isotope disequilibrium between feldspar phenocrysts and melt in large-volume silicic magma systems. J Volcanol Geotherm Res 87:7–13

Galipp K, Klügel A, Hansteen TH (2006) Changing depths of magma fractionation and stagnation during the evolution of an oceanic island volcano: La Palma (Canary Islands). J Volcanol Geotherm Res 155: 285–306

Garcia MO, Frey FA, Grooms DG (1986) Petrology of volcanic rocks from Kaula Island, Hawaii. Contrib Mineral Petrol 94:461–471

Gurenko AA, Hoernle KA, Hauff F, Schmincke HU, Han D, Miura YN, Kaneoka I (2006) Major, trace element and Nd-Sr-Pb-O-He-Ar isotope signatures of shield stage lavas from the central and western Canary Islands: Insights into mantle and crustal processes. Chem Geol 233:75–112

Hansteen TH, Troll VR (2003) Oxygen isotope composition of xenoliths from the oceanic crust and volcanic edifice beneath Gran Canaria (Canary Islands): consequences for crustal contamination of ascending magmas. Chem Geol 193:181–193

Hill DP, Pollitz F, Newhall C (2002) Earthquake-volcano interactions. Phys Today, pp 41–47

Hobson A, Bussy F, Hernandez J (1998) Shallow-level migmatization of gabbros in a metamorphic contact aureole, fuerteventura basal complex, Canary Islands. J Petrol 39:1025–1037

Hoernle K (1998) Geochemistry of jurassic oceanic crust beneath Gran Canaria (Canary Islands): implications for crustal recycling and assimilation. J Petrol 39:859–880

Hoernle K, Tilton G, Schmincke H-U (1991) Sr-Nd-Pb isotopic evolution of Gran Canaria: evidence for shallow enriched mantle beneath the Canary Islands. Earth Planet Sci Lett 106:44–63

Holloway MI, Bussy F (2008) Trace element distribution among rock-forming minerals from metamorphosed to partially molten basic igneous rocks in a contact aureole (Fuerteventura, Canaries). Lithos 102:616–639

Holloway M, Bussy F, Vennemann T (2008) Low-pressure, water-assisted anatexis of basic dykes in a contact metamorphic aureole, Fuerteventura (Canary Islands): oxygen isotope evidence for a meteoric fluid origin. Contrib Mineral Petrol 155:111–121

Huppert HE, Sparks RSJ (1988) The generation of granitic magmas by intrusion of basalt into continental crust. J Petrol 29:599–624

Kinman WS, Neal CR, Davidson JP, Font L (2009) the dynamics of kerguelen plateau magma evolution: new insights from major element, trace element and sr isotope microanalysis of plagioclase hosted in elan bank basalts. Chem Geol 264:247–265

Klügel A, Hansteen TH, Galipp K (2005a) Magma storage and underplating beneath Cumbre Vieja volcano, La Palma (Canary Islands). Earth Planet Sci Lett 236:211–226

Klügel A, Walter T, Schwarz S, Geldmacher J (2005b) Gravitational spreading causes en-echelon diking along a rift zone of Madeira Archipelago: an

experimental approach and implications for magma transport. Bull Volcanol 68:37–46
Krastel S, Schmincke H-U (2002) Crustal structure of northern Gran Canaria, Canary Islands, deduced from active seismic tomography. J Volcanol Geotherm Res 115:153–177
Longpré M-A, Troll VR, Hansteen TH (2008) Upper mantle magma storage and transport under a Canarian shield-volcano, Teno, Tenerife (Spain). J Geophys Res 113:B08203. doi:10.1029/2007jb005422
Longpré M-A, Troll VR, Walter TR, Hansteen TH (2009) Volcanic and geochemical evolution of the Teno massif, Tenerife, Canary Islands: Some repercussions of giant landslides on ocean island magmatism. Geochem Geophys Geosyst 10:Q12017. doi: 10.1029/2009gc002892
Manconi A, Longpré M-A, Walter TR, Troll VR, Hansteen TH (2009) The effects of flank collapses on volcano plumbing systems. Geology 37:1099–1102
Márquez A, López I, Herrera R, Martín-González F, Izquierdo T, Carreño F (2008) Spreading and potential instability of Teide volcano, Tenerife, Canary Islands. Geophys Res Lett 35:L05305. doi: 10.1029/2007GL032625
Marsh BD (2004) A magmatic mush column rosetta stone: the McMurdo dry valleys of antarctica. EOS Trans AGU 85:497–508
Marshall LA, Sparks RSJ (1984) Origin of some mixed-magma and net-veined ring intrusions (London, UK). J Geol Soc 141:171–182
Martí J, Mitjavila J, Araña V (1994) Stratigraphy, structure and geochronology of the Las Cañadas caldera (Tenerife, Canary Islands). Geol Mag 131:715–727
Martí J, Hurlimann M, Ablay GJ, Gudmundsson A (1997) Vertical and lateral collapses on Tenerife (Canary Islands) and other volcanic ocean islands. Geology 25:879–882
Meade FC, Chew DM, Troll VR, Ellam RM, Page LM (2009) Magma Ascent along a Major Terrane Boundary: Crustal Contamination and Magma Mixing at the Drumadoon Intrusive Complex, Isle of Arran, Scotland. J Petrol 50:2345–2374
Palacz ZA, Wolff JA (1989) Strontium, neodymium and lead isotope characteristics of the Granadilla Pumice, Tenerife: a study of the causes of strontium isotope disequilibrium in felsic pyroclastic deposits. Geol Soc Spec Publ 42:147–159
Petford N, Gallagher K (2001) Partial melting of mafic (amphibolitic) lower crust by periodic influx of basaltic magma. Earth Planet Sci Lett 193:483–499
Pinel V, Jaupart C (2000) The effect of edifice load on magma ascent beneath a volcano. Phil Trans R Soc Lond 358:1515–1532
Robertson AHF, Stillman CJ (1979) Submarine volcanic and associated sedimentary rocks of the Fuerteventura basal complex, Canary Islands. Geol Mag 116: 203–214
Simonsen SL, Neumann ER, Seim K (2000) Sr-Nd-Pb isotope and trace-element geochemistry evidence for a young HIMU source and assimilation at Tenerife (Canary Island). J Volcanol Geotherm Res 103:299–312
Spera FJ, Bohrson WA (2001) Energy-constrained open-system magmatic processes i: general model and energy-constrained assimilation and fractional crystallization (EC-AFC) formulation. J Petrol 42: 999–1018
Spera FJ, Bohrson WA (2002) Energy-constrained open-system magmatic processes 3. Energy-Constrained Recharge, Assimilation, and Fractional Crystallization (EC-RAFC). Geochem Geophys Geosyst 3, doi: 10.1029/2002gc000315
Spera FJ, Bohrson WA (2004) Open-system magma chamber evolution: an energy-constrained geochemical model incorporating the effects of concurrent eruption, recharge, variable assimilation and fractional crystallization (EC-E'RAFC). J Petrol 45:2459–2480
Staudigel H, Schmincke H-U (1984) The Pliocene seamount series of La Palma/Canary Islands. J Geophys Res 89, doi:10.1029/JB089iB13p11195
Stillman CJ, Fúster JM, Bennell Baker MJ, Muñoz M, Smewing JD, Sagredo J (1975) Basal complex of Fuerteventura (Canary Islands) is an oceanic intrusive complex with rift-system affinities. Nature 257:469–471
Stroncik N, Klügel A, Hansteen T (2009) The magmatic plumbing system beneath El Hierro (Canary Islands): constraints from phenocrysts and naturally quenched basaltic glasses in submarine rocks. Contrib Mineral Petrol 157:593–607
Sun SS (1980) Lead isotopic study of young volcanic rocks from mid-ocean ridges, ocean islands and island arcs. Philos Trans R Soc London, Ser A 297:409–445
Thompson G, Smith I, Malpas J (2001) Origin of oceanic phonolites by crystal fractionation and the problem of the Daly gap: an example from Rarotonga. Contrib Mineral Petrol 142:336–346
Troll VR, Chadwick JP, Ellam RM, McDonnell S, Emeleus CH, Meighan IG (2005) Sr and Nd isotope evidence for successive crustal contamination of Slieve Gullion ring-dyke magmas, Co. Armagh. Ireland. Geol Mag 142:659–668
Troll VR, Klügel A, Longpré MA, Burchardt S, Deegan FM, Carracedo JC, Wiesmaier S, Kueppers U, Dahren B, Blythe LS, Hansteen TH, Freda C, Budd DA, Jolis EM, Jonsson E, Meade FC, Harris C, Berg SE, Mancini L, Polacci M, Pedroza K (2012) Floating stones off El Hierro, Canary Islands: xenoliths of pre-island sedimentary origin in the early products of the October 2011 eruption. Solid Earth 3:97–110
Wellman TR (1970) The stability of sodalite in a synthetic syenite plus aqueous chloride fluid system. J Petrol 11:49–72
Wiesmaier S (2010) Magmatic differentiation and bimodality in oceanic island settings—implications for the petrogenesis of magma in Tenerife. Trinity College Dublin, Spain
Wiesmaier S, Troll VR, Carracedo JC, Ellam RM, Bindeman IN, Wolff JA (2012) Bimodality of Lavas

in the Teide–Pico Viejo succession in Tenerife—The role of crustal melting in the origin of recent phonolites. J Petrol. doi:10.1093/petrology/egs056
Wolff JA, Palacz ZA (1989) Lead isotope and trace element variation in Tenerife pumices—evidence for recycling within an ocean island volcano. Mineral Mag 53:519–525
Wolff JA, Grandy JS, Larson PB (2000) Interaction of mantle-derived magma with island crust? Trace element and oxygen isotope data from the Diego Hernandez Formation, Las Cañadas, Tenerife. J Volcanol Geotherm Res 103:343–366

Magma Mixing in the 1100 AD Montaña Reventada Composite Lava Flow: Interaction of Rift Zone and Central Complex Magmatism

11

Sebastian Wiesmaier, Frances M. Deegan, Valentin R. Troll, Juan Carlos Carracedo, and Jane P. Chadwick

Abstract

Zoned eruption deposits frequently show a lower felsic and an upper mafic member, thought to reflect eruption from a large, stratified magma chambers. In contrast, however, the Montaña Reventada composite flow in Tenerife consists of a lower basanite and a much thicker upper phonolite. A sharp interface separates the basanite and phonolite, and a chilled margin at this contact indicates the basanite was still hot upon emplacement of the phonolite, i.e. the two magmas erupted in very quick succession. Three types of mafic to intermediate inclusions are found in the phonolite, which comprise foamy quenched ones, inclusions with chilled margins and those that are physically mingled, reflecting progressive mixing with a decreasing temperature contrast between the end-member magmas involved. Analysis of basanite, phonolite and inclusions for majors, traces and Sr, Nd and Pb isotopes show the inclusions to be derived from binary mixing of basanite and phonolite end-members in ratios of 2:1–4:1. Although basanite and phonolite magmas were erupted in quick

S. Wiesmaier (✉)
Ludwig-Maximilians-Universität,
Geo- and Environmental Sciences,
Munich, Germany
e-mail: sebastian.wiesmaier@min.uni-muenchen.de

F. M. Deegan
Swedish Museum of Natural History, Laboratory for Isotope Geology, Stockholm, Sweden

V. R. Troll
Department of Earth Sciences, CEMPEG,
Uppsala University, 75236 Uppsala, Sweden

J. C. Carracedo
Department of Physics (Geology), GEOVOL,
University of Las Palmas, Gran Canaria, Spain

J. P. Chadwick
Science Gallery, Trinity College Dublin,
Dublin 2, Ireland

J. C. Carracedo and V. R. Troll (eds.), *Teide Volcano*, Active Volcanoes of the World,
DOI: 10.1007/978-3-642-25893-0_11, © Springer-Verlag Berlin Heidelberg 2013

succession, contrasting $^{206}Pb/^{204}Pb$ ratios show them to be genetically distinct. The Montaña Reventada basanite and phonolite first came into contact just prior to eruption and had seemingly limited interaction time. Montaña Reventada erupted from the transition zone between two plumbing systems, the phonolitic Teide-Pico Viejo complex and the basanitic Northwest rift zone. A rift zone basanite dyke most likely intersected a previously emplaced phonolite magma pocket, leading to eruption of geochemically and texturally unaffected basanite, followed by inclusion-rich phonolite that exploited the already established conduit.

11.1 Introduction

Magma mixing occurs when two liquid magmas of distinct composition interact with each other over a defined period of time. However, time-scales of magma mixing may be highly variable and range from hours to tens and hundreds of years. Large ignimbrite eruptions, for instance, have frequently been associated with voluminous stratified magma chambers, in which the compositionally distinct magmas formed over long time-scales from the same parent magma (e.g., Sparks et al. 1977; Blake 1981; Huppert et al. 1982; Wolff and Storey 1984; Blake and Ivey 1986; Freundt and Schmincke 1992; Calanchi et al. 1993; Kuritani 2001; Troll and Schmincke 2002). Alternatively, the origin of mixed magmas has also been explained by the forced intrusion or fountaining of a genetically distinct magma into another, whereby the newly arriving magma may trigger an eruption due to super-heating and re-mobilisation of the previously emplaced, already cooled pocket of magma (e.g., Turner 1980; Campbell and Turner 1986; Turner and Campbell 1986; Eichelberger et al. 2000; Izbekov et al. 2004; Troll et al. 2004). For example, Izbekov et al. (2004) suggested that in 1996, a mafic dyke had dissected a resident andesite magma chamber, triggering the intermittent eruption of a range of mixed products at Karymsky volcano, Kamchatka.

Various mechanisms for the mixing of magmas have been postulated, for example, (1) buoyant rise of mafic magma caused by a density decrease from strong vesiculation, (2) convective stirring or viscous coupling caused by thermal contrasts between mafic and felsic end-members, (3) forced intrusions of mafic magma into highly viscous and thus more competent felsic magma, and (4) mixing within the conduit during magma ascent (cf. Eichelberger 1980; Bacon 1986; Coombs et al. 2003; Troll et al. 2004; De Campos et al. 2008). The circumstances of magma mixing for a given deposit are thus a result of the dynamics and compositional controls of the related magmatic plumbing system. Key issues for understanding magma mixing are therefore: (a) how long have the compositionally distinct magmas interacted with each other? (b) are these compositionally distinct magmas co-genetic, i.e., derived from the same parental magma? or (c) are they genetically unrelated and met only prior to eruption? This chapter will try to answer these questions through a summary of detailed studies of the compositionally mixed Montaña Reventada lava flow (Araña et al. 1994; Wiesmaier et al. 2011), one of the most recent deposits within the Teide–Pico Viejo succession in Tenerife. At Reventada, a basanite lava flow erupted and was shortly followed by the eruption of phonolite lava from the same vent, thus forming a composite flow or cooling unit. The phonolite part of the flow contains abundant dark inclusions that appear to be related to the basanite part. Earlier studies on Montaña Reventada (Araña et al. 1989, 1994) provided mass balance calculations that, combined with mineral abundances, allowed them to exclude continuous closed system fractional crystallisation as the origin of these inclusions, but supported a hybrid (mixing) origin instead. The lack of mingling textures within the phonolite matrix led the

Fig. 11.1 Picture of Montaña Reventada. Montaña Reventada, the edifice of a complex eruption located in the zone of interaction between the NW rift zone and the central felsic volcanoes Teide and Pico Viejo. This eruption involved mixing of mafic and felsic magmas

authors to believe that the intermediate inclusions formed exclusively through diffusional hybridisation, which would require a long-lived, diffusional interface between basanite and phonolite in the Reventada magma chamber. A long-lived stratified magma chamber, however, is inconsistent with the relatively small volume of the Reventada eruption (0.1 km^3) as thermodynamic considerations indicate that comparably small magma chambers are likely to solidify completely before significant diffusional gradients can develop (cf. Hawkesworth et al. 2000). Equally, the eruption order, i.e., basanite before phonolite, does not agree with common models of stratified magma chambers (Hildreth 1979; Blake 1981; Troll and Schmincke 2002), which are believed to hold the denser magma (here basanite) below the less dense one (here phonolite).

In the study of Wiesmaier et al. (2011), new isotope and geochemical data have been merged with those of Araña et al. (1994), and combined with a detailed textural analysis of inclusion types. This approach resulted in a refined model of magma mixing that is consistent with field and textural constraints and allows for a substantial revision of the magmatic processes ongoing during the Montaña Reventada eruption, with implications for the interaction of distinct magma plumbing systems in Tenerife.

11.2 The Montaña Reventada Lava Flow

Montaña Reventada consists of a small group of vents and associated flow lobes, which have been radiometrically dated at 895 ± 155 a BP (Carracedo et al. 2007). Two exceptional roadcuts at 330437/3128642 (UTM 28R ± 15 m) at either side of the road TF-38 (locally referred to as "Carretera Boca Tauce–Chío") provide a cross-section through the complete stratigraphy of this eruption, including the bottom contact with older lavas (Fig. 11.2). This roadcut, at the base of Pico Viejo in the NWRZ, has been previously described by Araña et al. (1994).

The Montaña Reventada stratigraphy comprises the following components from bottom to top: (1) A red basal breccia of about 10–20 cm thickness, composed of scoriaceous basanite, which is scarcely porphyritic and shows flow banding in some clasts. (2) A lower basanite layer of variable thickness (20–200 cm), composed of massive, dark, mainly aphyric lava with flow banding that is frequently folded (Fig. 11.2a). Laterally, the massive parts grade into welded scoria, where the scoria clasts are of variable vesicularity. At 1–2 km downhill from the outcrop described here, the basanite contains abundant plagioclase. (3) An upper phonolite layer of

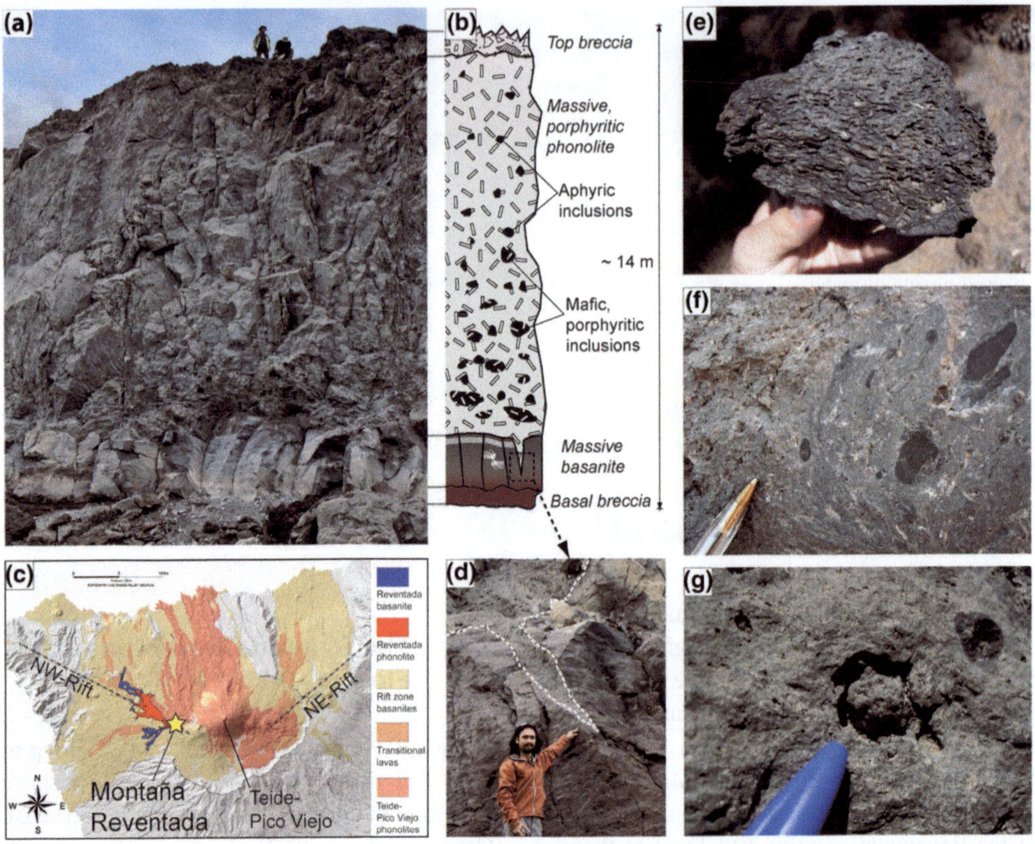

Fig. 11.2 Picture of outcrop and map inset. **a** A photograph of the main outcrop of the Montaña Reventada composite flow with people for scale. **b** A simplified stratigraphic column of this main outcrop. **c** A location map after Carracedo et al. (2007). **d** An opened fracture within the basanite that has been filled with phonolite. **e** Vesicle-rich and plagioclase bearing basanite can be found at the flow front. **f** Mingled appearance of a light-coloured inclusion. **g** Degassing halo around an inclusion in host phonolite

10–12 m thickness that is massive, light-coloured and porphyritic. The contact between the basanite and the overlying phonolite is sharp and undulating, lacking both top and basal brecciation. In places, the phonolite intrudes the lower basanite or appears to "lift out" basanite blocks (up to 50 cm). At one location, the intruding phonolite caused a chilled margin in the underlying basanite, indicating a considerable temperature contrast (Fig. 11.2d). Within the first metre above the basanite-phonolite contact, vesicles up to 10 cm are abundant. These are elongated parallel to the contact and grade into equant shapes some 40 cm above the contact. The phonolite hosts frequent dark inclusions of varying texture that range in size from a few cm to 50 cm across and appear to gradually decrease in abundance up-section. The phonolite becomes pink in the uppermost half metre (oxidised top). (4) A top breccia to the phonolite of up to 1.5 m in thickness that consists of large clinker fragments and glassy blocks.

11.3 Research Techniques

To define the lithological units and constrain the processes that gave rise to the Montaña Reventada composite eruption, Wiesmaier et al. (2011) analysed 20 samples from the outcrop for their major- and trace element concentrations as well as for their Sr, Nd and Pb isotopes. The compositional data are complemented by field and petrographical evidence from the outcrop as well as hand-specimen samples and thin-sections. The sample set comprises 14 whole-rock and six groundmass samples collected from the two road sections at TF-38. Of the whole-rock samples, three basanite, seven phonolite and four inclusion samples were selected for whole-rock analyses and groundmass measurements included two basanite and four phonolite samples.

Results are listed in Table 11.1 with all errors reported as 2SD. A detailed documentation of the analytical methods applied can be found in Wiesmaier et al. (2011).

11.4 Petrological and Geochemical Observations

11.4.1 Petrography

The petrographic description of the samples allows the distinction of basanite, phonolite and in total four different types of mafic inclusions.

11.4.1.1 Basanite

Plagioclase phenocrysts and vesicles are abundant in the flow front of the basanite lava. In contrast, at the outcrop described here, Reventada basanite is essentially aphyric and vesicle-free with only scarce plagioclase phenocrysts. The microcrystalline, melanocratic groundmass consists of lath-shaped plagioclase, pyroxene microlites with high birefringence colours and opaque Fe/Ti-oxides. The groundmass shows abundant flow lamination, which is frequently folded (Fig. 11.3a).

11.4.1.2 Phonolite

The overlying phonolite contains 10 % alkali feldspar, 3 % opaque minerals and scarce clinopyroxene and amphibole with dehydration rims. Feldspar may be intergrown with opaque minerals and/or pyroxene. Feldspar also often displays sieve textures and occurs as single, euhedral crystals with rounded corners and abundant Carlsbad twinning or, less often, as glomerocrysts of up to 10 mm across.

The microcrystalline, leucocratic phonolite groundmass is holocrystalline and consists mainly of feldspar and opaque minerals. Vesicles are abundant and make up $\sim$10 vol.% of the rock close to the contact with the lower basanite, but this decreases to $\sim$1 vol.% farther away from the basanite (Fig. 11.3c).

11.4.1.3 Inclusions

Inclusion textures range from frothy and vesicle-rich through scarcely porphyritic and banded to porphyritic and mingled. Four major types are distinguished. *Type I*: finely vesicular with a cryptocrystalline groundmass (diktytaxitic texture, cf. Bacon 1986), sometimes containing alkali feldspars with an anhedral relict appearance. This type of inclusion has angular outlines and is occasionally intruded by phonolite and thus appears to have behaved competently against the liquid phonolite magma (Fig. 11.3d, e). *Type II*: dark-coloured, feldspar-bearing inclusions with a lobate, sometimes chilled margin that indicates fluidal behaviour at the time of formation. Type II inclusions contain nodules of darker material (Fig. 11.3f, g). *Type III* inclusions are lighter coloured than type II, are feldspar-bearing and show a coarser-grained groundmass of microlites, feldspars and amphiboles. These inclusions have lobate and diffuse margins (blob-like), and filaments and blobs of dark magma are visible within them. Inclusions of about 1 cm or less in size may show a sharp, well-defined contact, or a diffuse transition between inclusion and phonolite host (Fig. 11.3g, h). Glomerocrysts of feldspar intergrown with opaque oxides, clinopyroxene and amphibole occur. *Type IV:* dense inclusions with scarce feldspar that show flow-banding. Phenocryst orientations generally appear to follow the observed groundmass lamination. The contact to the host phonolite is sharp and angular (Fig. 11.3i).

Feldspars within the inclusions show anorthoclase compositions but also a range of labradorite to oligoclase (Wiesmaier 2010).

Table 11.1 XRF major element and ICP-MS trace element data for Montaña Reventada basanite, inclusion and phonolite samples.

	Basanite					Inclusions				Phonolite
sample:	205-1	205-2	205-3	205-1 gm	205-2 gm	E 206A	E 206B	E 206D	E 204F	206 Cont
SiO_2 (wt%)	46.63	46.2	46.19	46.86	46.71	50.08	50.12	50.44	48.4	57.46
TiO_2	3.31	3.35	3.33	3.32	3.36	2.62	2.64	2.6	2.92	1.35
Al_2O_3	17.16	17.13	17.14	17.18	17.17	17.65	17.68	17.74	17.65	18.49
Fe_2O_3	11.13	11.22	11.21	11.09	11.19	9.04	9.04	9.02	9.84	5.3
MnO	0.18	0.18	0.18	0.18	0.18	0.18	0.18	0.18	0.21	0.17
MgO	4.42	4.53	4.55	4.48	4.58	3.35	3.44	3.39	3.75	1.46
CaO	9	9.15	9.12	9.06	9.14	6.91	7.01	6.89	7.72	2.89
Na_2O	4.94	4.97	4.93	4.83	4.86	6.3	6.05	6.07	5.67	7.59
K_2O	1.92	1.85	1.88	1.91	1.91	2.46	2.62	2.66	1.75	4.32
P_2O_5	1.26	1.29	1.3	1.29	1.29	0.99	1	1	1.17	0.4
H_2O	0.08	0.09	0.09	–	–	0.12	0.14	0.08	0.24	0.17
CO_2	0.02	0.02	0.02	–	–	0.02	0.02	0.01	0	0.04
Sum	100.28	100.17	100.13	100.51	100.7	99.91	100.1	100.29	99.45	99.69
Ba (ppm)	581.9	728.4	528.8	526.1	616.2	780.3	594.6	708.7	1053.7	668.5
Sr	982.1	1327.3	936.7	910.3	1068.2	968.5	767.4	885.2	1075.5	239.6
Hf	7.04	8.53	5.95	6.29	7.32	8.28	6.65	8.43	6.53	7.68
Th	6.48	8.59	6.32	5.59	7.21	11.07	7.68	11.09	6.36	8.97
U	1.90	2.19	1.55	1.68	1.86	2.43	2.00	2.49	1.59	2.79
Nb	82.3	115.3	77.5	81.3	99.8	113.4	93.3	110.3	110.8	113.8
Ta	5.75	7.20	4.63	4.87	5.54	7.07	5.64	6.74	6.05	6.41
Rb	39.50	36.31	29.89	36.00	33.18	58.16	47.98	53.57	25.93	64.70
Pb	3.52	4.09	2.88	3.18	3.62	5.43	4.12	5.32	3.53	5.53
$^{206}Pb/^{204}Pb$	19.7418(16)	19.7401(10)	19.7355(7)	19.7377 (9)	19.7193(10)	19.7641(12)	19.7528(7)	19.7594(7)	19.7660(8)	19.7671(9)
$^{207}Pb/^{204}Pb$	15.6122(17)	15.6163(9)	15.6213(9)	15.6173(16)	15.6146(17)	15.6196(15)	15.6117(8)	15.6175(8)	15.6142(9)	15.6168(15)
$^{208}Pb/^{204}Pb$	39.5607(31)	39.5673(20)	39.5720(14)	39.5638(18)	39.5423(22)	39.5858(23)	39.5603(15)	39.5786(14)	39.5701(16)	39.5769(19)

(continued)

Table 11.1 (continued)

	Basanite					Inclusions				Phonolite
	Phonolite									
sample:	206-2	206-3	206-5	206-2 gm	206-5 gm	207-4	207-5	207-6	207-4 gm	207-6 gm
SiO_2 (wt%)	58.68	59.12	57.65	59.31	57.82	58.88	58.17	58.75	59.16	58.86
TiO_2	1.08	1.03	1.28	1.06	1.28	1.1	1.21	1.12	1.08	1.11
Al_2O_3	18.53	18.61	18.58	18.63	18.52	18.58	18.5	18.51	18.55	18.47
Fe_2O_3	4.56	4.41	5.09	4.54	5.09	4.54	4.99	4.57	4.53	4.76
MnO	0.16	0.16	0.17	0.16	0.17	0.17	0.16	0.16	0.17	0.17
MgO	1.05	1	1.33	1.04	1.35	1.09	1.28	1.12	1.06	1.12
CaO	1.99	1.87	2.65	1.97	2.63	1.98	2.35	2.1	1.96	2.08
Na_2O	7.91	7.85	7.7	7.9	7.64	7.67	7.67	7.81	7.73	7.88
K_2O	4.75	4.81	4.42	4.82	4.52	4.73	4.57	4.66	4.83	4.74
P_2O_5	0.29	0.28	0.37	0.29	0.38	0.29	0.35	0.31	0.29	0.32
H_2O	0.09	0.1	0.09	–	–	0.16	0.26	0.2	–	–
CO_2	0	0	0.01	–	–	0.02	0.01	0.02	–	–
Sum	99.24	99.4	99.5	99.98	99.67	99.29	99.52	99.37	99.6	99.76
Ba (ppm)	996.9	1289.7	881.1	789.1	796.4	975.8	1055.4	1032.9	839.8	694.3
Sr	186.7	218.4	274.0	154.2	270.4	171.3	244.7	203.0	159.1	155.6
Hf	10.86	13.18	9.35	10.79	10.71	9.59	10.07	9.31	11.10	7.83
Th	14.90	19.87	12.27	12.45	12.57	13.99	15.06	11.83	14.87	10.63
U	3.71	4.68	3.28	3.29	3.49	3.58	3.73	2.75	3.75	2.33
Nb	157.8	192.1	134.4	157.3	164.4	146.6	157.8	162.0	165.1	121.8
Ta	8.75	10.39	7.51	8.14	8.20	8.28	8.17	7.85	8.43	6.01
Rb	101.62	116.55	79.90	91.25	95.38	90.05	97.71	86.14	91.28	62.88
Pb	8.91	10.93	7.34	8.18	7.93	8.27	8.57	6.75	8.67	5.25
$^{206}Pb/^{204}Pb$	19.7807(11)	19.7762(6)	19.7746(6)	19.7723(6)	19.7708(10)	19.7767(8)	19.7761(7)	19.7802(7)	19.7723(10)	19.7750(12)
$^{207}Pb/^{204}Pb$	15.6232(14)	15.6175(9)	15.6189(8)	15.6178(8)	15.6195(15)	15.6219(10)	15.6210(9)	15.6210(9)	15.6203(16)	15.6209(17)
$^{208}Pb/^{204}Pb$	39.5997(23)	39.5845(14)	39.5882(13)	39.5835(14)	39.5843(22)	39.5980(18)	39.5929(14)	39.5983(15)	39.5873(22)	39.5931(25)

Major and trace element and Pb isotope data for Montaña Reventada. Samples with number 205 are from the basanite. 206 Cont is the phonolite sample just above the contact between basanite and phonolite. Samples with number 207 are from the top of the phonolite layer and 206 samples are from the bottom of the phonolite. "gm" in a sample name denotes a groundmass sample

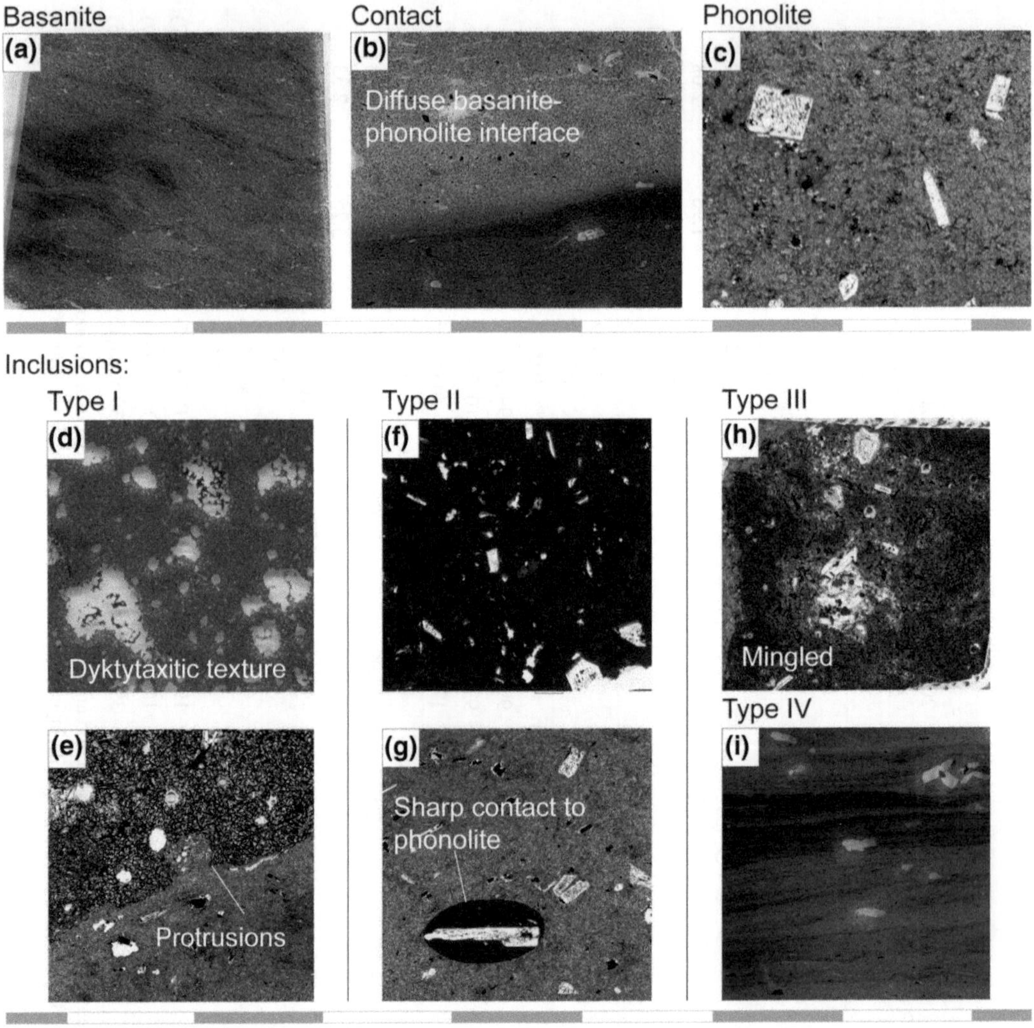

Fig. 11.3 Photomicrographs and scans of thin-sections. Thin-section images of Montaña Reventada rocks (scans: a, b, d, i; photomicrographs: c, e, f, g, h). **a** basanite, **b** diffuse contact between basanite and phonolite, **c** phonolite, **d–e** type I inclusions, frothy and vesicle-rich, **f–g** type II inclusions, crystal-rich and possessing a chilled margin, **h** type III inclusions massive, crystal-rich, and mingled, **i** type IV inclusions, flow-banded. Scale bar in 1 cm divisions. Sieve-textured feldspar occurs in all samples

11.4.2 Whole-Rock and Groundmass Composition

11.4.2.1 Major Elements

In the Total Alkali versus Silica diagram (TAS; Le Bas et al. 1986), the lower lava layer classifies as basanite and the upper one as phonolite, while inclusions contained within the phonolite represent variable compositions between the two, plotting as either basanite, phonotephrite or tephriphonolite (Table 11.1, Fig. 11.4). Inclusion data from Araña et al. (1994) plot in the same linear array between basanite and phonolite, with higher alkali element and silica concentrations. In fact, all major element data form linear trends between basanite and phonolite and the gap between the two principal lava types is always bridged by inclusions of intermediate composition from both data sets (Fig. 11.5).

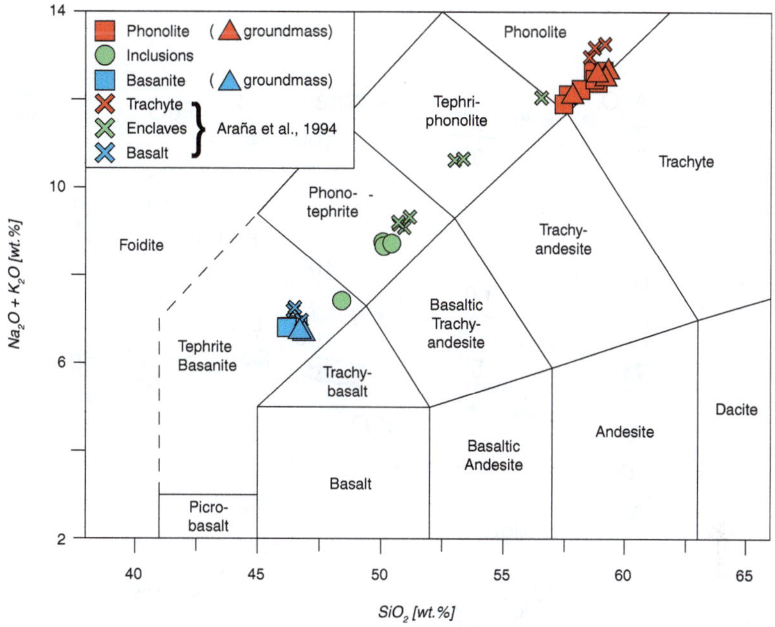

Fig. 11.4 TAS diagram. Total alkali versus silica diagram after Le Bas et al. (1986). The two principal lava types, basanite and phonolite, are end-members, while the inclusions are of variable intermediate compositions. The data of Araña et al. (1994) (crosses) plot on the same linear trend as the samples from Wiesmaier et al. (2011), between the two principal lava compositions

11.4.2.2 Trace Elements

In a multi-element variation diagram normalised to primitive mantle, basanite and inclusion samples show comparable patterns, apart from the Large Ion Lithophile Elements (LILE) Cs, Rb and Ba and the High Field Strength Elements (HFSE) Th and U, in which inclusions appear more enriched. Phonolite samples are more enriched than basanite in the LILE, but display a pronounced negative Sr and positive Zr anomaly and an overall depletion in MREE (Table 11.1).

When whole-rock trace element data are plotted against Zr concentration as an index of differentiation (cf. Wolff et al. 2000), basanite and phonolite samples again plot as end-members, with the inclusions generally filling the space in-between. However, phonolites show a rather widespread array in several incompatible trace elements, while the basanites appear closely spaced.

11.4.2.3 Isotope Data

Basanite whole-rocks have $^{87}Sr/^{86}Sr$ values of between 0.703032(9) and 0.703040(7) (groundmass: 0.703024(10) to 0.703046(9)). The phonolite whole-rocks range from 0.703032(7) to 0.703062(9) (groundmass: 0.703032(9) to 0.703082(7)). The inclusions display values from 0.703032(7) to 0.703059(9).

Basanites show $^{143}Nd/^{144}Nd$ ratios from 0.512855(38) to 0.512896(46) and phonolites from 0.512848(42) to 0.512910(46). Inclusions show a range in Nd ratios between 0.512871(46) and 0.512899(42). All Nd ratios are within error of each other.

The $^{206}Pb/^{204}Pb$ ratios range from 19.7193(21) to 19.7418(31) versus 19.7528(14) to 19.7660(16) versus 19.7671(18) to 19.7807(23), for basanite, inclusions and phonolite, respectively, with significant differences among these three groups. In contrast however, basanite, inclusion and phonolite samples overlap in their $^{207}Pb/^{204}Pb$ ratios (15.6122(34) to 15.6213(17) versus 15.6117(17) to 15.6196(29) versus 15.6168(30) to 15.6232(28), respectively). The $^{208}Pb/^{204}Pb$ ratios partially overlap between basanite, inclusions and phonolite (39.5423(43) to 39.5720(29) versus 39.5603(29) to 39.5858(46) versus 39.5769(39) to 39.5997(45), respectively), but with each group reaching higher values. The results for Sr, Nd and Pb isotopes agree well with existing data for Tenerife igneous rocks (Palacz

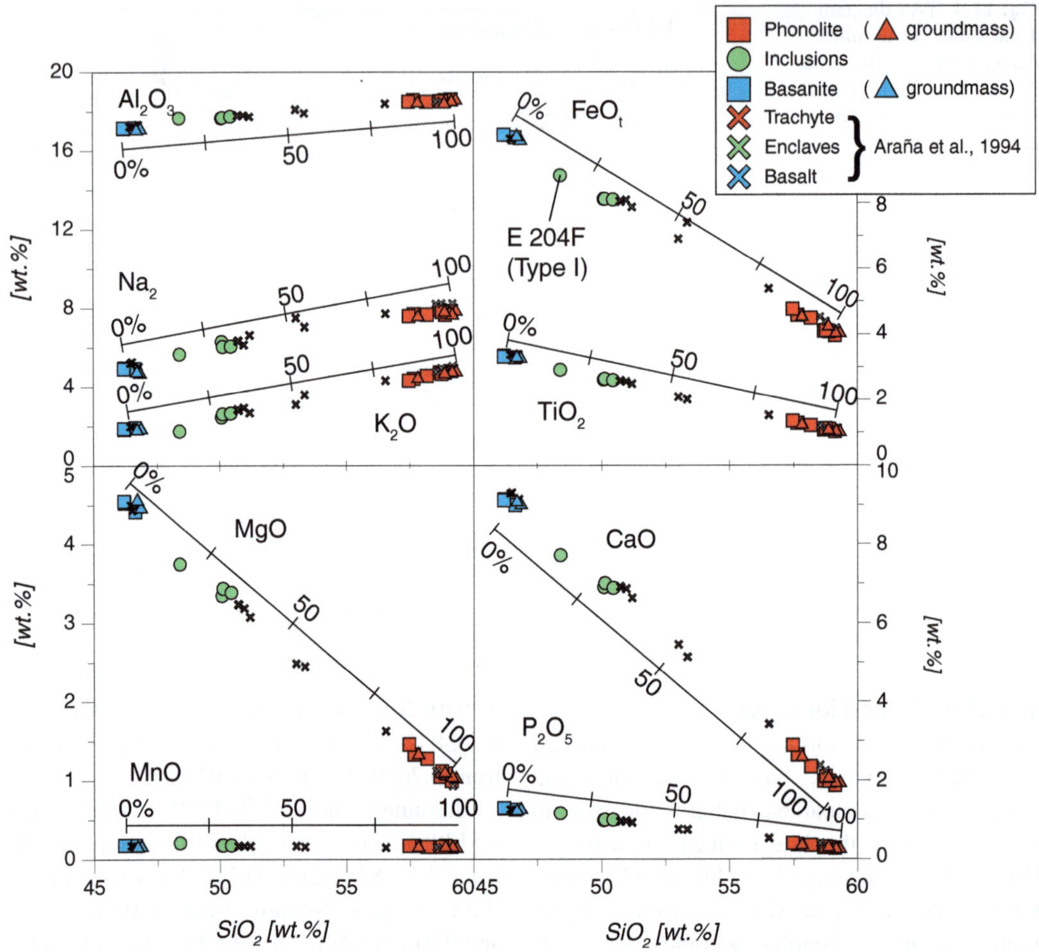

Fig. 11.5 Harker diagrams with graphical mixing solution. Whole-rock major element composition of the Montaña Reventada eruption. Fe data recalculated to FeO_{tot} using the formula $FeO_{tot} = FeO + 0.899\ Fe_2O_3$ (Bence and Albee 1968). All major elements define straight trends when correlated to SiO_2, which indicates the origin of the inclusions to be mixing of the two principal components basanite and phonolite, rather than by fractional crystallisation. Note graphical mixing lines that indicate the percentage of phonolite material for intermediate compositions

and Wolff 1989; Simonsen et al. 2000; Abratis et al. 2002; Gurenko et al. 2006). All errors are reported as 2SD (Table 11.1).

11.5 Emplacement and Formation of the Montaña Reventada Lava Flow

11.5.1 Subaerial Emplacement of Lava

Several lines of evidence allow us to establish that the basanite and phonolite were part of the same eruption. Firstly, the focus will be on the eruption dynamics at the surface.

The basanite shows a chilled margin where the phonolite intruded (Fig. 11.2d), implying that the basanite was still hot at the time the much cooler phonolite came in contact with it. Further evidence for this is the vesiculation of the phonolite, which is limited to a zone of one metre upwards from the basanite-phonolite contact. This localised zone of vesicles is probably the result of inclusions that were reheated at atmospheric pressure within the phonolite and which consequently liberated volatiles (see inclusion degassing halos in Fig. 11.2g). The basanite and the phonolite are thus effectively contemporaneous. Only an underlying basanite that was still hot can have caused the inclusions to decompose and develop a chilled margin.

The sharp contact in-between basanite and phonolite holds further clues to the emplacement process. Common ‘a’ā lava flows form by developing a chilled crust against the air which brecciates and gives an ‘a’ā flow its characteristic rugged surface appearance. During flow, this brecciated top is continuously transported to the flow front, where it rolls onto the ground (Merle 1998) and is consequently run over by the proceeding lava. As a result, the standard stratigraphy of an ‘a’ā lava flow consists of a basal breccia, a massive inner part and a top breccia (Cas and Wright 1987).

However, at Montaña Reventada, the contact between basanite and phonolite is sharp, i.e., the basanite lacks a top breccia and the phonolite lacks a bottom breccia. The missing breccias are either not preserved or have never formed. However, the type IV inclusions (massive, flow-banded and angular) closely resemble the laminated texture of the underlying basanite, hence, they have most likely been picked up by the phonolite while overriding the basanite. It thus seems likely that the phonolite rafted on top of the ductile basanite, at least in the area of the outcrop. Rafting would allow it to erode the basanite top breccia and pick up the breccia clasts while not developing its own bottom breccia. The lack of a top basanite breccia and bottom phonolite breccia therefore provides further evidence in support of the two magmas of Montaña Reventada having been emplaced almost simultaneously.

11.5.2 Origin of Inclusions

11.5.2.1 Major and Trace Element Constraints

Whole-rock major element trends are linear for all oxide data from both Araña et al. (1994) and Wiesmaier et al. (2011). Together they form a semi-continuous compositional sequence, with intermediate inclusions bound by the basanite and phonolite end-member compositions (Fig. 11.5). As the major element patterns are exclusively straight, lacking the typical kinks expected from fractional crystallisation (cf. Geldmacher et al. 1998), physical and chemical mingling and mixing are thought to be the dominant processes active during formation. Such straight trends should not be confused with the much more complex and fluctuating geochemical trends that are produced by the diffusive gradients between two magmas. Although these diffusive gradients are the fundamental driver of mixing and homogenisation, they occur on diffusion distances of sub-cm scale and in single samples only (De Campos et al. 2008). However, for analysis each sample has to be homogenised, including those that show various degrees of magma mixing. As a result complex S-shaped diffusional trends are not preserved and the analysis of a sample suite will produce straight mixing trends, such as observed at Reventada. Alternatively, inclusions may have been thoroughly hybridised when still liquid, so that straight geochemical trends may also be interpreted to reflect an advanced stage of mixing. The straight trends observed at Montaña Reventada thus point towards a mixing origin for the mafic inclusions that are found in the phonolite.

These straight mixing trends allow calculation of the proportions of each component involved during the mixing process. Trace element and major element oxide concentrations in inclusions were modelled as two-component bulk mixtures of basanite and phonolite and the respective maximum and minimum concentrations found for each major and trace element were used.

For most major and trace elements, the inclusions can be equated to mixtures of between 66:34 basanite to phonolite (E206A, E206B and E206D) and 80:20 basanite to phonolite (E204F). This agrees well with graphical mixing solutions in the Harker diagrams, where the three inclusions (E206A, E206B and E206D) cluster together but the latter (E204F) shows a slightly more mafic composition (Fig. 11.5).

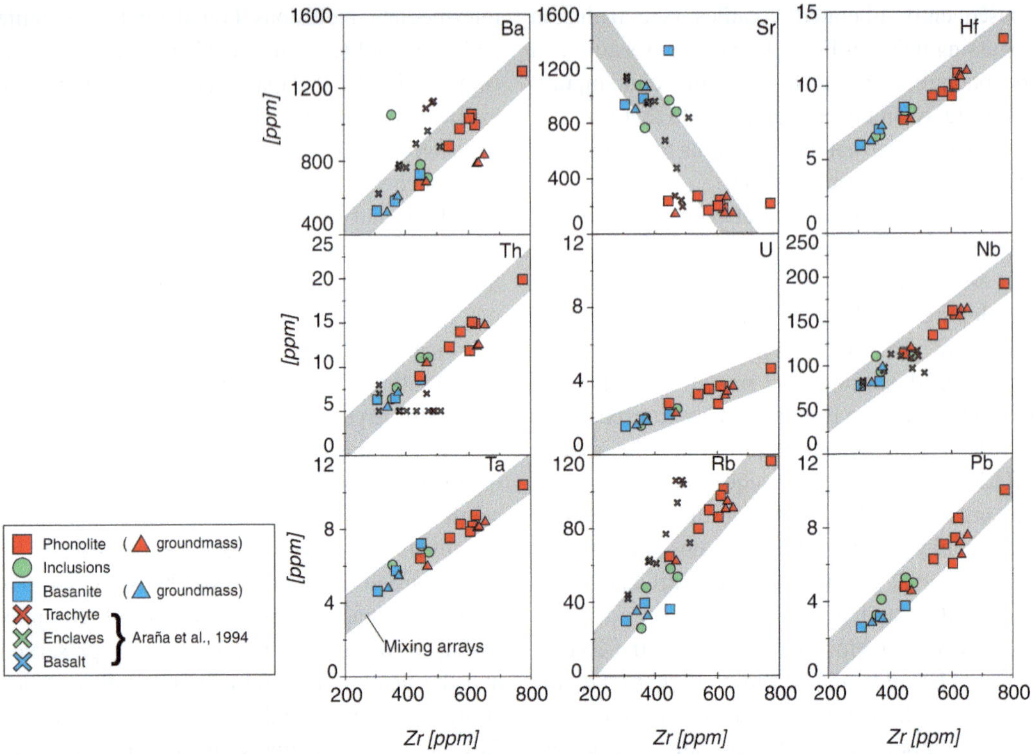

Fig. 11.6 Matrix of trace element plots. Selected trace elements are plotted versus Zr concentration. The crosses denote data from Araña et al. (1994) for comparison. Note the linear variation among the sample suite in most trace elements. Ba, Sr and Rb may be affected by crystal transfer of feldspar between the end-member magmas, the dominant mineral phase at Montaña Reventada. Phonolite samples show a wide spread in trace element concentrations, which is possibly a result of diffusional hybridisation

Two-component bulk mixing of basanite and phonolite yields matches for the major oxides SiO_2, MgO, Fe_2O_3 and TiO_2. The two major element oxides MnO and P_2O_5 are within 0.01 wt% of the model limits, which we deem a satisfactory fit. Most trace elements are modelled satisfactorily too.

Deviations in element concentrations from ideal mixing behaviour are few and can be well explained by diffusive phenomena. Slightly lower K_2O concentrations than expected may indicate uphill diffusion of K_2O towards the phonolite (cf. Watson and Baker 1991; Bindeman and Perchuk 1993; Araña et al. 1994; Bindeman and Davis 1999), which would be in line with the enhanced diffusivities of this element (Walker et al. 1981; Watson 1982; Walker and DeLong 1982; Lesher 1986; Lesher and Walker 1986). The higher than expected concentration of Al_2O_3 and Na_2O in all inclusions is, in turn, suggestive of anorthoclase added from the phonolite magma (Fig. 11.6).

Trace elements that deviate by more than 10 % from the linear two-component bulk mixing interval are mainly the lithophile elements Li, Sc, Cs, Rb and U. These elements are depleted in inclusions, which can also be explained by uphill diffusion, previously recognised for Li, Cs and Rb in basalt–rhyolite systems (Bindeman and Davis 1999). The siderophile element Ni and the chalcophile element Cu are also depleted in the inclusions with respect to the mixing calculation (Table 11.2). Barium, in turn, is enriched in the type I inclu-

Table 11.2 Modelling of inclusion compositions.

Inclusion	Basanite (%)	Phonolite (%)	Compared with calculated mixture	
			Enriched in:	Depleted in:
E206A	66	34	–	Li, Cu
E206B	66	34	–	Sc, Cu
E206D	66	34	–	–
E204F	80.4	19.6	Ba	Ni, Cu, Cs, Rb, U

Percentages of two-component bulk mixtures between basanite and phonolite that reproduce inclusion compositions. Some trace elements were enriched or depleted in the real samples compared to the theoretical mixture, but variations remain unique to each sample

sion E204F, which may be explained by the addition of Ba-rich anorthoclase from Montaña Reventada phonolite.

In summary, the majority of major and trace elements agree with the ideal mixing trend and apparent deviations are in line with well-documented diffusive or accumulative mechanisms.

11.5.2.2 Isotope Fingerprinting of Basanite, Phonolite and Inclusions

Although the inclusions occupy intermediate values between basanite and phonolite, all three rock types overlap within their analytical uncertainty in Sr and Nd isotopes (not shown) and are thus indistinguishable from each other for these components. The Pb isotopes, however, show distinct $^{206}Pb/^{204}Pb$ signatures for basanite and phonolite. The basanites, phonolites and inclusions show similar $^{207}Pb/^{204}Pb$ ratios, but systematic variation in $^{206}Pb/^{204}Pb$. An implication of the significant difference in the observed Pb isotope ratios between basanite and phonolite is that they cannot be co-genetic. This is consistent with the recent phonolite eruptions from the Teide-Pico Viejo central complex incorporating variable amounts of crustal components, i.e. they are likely isotopically distinct from the rift zone basanites that are comparatively free from crustal assimilation (Chap. 10).

The distinct Pb isotope signatures of basanite and phonolite also allow testing whether or not the inclusions are intermediate to these end-members. From the dataset (n = 20), all inclusion samples can be explained as isotopic mixtures between the basanite and phonolite end-members. This divides the components of Montaña Reventada into three arrays, with the mafic to intermediate inclusions placed in-between the basanite and phonolite (Fig. 11.7). The magmatic inclusions are therefore direct evidence of basanite–phonolite interaction at depth. In the next section, a detailed textural analysis of these inclusions will help to decipher the magma chamber configuration and the mechanism that led to mixing of the two end-member magmas.

11.5.3 Subsurface Dynamics

Altogether four types of inclusions were recorded in the Reventada phonolite. Since the following discussion is concerned with processes at depth prior to and during eruption, type IV inclusions, which have been identified as pick-up clasts, i.e. as subaerially solidified rocks, will not be mentioned here. The remaining inclusion types, I through III, record a progressive sequence of basanite-phonolite interaction, reflecting a continuous interaction while temperature differences between the end members gradually dissappear.

11.5.3.1 Type I Inclusions

Type I inclusions in the Montaña Reventada phonolite are angular fragments of a vesicle-rich "mafic foam" (a term coined by Eichelberger 1980), indicating that a large temperature contrast must have led to rapid crystallisation of the hotter, mafic magma (cf. Bacon and Metz 1984; Bacon 1986). The resultant exsolution of volatiles in the residual melt raises the melt's

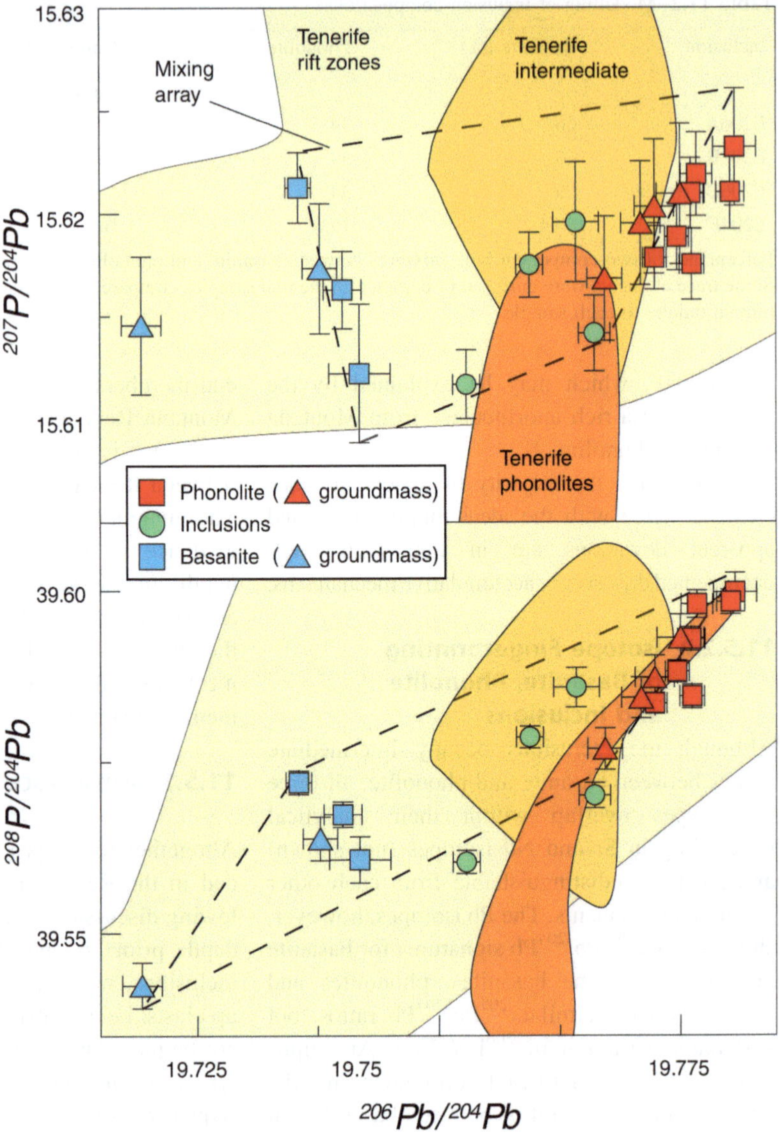

Fig. 11.7 Pb isotope plot. Pb-Pb isotope systematics of the Montaña Reventada eruption. Errors are 2SD. Fields denote existing data from the Tenerife Teide-Pico Viejo complex and rift zones (Wiesmaier 2010): in yellow primitive rift zone basanites, in orange intermediate rocks and in red phonolites. Basanite and phonolite data from this study define independent sub-vertical trends. Mafic to intermediate inclusions that are found in the phonolites show a similar range in $^{207}Pb/^{204}Pb$ but bridge the gap in $^{206}Pb/^{204}Pb$ between the basanite and phonolite, consistent with a mixing origin. The basanite and phonolite end-members, in turn, define two parallel trends that do not overlap, characterising them as two genetically distinct magmas that define a mixing array

solidus and hence enhances solidification in the residual melt (cf. Sparks et al. 1977; Hammer et al. 2000). The result is a boundary layer of foamy, vesiculated material that is interpreted to form upon initial contact of hot, mafic with cool, felsic magma (Eichelberger 1980) and we interpret type I inclusions accordingly. By implication, the phonolite was probably rather cool when the basanite first arrived, but must have been conductively heated by the basanite. The initially quenched and almost solid, vesicular boundary zone was subsequently disrupted, creating the angular fragments of type I inclusions and allowing for direct contact between the liquid portions of the basanite with the heated phonolite magma.

11.5.3.2 Type II Inclusions

In comparison to type I inclusions, type II inclusions are indicative of a lesser, but still considerable temperature contrast. Instead of the angular outlines characteristic of type I inclusions, the smooth and undulate contacts of type II inclusions resemble liquid blobs and show that the basanite was at some point able to undergo ductile deformation. The temperature difference between the two bodies of magma was still sufficiently large to cause chilled margins (cf. Sparks et al. 1977; Eichelberger 1980; Marshall and Sparks 1984). The two magmas were therefore far from thermal equilibrium, which implies a close temporal relationship between the formation of type I (mafic foam) and type II inclusions. Type II inclusions are interpreted to result from entrainment of mafic magma into reheated and hence re-mobilised phonolite magma. Type II inclusions have thus likely formed after the initial contact zone had been disrupted to form type I inclusions (cf. Troll et al. 2004).

11.5.3.3 Type III Inclusions

Type III inclusions are a lighter colour, with filaments and blobs of darker, mafic magma within them (see Fig. 11.2f). Physical mixing (mingling) of two liquids results in active regions of intense mingling (filaments), and coherent regions largely unaffected by mingling (blobs) (Perugini et al. 2003). The banded textures of type III inclusions demonstrate that a phase of mingling occurred at some point during basanite-phonolite interaction. To allow such intimate mingling, a reduced viscosity contrast must have prevailed between basanite and phonolite (cf. Jellinek et al. 1999), otherwise all other inclusion types ought to show comparable filament textures. The filaments and blobs in type III inclusions are thus most probably the result of intense physical mingling of cooled basanite and heated-up phonolite, i.e. magmas of similar or near-identical viscosities (cf. Perugini et al. 2003). Further evidence for reheating of phonolite is provided by sieve textures in anorthoclase crystals, which are interpreted to originate by remelting of the crystal (cf. Hibbard 1995; Stewart and Pearce 2004). As such, type III inclusions are indicative of advanced thermal equilibration between basanite and phonolite, which makes them the youngest inclusions to have formed during interaction of these two magmas.

This mirrors the results of several studies that have suggested that magma mixing is a progressive interplay of initially dominant mingling and successively more important diffusion (e.g., Kouchi and Sunagawa 1985; Perugini et al. 2003; Zimanowski et al. 2004).

To achieve hybridisation of two magmas within a short timeframe, it is vital that intense physical mingling takes place first. Type III inclusions are thus interpreted to be the result of physical interaction between the basanite and the phonolite magma, suggesting that the interaction was not limited to diffusive hybridisation as suggested by Araña et al. (1994), but was locally associated with intense physical mixing. The succession of type I through III inclusions thus indicates the progressive interaction of two magmas that were initially distinct in composition, temperature and viscosity. Upon interaction, they began to thermally equilibrate and approached each other in their viscosities to allow progressive hybridisation.

11.5.4 Timescale of Basanite-Phonolite Interaction

The preservation of textural transitions in the inclusions from initial formation of mafic foam through quenching and chilled margin development to final liquid–liquid interaction indicates a rapid succession of events. The formation of chilled margins in type II inclusions must have swiftly followed the initial quench-type inclusions in type one inclusions, for the thermal contrast between basanite and phonolite to still be strong enough to allow chilled margins to form. The transition between type II and type III inclusions is less clear, but a close temporal relationship is likely. By using the MELTS algorithm in combination with cooling and decompression experiments, Coombs et al. (2003) temporally constrained the formation of inclusions and chilled margins between an andesite and a dacite to be on the order of hours only. The duration of mixing at Montaña Reventada was therefore probably on a similar order of magnitude (hours to days).

This has implications for the configuration of the magma chamber at depth, because short interaction between basanite and phonolite on the order of few hours is inconsistent with the concept of a long-lived, stratified magma chamber, in which compositionally distinct magmas develop by magmatic differentiation for decades prior to eruption as postulated by Araña et al. (1994). Such a compositionally stratified body of magma is thought to possess stable, diffusive gradients between distinct magmas. In such a magma chamber, quenched material, like that found at Montaña Reventada, is less likely to form because thermal gradients are smooth.

Furthermore, the groundmass of both basanite and phonolite appear to be relatively free of physical mixing; it is only the inclusions that are the result of magma mixing. The total volume of mixed inclusions therefore is restricted at Reventada, with inclusions estimated to amount to less than 1 vol.% of the total deposit volume (Araña et al. 1994). Mingling of basanite and phonolite appears to have been confined to a spatially small zone of interaction, and took place after an initial carapace of quenched basanite (mafic foam) had been disaggregated. In order to efficiently mix two magmas, their viscosities need to be comparable and it has been suggested that only thermal equilibration may reasonably cause approaching viscosities (Campbell and Turner 1986). According to these authors, this may occur either in long-lived, stratified magma chambers or as a spatially restricted phenomenon during fountaining or forced intrusion, creating a small-volume, hybrid boundary layer between mafic and felsic magma. Since at Montaña Reventada a rather short period of interaction is indicated and a relatively small volume of hybrid inclusions is observed within the host phonolite, Wiesmaier et al. (2011) postulated a forced intrusion of basanite into an ambient body of phonolite magma.

11.5.5 Mixing Mechanism

In the previous section it was shown that basanite and phonolite likely interacted over a short time-scale by means of a forced intrusion of basanite into phonolite. The following discussion will now establish the detailed mechanism(s) by which the basanite and phonolite mingled to produce the observed range of hybrid inclusions. Viscous coupling has been suggested in a comparable case in which the mafic member of an eruption was emplaced before the felsic one. Pinatubo erupted andesite before dacite in 1991, the latter of which comprises the bulk of the final deposit (Pallister et al. 1992). Snyder and Tait (1996) tested the Pinatubo scenario experimentally by using the viscous coupling of magmas driven by thermal convection (after Huppert et al. 1983, 1984). It was found that a strong temperature contrast between mafic and felsic magma may trigger local convection within the felsic member, thereby entraining mafic liquid by viscous coupling. Their mafic magma analogue liquid reached the roof of the chamber as a mixed layer, thus providing a model for the eruption of mixed andesite erupting before pristine dacite (as at Pinatubo in 1991).

The Pinatubo model, however, does not satisfactorily explain the situation at Reventada. First of all, at Pinatubo hardly any quenched inclusions have been found, whereas at Reventada these type I and type II inclusions are ubiquitous, indicating a larger temperature contrast between basanite and phonolite compared to andesite and dacite. Furthermore, the first-erupting andesite at Pinatubo is of hybrid origin followed by a pristine dacite, while at Reventada the situation is reverse; the basanite appears texturally and compositionally pristine. Viscous coupling may thus not be the driving mechanism for magma mixing at Montaña Reventada.

In turn, Reventada phonolite, i.e., the later erupted magma, is indeed affected by mixing demonstrated by the inclusions that originate from initial contact between the two magmas. It appears that the phonolite collected the leftover basanite material that had initially quenched against the phonolite. Two possible configurations of interaction were thus suggested. Either the basanite was able to largely bypass the phonolite chamber at its side and only tap it peripherally, or the basanite formed a dyke through the phonolite, being shielded from interaction by the early-formed quench horizon.

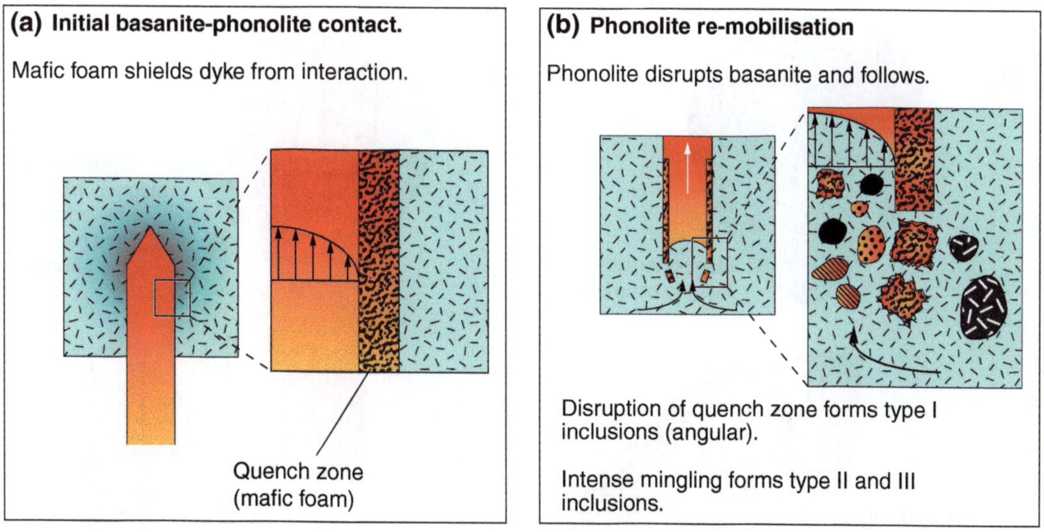

Fig. 11.8 Quench sequence of basanite and phonolite. Schematic representations of the immediate contact between basanite and phonolite. **a** At first contact between basanite and phonolite, the basanite develops a vesicular, solid layer of quenched material (mafic foam) that isolates the bulk of the basanite from interaction with the phonolite. However, the lifespan of this screen of quenched material may be short. **b** Both magmas equilibrate thermally, thus remobilising the phonolite. After the basanite eruption wanes, the phonolite exploits the pre-established conduit, collecting angular fragments of mafic foam and mingling with the remains of liquid basanite on the way to the surface

In both scenarios, the phonolite must have followed the basanite through its conduit. A comparable mechanism has been found in the Katmai region, Alaska. There, the default type of eruption has been described as coming from small, andesite magma chambers that experience mafic recharge. At low recharge vigour, the basalt mixes with the andesite. However, when unmixed mafic scoria is erupted, this is interpreted as basalt magma passing through the andesite chamber with limited interaction only (Coombs et al. 2000; Eichelberger and Izbekov 2000). At Montaña Reventada, the phonolite was probably rather cool prior to interaction, indicating a high viscosity body into which the basanite intruded, a Katmai-type scenario is therefore highly conceivable. The basanite would initially quench at the interface with the phonolite (mafic foam, type I inclusions, Fig. 11.8a). Type II inclusions would form when the reheated and partly re-mobilised phonolite would have started to enter the established basanite conduit, collecting the type I fragments (the former quench zone) and commencing interaction with the liquid basanite magma that is left within the conduit. Because of the ongoing thermal interaction between the phonolite and the basanite, the temperature contrast progressively lowered, allowing for converging viscosities that increasingly permitted mingling to form the type III inclusions (Fig. 11.8b).

Equally plausible, however, is the notion that the basanite was blocked by the phonolite chamber, thereby partly intruding it, but eventually continuing to ascend to the side of it. Again, type I inclusions would have formed at initial contact, while type II and type III would have been generated when the re-heated phonolite exploited the basanite conduit afterwards. Examples for a similar scenario have been found at Karymsky (Kamchatka), Katmai/Novarupta centre (Alaska) and also the 2010 Eyjafjallajökull/Fimmvörðuháls eruption in Iceland, where mafic dykes first opened a fissure at the flank of these volcanoes before triggering more silicic eruptions from central vents (Eichelberger and Izbekov 2000; Gertisser 2010; Gudmundsson et al. 2010).

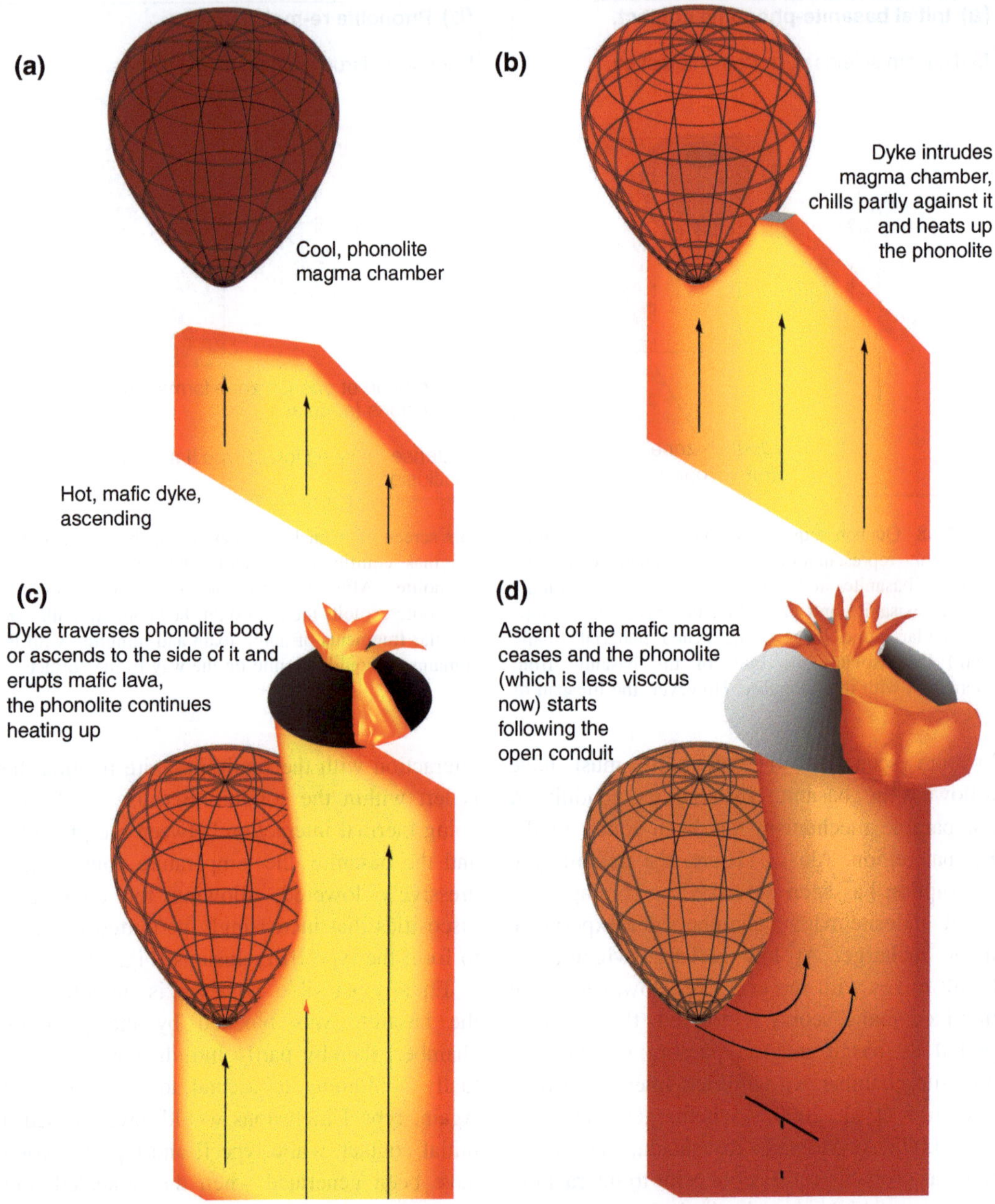

Fig. 11.9 3D Sketch of subsurface interaction between phonolite and basanite. Sketch of the envisaged subsurface dyke ascent and magma chamber dynamics. At Montaña Reventada, two possibilities of dyke ascent are conceivable; a basanite dyke taps a phonolite body peripherally, or a basanite dyke cuts through a cool phonolite magma chamber. **a** A mafic dyke encounters a phonolite body in its ascent path. **b** The dyke taps the phonolite magma and initially quenches against it due to a large temperature contrast. **c** The dyke either intrudes the phonolite or cuts it peripherally and erupts before the phonolite. **d** The phonolite follows the basanite into its conduit after the basanite eruption wanes

For the reasons outlined, the interaction of a relatively small pocket of phonolite magma with a basanite dyke is envisaged at Montaña Reventada (Fig. 11.9). The thermal contrast between cold phonolite and hot basanite inhibited hybridisation, and mixing only initiated after some thermal equilibration had occurred. This interaction was limited to the remainders of the basanite dyke, which must have ceased erupting at that point.

11.6 Eruption Sequence

At Montaña Reventada, hybridisation remained incomplete as mixing was interrupted by eruption, and the inclusions reflect only short-term interaction between basanite and phonolite. However, prolonged interaction of basanite and phonolite would likely lead to homogenisation of the liquid magma portions and may be one of the processes responsible for producing intermediate magmas in the Canary archipelago and ocean islands elsewhere.

Most likely, a pre-existing phonolite magma pocket of the central Teide-Pico Viejo complex was cut by an ascending mafic dyke of the NW rift zone (Fig. 11.8a, b). From the isotopes it is evident that the phonolite had formed by processes that are unrelated to the basanite and the two magmas must have met just prior to eruption. The distinct Pb isotope signatures of basanite and phonolite magmas support the view that the two magmas were co-eruptive, but not co-genetic. When the basanite dyke intruded the phonolite magma chamber, it partly quenched against it, forming solidified, vesicular mafic inclusions within the phonolite. The subsequent entrainment of liquid basanite magma into the phonolite liquid gave rise to type II and III mafic inclusions that were locally hybridised by mingling and mineral exchange along with diffusion. Apart from the resulting hybrid inclusions, both end-members remained largely mechanically and chemically distinct.

The Montaña Reventada composite flow, therefore, is a direct manifestation of the petrogenetic bimodality in recent Tenerife activity. Reventada is located above the assumed boundary of the central Teide-Pico Viejo complex with the NW rift zone. In recent times, Teide and Pico Viejo have erupted phonolite from shallow magma chambers (below sea level, e.g., Ablay et al. 1998), whereas the rift zones have continued to produce lavas of primitive composition that ascended in dykes from upper mantle or lower crustal levels (Carracedo et al. 2007). In the border zone between these two plumbing systems, not only Montaña Reventada shows a lower mafic and an upper felsic member, but so do the lavas of Cuevas Negras which also erupted successively during a single event (Carracedo et al. 2008). The Reventada eruption thus occurred in the transition zone between the central, phonolite-erupting Teide-Pico Viejo complex and the basanite-erupting NW rift zone, implying that two genetically distinct magmas have accidentally met to form this composite eruption.

References

Ablay GJ, Carroll MR, Palmer MR, Martí J, Sparks RSJ (1998) Basanite-phonolite lineages of the teide-pico viejo volcanic complex, Tenerife, Canary Islands. J Petrol 39:905–936

Abratis M, Schmincke H-U, Hansteen T (2002) Composition and evolution of submarine volcanic rocks from the central and western Canary Islands. Int J Earth Sci 91:562–582

Araña V, Aparicio A, Garcia Cacho L, Garcia Garcia R (1989) Mezcla de magmas en la región central de Tenerife. In: Araña V, Coello J (eds) Los Volcanes y La Caldera del Parque Nacional del Teide (Tenerife, Islas Canarias). Serie Técnica, ICONA, pp 269–298

Araña V, Martí J, Aparicio A, García-Cacho L, García-García R (1994) Magma mixing in alkaline magmas: an example from Tenerife, Canary Islands. Lithos 32:1–19

Bacon CR (1986) Magmatic inclusions in silicic and intermediate volcanic rocks. J Geophys Res 91:6091–6112

Bacon CR, Metz JM (1984) Magmatic inclusions in rhyolites, contaminated basalts, and compositional zonation beneath the Coso volcanic field, California. Contrib Mineral Petrol 85:346–365

Bence AE, Albee AL (1968) Empirical correction factors for the electron microanalysis of silicates and oxides. J Geol 76:382–403

Bindeman IN, Perchuk LL (1993) Experimental studies of magma mixing at high pressures. Int Geol Rev 35:721–733

Bindeman IN, Davis AM (1999) Convection and redistribution of alkalis and trace elements during the mingling of basaltic and rhyolitic melts. Petrol 7:91–101

Blake S (1981) Eruptions from zoned magma chambers. J Geol Soc (London, UK) 138:281–287
Blake S, Ivey GN (1986) Magma-mixing and the dynamics of withdrawal from stratified reservoirs. J Volcanol Geotherm Res 27:153–178
Calanchi N, Rosa R, Mazzuoli R, Rossi P, Santacroce R, Ventura G (1993) Silicic magma entering a basaltic magma chamber: eruptive dynamics and magma mixing—an example from Salina (Aeolian islands, Southern Tyrrhenian Sea). Bull Volcanol 55:504–522
Campbell IH, Turner JS (1986) The influence of viscosity on fountains in magma chambers. J Petrol 27:1–30
Carracedo JC, Rodríguez Badiola E, Guillou H, Paterne M, Scaillet S, Pérez Torrado FJ, Paris R, Fra-Paleo U, Hansen A (2007) Eruptive and structural history of Teide Volcano and Rift Zones of Tenerife, Canary Islands. Geol Soc Am Bull 119:1027–1051
Carracedo JC, Rodríguez Badiola E, Guillou H, Paterne M, Scaillet S, Pérez Torrado FJ, Paris R, Rodríguez González A, Socorro S (2008) El Volcán Teide–Volcanología, Interpretación de Pasajes y Iterinarios Comentados. Caja Generál de Ahorros de Canarias
Cas RAF, Wright JV (1987) Volcanic successions–modern and ancient. Allen & Unwin Ltd, London
Coombs ML, Eichelberger JC, Rutherford MJ (2000) Magma storage and mixing conditions for the 1953–1974 eruptions of Southwest Trident volcano, Katmai National Park, Alaska. Contrib Mineral Petrol 140:99–118
Coombs ML, Eichelberger JC, Rutherford MJ (2003) Experimental and textural constraints on mafic enclave formation in volcanic rocks. J Volcanol Geotherm Res 119:125–144
De Campos CP, Dingwell DB, Perugini D, Civetta L, Fehr TK (2008) Heterogeneities in magma chambers: Insights from the behavior of major and minor elements during mixing experiments with natural alkaline melts. Chem Geol 256:131–145
Eichelberger JC (1980) Vesiculation of mafic magma during replenishment of silicic magma reservoirs. Nature 288:446–450
Eichelberger JC, Izbekov PE (2000) Eruption of andesite triggered by dyke injection: Contrasting cases at Karymsky Volcano, Kamchatka and Mt Katmai, Alaska. Philos Trans R Soc London, Ser A 358:1465–1485
Eichelberger JC, Chertkoff DG, Dreher ST, Nye CJ (2000) Magmas in collision: rethinking chemical zonation in silicic magmas. Geology 28:603–606
Freundt A, Schmincke H-U (1992) Mixing of rhyolite, trachyte and basalt magma erupted from a vertically and laterally zoned reservoir, composite flow P1, Gran Canaria. Contrib Mineral Petrol 112:1–19
Geldmacher J, Haase KM, Devey CW, Garbe-Schönberg CD (1998) The petrogenesis of tertiary cone-sheets in Ardnamurchan, NW Scotland: petrological and geochemical constraints on crustal contamination and partial melting. Contrib Mineral Petrol 131:196–209
Gertisser R (2010) Eyjafjallajökull volcano causes widespread disruption to European air traffic. Geol Today 26:94–95
Gudmundsson MT, Pedersen R, Vogfjörd K, Thorbjarnardóttir B, Jakobsdóttir S, Roberts MJ (2010) Eruptions of Eyjafjallajökull Volcano, Iceland. Eos Trans AGU 91:190–191
Gurenko AA, Hoernle KA, Hauff F, Schmincke H-U, Han D, Miura YN, Kaneoka I (2006) Major, trace element and Nd-Sr-Pb-O-He-Ar isotope signatures of shield stage lavas from the central and western Canary Islands: Insights into mantle and crustal processes. Chem Geol 233:75–112
Hammer JE, Cashman KV, Voight B (2000) Magmatic processes revealed by textural and compositional trends in Merapi dome lavas. J Volcanol Geotherm Res 100:165–192
Hawkesworth CJ, Blake S, Evans P, Hughes R, Macdonald R, Thomas LE, Turner SP, Zellmer G (2000) Time scales of crystal fractionation in magma chambers–integrating physical, isotopic and geochemical perspectives. J Petrol 41:991–1006
Hibbard MJ (1995) Petrography to petrogenesis. Prentice Hall, Upper Saddle River
Hildreth EW (1979) The bishop tuff: evidence for the origin of compositional zonation in silicic magma chambers. Geol Soc Spec Publ 180:43–75
Huppert HE, Turner JS, Sparks RSJ (1982) Replenished magma chambers: effects of compositional zonation and input rates. Earth Planet Sci Lett 57:345–357
Huppert HE, Sparks RSJ, Turner JS (1983) Laboratory investigations of viscous effects in replenished magma chambers. Earth Planet Sci Lett 65:377–381
Huppert HE, Sparks RSJ, Turner JS (1984) Some effects of viscosity on the dynamics of replenished magma chambers. J Geophys Res 89:6857–6877
Izbekov PE, Eichelberger JC, Ivanov BV (2004) The 1996 Eruption of Karymsky Volcano, Kamchatka: historical record of basaltic replenishment of an andesite reservoir. J Petrol 45:2325–2345
Jellinek AM, Kerr RC, Griffiths RW (1999) Mixing and compositional stratification produced by natural convection 1. Experiments and their application to Earth's core and mantle. J Geophys Res Solid Earth 104:7183–7201
Kouchi A, Sunagawa I (1985) A model for mixing basaltic and dacitic magmas as deduced from experimental data. Contrib Mineral Petrol 89:17–23
Kuritani T (2001) Replenishment of a mafic magma in a zoned felsic magma chamber beneath Rishiri Volcano, Japan. Bull Volcanol 62:533–548
Le Bas MJ, Maitre RWL, Streckeisen A, Zanettin B, ISotSoI Rocks (1986) A chemical classification of volcanic rocks based on the total alkali-silica diagram. J Petrol 27:745–750
Lesher CE (1986) Effects of silicate liquid composition in mineral-liquid element partitioning from soret diffusion studies. J Geophys Res 91:6123–6141
Lesher CE, Walker D (1986) Solution properties of silicate liquids from thermal diffusion experiments. Geochim Cosmochim Acta 50:1397–1411
Marshall LA, Sparks RSJ (1984) Origin of some mixed-magma and net-veined ring intrusions. J Geol Soc (London, UK) 141:171–182

Merle O (1998) Internal strain within lava flows from analogue modelling. J Volcanol Geotherm Res 81:189–206
Palacz ZA, Wolff JA (1989) Strontium, neodymium and lead isotope characteristics of the Granadilla Pumice, Tenerife: a study of the causes of strontium isotope disequilibrium in felsic pyroclastic deposits. Geol Soc Spec Publ 42:147–159
Pallister JS, Hoblitt RP, Reyes AG (1992) A basalt trigger for the 1991 eruptions of Pinatubo volcano? Nature 356:426–428
Perugini D, Poli G, Mazzuoli R (2003) Chaotic advection, fractals and diffusion during mixing of magmas: evidence from lava flows. J Volcanol Geotherm Res 124:255–279
Simonsen SL, Neumann ER, Seim K (2000) Sr-Nd-Pb isotope and trace-element geochemistry evidence for a young HIMU source and assimilation at Tenerife (Canary Island). J Volcanol Geotherm Res 103:299–312
Snyder D, Tait S (1996) Magma mixing by convective entrainment. Nature 379:529–531
Sparks SRJ, Sigurdsson H, Wilson L (1977) Magma mixing: a mechanism for triggering acid explosive eruptions. Nature 267:315–318
Stewart ML, Pearce TH (2004) Sieve-textured plagioclase in dacitic magma: Interference imaging results. Am Mineral 89:348–351
Troll VR, Schmincke H-U (2002) Magma mixing and crustal recycling recorded in ternary feldspar from compositionally zoned peralkaline ignimbrite 'A', Gran Canaria, Canary Islands. J Petrol 43:243–270
Troll VR, Donaldson CH, Emeleus CH (2004) Pre-eruptive magma mixing in ash-flow deposits of the Tertiary Rum Igneous Centre Scotland. Contrib Mineral Petrol 147(6):722–739
Turner JS (1980) A fluid-dynamical model of differentiation and layering in magma chambers. Nature 285:213–215
Turner JS, Campbell IH (1986) Convection and mixing in magma chambers. Earth Sci Rev 23:255–352
Turner SP, Platt JP, George RMM, Kelley SP, Pearson DG, Nowell GM (1999) Magmatism associated with orogenic collapse of the betic-alboran domain, SE Spain. J Petrol 40:1011–1036
Walker D, DeLong SE (1982) Soret separation of mid-ocean ridge basalt magma. Contrib Mineral Petrol 79:231–240
Walker D, Lesher CE, Hays JF (1981) Soret separation of lunar liquids. Paper presented at the lunar and planetary science XII, 16–20 March
Watson EB (1982) Basalt contamination by continental crust: Some experiments and models. Contrib Mineral Petrol 80:73–87
Watson EB, Baker DR (1991) Chemical diffusion in Magmas: an overview of experimental results and geochemical applications. In: Perchuk LL, Kushiro I (eds) Advances in physical geochemistry, vol 6. Springer, New York, pp 120–151
Wiesmaier S (2010) Magmatic differentiation and bimodality in oceanic island settings—implications for the petrogenesis of magma in Tenerife, Spain. PhD Thesis, Trinity College Dublin, Dublin
Wiesmaier S, Deegan F, Troll V, Carracedo J, Chadwick J, Chew D (2011) Magma mixing in the 1100 AD Montaña Reventada composite lava flow, Tenerife, Canary Islands: interaction between rift zone and central volcano plumbing systems. Contrib Mineral Petrol 162:651–669
Wolff JA, Storey M (1984) Zoning in highly alkaline magma bodies. Geol Mag 121:563–575
Wolff JA, Grandy JS, Larson PB (2000) Interaction of mantle-derived magma with island crust? Trace element and oxygen isotope data from the Diego Hernandez Formation, Las Canadas, Tenerife. J Volcanol Geotherm Res 103:343–366
Zimanowski B, Büttner R, Koopmann A (2004) Experiments on magma mixing. Geophys Res Lett 31:L09612

Eruptive Styles at the Teide Volcanic Complex

12

Francisco J. Perez-Torrado, Juan Carlos Carracedo, Alejandro Rodriguez-Gonzalez, Eduardo Rodríguez-Badiola, Raphaël Paris, Valentin R. Troll, Hilary Clarke, and Sebastian Wiesmaier

Abstract

The wide variety of volcanic products composing the Teide Volcanic Complex (TVC) reflects an unusual assemblage of eruptive styles, with a wide range of phenomena represented and only plinian and phreatoplinian styles truly lacking. This diversity is due to spatial and temporal variations in magma composition (mafic magmas of the rift zones and felsic magmas of the central edifice), variable magmatic volatile contents and the interaction of magma with external water (snow, groundwater, etc.). Overall, strombolian eruptions are the most frequent eruptive style at the TVC. Explosive eruptions of felsic material tend to be of low volume, for example, the largest explosive event during the Holocene,

F. J. Perez-Torrado (✉) · J. C. Carracedo · A. Rodriguez-Gonzalez
Departamento de Física (GEOVOL), Universidad de Las Palmas de Gran Canaria, Las Palmas de Gran Canaria, 35017 Canary Islands, Spain

E. Rodríguez-Badiola
Museo Nacional de Ciencias Naturales, CSIC, 28006 Madrid, Spain

R. Paris
Laboratoire Magmas et Volcans UMR 6524 CNRS-UBP-IRD, 5 rue Kessler, 63038 Clermont-Ferrand, France

V. R. Troll
Department of Earth Sciences, CEMPEG, Uppsala University, 75236 Uppsala, Sweden

H. Clarke
Serengeti Resources Inc. SIR: TSX-V, Vancouver, 34S:FSE, Canada

S. Wiesmaier
Department of Earth and Environmental Sciences, Ludwig-Maximilians Universität (LMU), Munich, Germany

J. C. Carracedo and V. R. Troll (eds.), *Teide Volcano*, Active Volcanoes of the World,
DOI: 10.1007/978-3-642-25893-0_12,

Montaña Blanca (ca. 2 ka), produced ~0.2 km^3 DRE of phonolitic pumice during an eruptive sequence that reached explosivity of sub-plinian magnitude. Examples of phreatomagmatic activity (surge deposits) have been described both on the northern flanks of Teide volcano as well as from the summit area of Pico Viejo volcano. Until now most studies on volcanic hazard assessment have focussed on ash fall and lava flow hazards in the Canary Islands, but phreatomagmatic eruptions and their potential effects may have to be seriously considered as well.

12.1 Introduction

The definition and classification of volcanic eruptions is a challenging task because of the inherent complexity of the eruptive process and the number of parameters involved (e.g. temperature, silica content, and volatile content of the magma, the presence of external water and the structural state of an edifice). During its evolution, the eruptive activity of a volcano can hence display a variety of styles, some of them very different from previous eruptive stages. These styles may change drastically even in the course of a single eruption, and in the Teide Volcanic Complex (TVC) perhaps the best examples of these changes are derived from the interaction of magma with meteoric water (including snow) and groundwater.

Attempts to classify eruptions using different characteristics (genetic, descriptive, etc.) have resulted in an abundance of terms, many of them redundant and of unclear significance. The simplest classification, however, distinguishes "magmatic" and "phreatomagmatic" eruptions, and depends on the presence or absence of external water (e.g. from the sea, lakes or groundwater) during the eruptive process.

Magmatic eruptions are driven essentially by the magma and its contained gases, and the different eruptive styles which then ensue from variations in magmatic composition and rheology. Following Walker (1973), a widely accepted classification of magmatic eruptions defines types (with progressively increasing explosivity), such as Hawaiian, Strombolian, Vulcanian and Plinian eruption styles. All except the Plinian type are represented in the volcanic succession of the TVC. Hawaiian and Strombolian eruptive styles, commonly grouped as effusive eruptions, are characterised by mafic and intermediate lava compositions, and on Tenerife frequently form multiple vents along fissures parallel to the rift zones (fissure eruptions) (Fig. 12.1). Vulcanian (narrow conduit) and Plinian (wide conduit) eruptive styles are classified as explosive eruptions, and are commonly associated with felsic magma compositions on Tenerife (Wolff and Storey 1983).

Phreatomagmatic eruptions (e.g. Wohletz 1983; Lorenz 1987) occur when large volumes of hot magma mix with water (e.g. groundwater, sea water, lake water and snow and ice melt water). Highly explosive eruptions occur if the magma/water interaction is efficient, i.e. if the balance between fuel and coolant allows for explosive energy release. Transfer of heat (fuel) to water (coolant) becomes more effective as the contact area between hot magma and water increases, for example with initially fragmenting magma. This is because the intensity of phreatomagmatic explosions results from the expansion of heated water and is therefore proportional to the area of water/magma contact during the eruption (Büttner et al. 1999; Morrissey et al. 2000). In this context, highly fragmented magma and an intermediate magma/water mass ratio produce the most explosive eruptions. In general, felsic magmas yield more efficient "fuel–coolant" interactions than mafic magmas, so phreatomagmatic eruptions vary from the lower end of the explosivity range, for example in Surtseyan types (interaction of water with mafic magmas) to the very explosive phreatoplinian types, caused by interaction of water with felsic magmas, which are often an

Fig. 12.1 Panoramic photograph of the Chinyero volcano (Nov. 1909), the latest eruption on the island of Tenerife (Maximilian Löhr, Fotografía Alemana)

order of magnitude more energetic (Morrissey et al. 2000).

Explosive eruptions are in general relatively scant in the TVC, despite the abundance of felsic volcanism (Araña et al. 1989a; Ablay et al. 1998; Rodríguez-Badiola et al. 2006; and Chaps. 9 and 10). In contrast, highly explosive (Plinian) events are frequent in the older volcanic phases of the Las Cañadas Volcano. The total volume of these older explosive, predominantly phonolitic, eruptions has been estimated at ca. 150 km^3, with single events of about 20 km^3 e.g. Edgar et al. (2007). Martí et al. (2008) defined this difference as follows "*Pre-Teide central activity is mostly characterised by large-volume* (1 –> 20 km^3, *DRE*) *eruptions of phonolitic magmas, while Teide–Pico Viejo is dominated by effusive eruptions*". The lower volatile contents (particularly H_2O) and lower viscosity of the peralkaline TVC magmas may account for their lower explosivity (Araña et al. 1989a; Albert-Beltran et al. 1990; Ablay et al. 1995; Rodríguez-Badiola et al. 2006; see also Chaps. 9 and 10). Even mixing of mafic and felsic magmas, a relatively frequent event in the TVC, tends to produce effusive eruptions (Araña et al. 1989b; Rodríguez-Badiola et al. 2006; Wiesmaier et al. 2011; and Chap. 11), with the exception of the Montaña Blanca subplinian event (Ablay et al. 1995). The explosive eruptions that have occurred have mainly been of phreatomagmatic style (Pérez Torrado et al. 2004, 2006; del Potro et al. 2009). From the geological record it thus appears that Strombolian eruptions, producing both mafic and felsic magmas, are the most frequent eruptive style found in the TVC.

12.2 Effusive Eruptions in the TVC

Magmas contain a significant proportion of magmatic gases at high pressure, which are usually liberated to the atmosphere in the initial stages of an eruption. Consequently, eruptions tend to be mildly more explosive at the onset (e.g. Strombolian), fragmenting the lava and producing pyroclasts such as lapilli (Fig. 12.2a), spatters and bombs, some of which reach a considerable size (Fig. 12.2b). After most of the gas has been released, the eruptive style may change to a more effusive one where lava flows are predominant.

In the TVC, the vents of effusive eruptions are largely clustered along the rift zone axes. Therefore, pyroclasts abound at the crest of the rifts, whereas lava flows are more prevalent on the rift flanks. Lava predominantly forms 'a'ā type flows, whereas pāhoehoe morphologies are less abundant, except in the oldest sequences of Teide and Pico Viejo stratocones. There, abundant, smooth to ropy surfaced plagioclase basalt flows occur (Fig. 12.2c). Phonolitic lavas, that in turn frequently form blocky flows, present a wide range of increasingly rough and irregular morphologies (Fig. 12.2d). 'A'ā and blocky lava morphologies are known in the Canary Islands as "*malpaíses*" (badlands).

(a)
(b)
(c)
(d)
(e)
(f)
(g)
(h)

◀**Fig. 12.2** Examples of characteristic features resulting from effusive eruptions in the TVC. **a** Basaltic lapilli fall deposit from the Montaña de Garachico eruption. **b** Basaltic bomb from the area of Montaña Reventada. **c** Pāhoehoe lavas from the initial phase of Pico Viejo. Road near the town of Chío, south Tenerife. **d** Characteristic rough surface 'a'ā lava flow from Boca Cangrejo volcano. **e** Internal structure of a channelled basaltic lava flow lobe with radial cooling joints. **f** Interior of the Cueva del Viento lava tube, formed in pāhoehoe lavas during the initial stages of Pico Viejo eruptive activity that flow all the way to the sea. This lava tube is the 5th longest volcanic cavity in the world (after Hawaiian lava tubes). **g** Strombolian cinder cone of Montaña de Arafo (1705 A.D.) **h** Hornito in deposits of the Montaña de Garachico eruption (1706 A.D.)

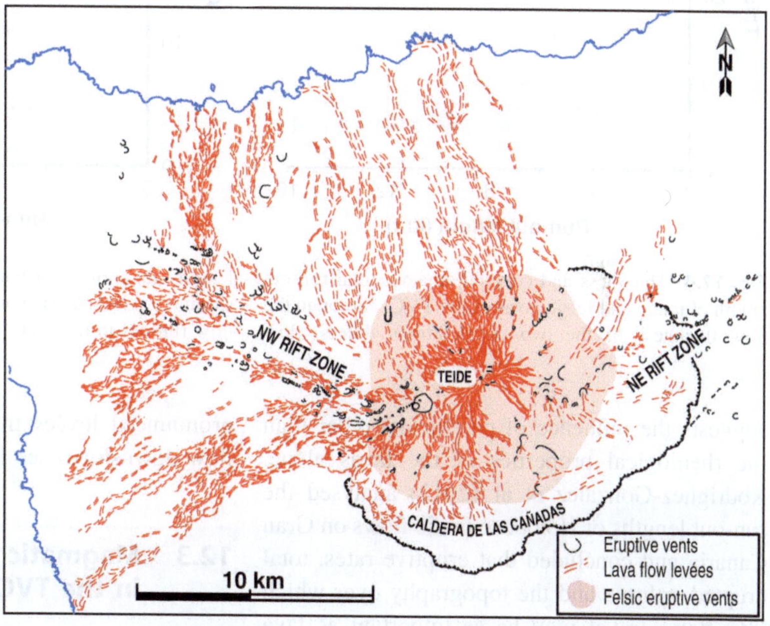

Fig. 12.3 Eruptive vent distribution and flow run-out length in the TVC. Note that a number of felsic as well as mafic flows reach the sea in north and west Tenerife

Associated with effusive eruptions in the TVC are spectacular features like accretionary lava balls (Fig. 12.2e), lava tubes (Fig. 12.2f), cinder cones (Fig. 12.2g), and hornitos (Fig. 12.2h). Detailed morphological descriptions of volcanic cones and lava flows found in the TVC are presented in Chap. 3.

12.2.1 Eruptive Vent Distribution

The distribution of mafic and felsic eruptive centres shows a clear pattern (Fig. 12.3). Mafic eruptions tend to group in the rift zones and form multiple vents from fissures broadly aligned with the main direction of the rift (see Chap. 4). Felsic vents, however, are restricted to the interior of the Las Cañadas Caldera, and often show concentric patterns, forming domes located around the base of the Teide and Pico Viejo stratocones (see also Chaps. 9 and 10). As a result, felsic flows are often thicker than mafic flows on Tenerife but are comparable in total surface area and run-out length, accommodating the larger eruption volumes of the felsic TPV events.

12.2.2 Lava Run-Out Lengths

Mafic and felsic lavas in the TVC differ considerably in thickness, but their total run-out lengths are seemingly similar (Fig. 12.4a). Generally, felsic flow lengths are significantly shorter in other volcanic settings because of their higher viscosity (Fisher and Schmincke 1984; Cas and Wright 1987). Therefore, the total run-out length of felsic (phonolitic) lavas associated with the Teide and Pico Viejo volcanoes

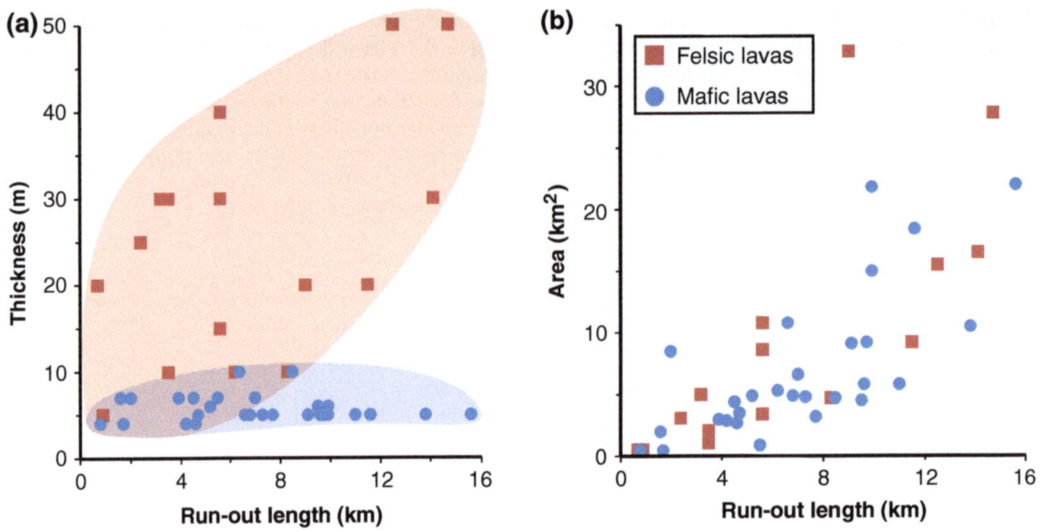

Fig. 12.4 Thickness and covered area vs. total run-out length of mafic and felsic lavas of the TVC. Although the flow thickness of the two compositions differs, flow lengths are similar for both types indicating that composition (i.e. viscosity) is not the sole factor controlling lava flow run-out in the TVC

suggests the influence of parameters other than the rheological properties of the lavas alone. Rodriguez-Gonzalez et al. (2012) analysed the run-out lengths of Holocene mafic flows on Gran Canaria and concluded that eruptive rates, total erupted volume and the topography over which lava flows travel may be as important as lava composition for determining their final run-out lengths. Another aspect affecting the total run-out length of phonolitic lava flows of the TVC is the formation of an external lava carapace (see Fig. 8.21), which generates effective thermal isolation of the lava, and allows flows to travel longer distances producing run-outs approaching those of mafic lavas.

Notably, when run-out lengths are compared with total surface covered by the lava flows, mafic and felsic lava flows frequently show different patterns (Fig. 12.4b). The explanation probably lies in the compositional differences of the two types of lavas. Although felsic lava flows are volumetrically superior (even by some orders of magnitude) to mafic lava flows, they are unable to expand laterally due to their somewhat higher viscosities and thus form pronounced levées that effectively channel the flow downslope (see Fig. 12.5h).

12.3 Magmatic Explosive Eruptions in the TVC

In contrast to the Hawaiian Islands, where eruptions of felsic magmas are scarce to absent, Tenerife and the TVC display a wide variety of these products and associated features (Fig. 12.5), representing probably one of the best examples of felsic volcanism in an oceanic island setting. However, highly explosive eruptive events related to the felsic magmas in the TVC are relatively infrequent and tend to be of low volume. Occasional collapses of asymmetrical domes and phonolitic lava flow fronts cause Vulcanian events, but Plinian eruptions, very common in the older Las Cañadas Volcano, are not preserved in the TVC geological record. So far, the subplinian Montaña Blanca phonolitic event, some 2,000 years ago, appears to have deposited the most explosive eruption sequence within the TVC succession (Ablay et al. 1995).

(a)
(b)
(c)
5 cm
(d)
(e)
(f)
(g)
(h)

◀ **Fig. 12.5** Characteristic felsic products and features of the TVC. **a** Strombolian phonolitic cone of Montaña Majúa. **b** Montaña Blanca-derived airfall pumice in the saddle between Teide and Pico Viejo. The red colour is caused by oxidation of the frothy pumice. **c** Bread-crust phonolitic bomb, Montaña Blanca. **d** Large accretionary lava ball (locally known as Huevos del Teide or Teide's Eggs). This one detached from the front of the Lavas Negras flow (ca. 1240 B.P.), northern slope of Montaña Blanca. **e** Very rough surface typical of phonolitic blocky lava flows, Lavas Negras. Encircled people for scale. **f** Chaotic assemblage of angular blocks of obsidian phonolites from the Montaña Blanca group (El Tabonal Negro). Encircled is a person for scale. **g** An example of a short-length phonolitic flow: Los Gemelos, north of Pico Viejo (image is 600 m across). **h** Pronounced levées in phonolitic flows of Roques Blancos (image is 150 m across). Note that the best growth conditions for the pine trees are found inside the former flow channels, where pumice and rubble accumulate and preserve humidity

12.3.1 The Montaña Blanca Subplinian Event

Montaña Blanca is a large volcanic dome complex formed by phonolitic lavas on the eastern flank of Teide stratocone, nearby the Las Cañadas Caldera region influenced by the NERZ (see Chaps. 7 and 8). The Montaña Blanca eruption can be divided into multiple events, which have been dated at ca. 2 ka (Ablay et al. 1995; Carracedo et al. 2007). A map of this volcanic complex and stratigraphic relationships is provided in (Fig. 8.23). Ablay et al. (1995) also made a detailed study of the different eruptive styles involved in this eruption. The eruption began with lava extrusion, followed by subplinian explosive activity, which produced large volumes of pumice and ash. The eruption ended with lower energy vulcanian and, eventually, dome-building activity. The bulk volume of the pumice and ash deposit from Montaña Blanca is estimated at about 0.815 km^3 ($\approx$0.17–0.25 km^3 DRE), emitted from a NW–SE fissure located at the most elevated area of the Montaña Blanca dome complex. Except at some proximal localities, the pumice deposit consists of a single, well-sorted, massive bed of pale green, crystal-poor, phonolitic, angular pumice lapilli (Fig. 12.6a). The coarseness, good sorting, angularity and grain size characteristics clearly indicate its fallout origin. Isopach and isopleth maps show the deposit to be elongated in SW-NE direction. Ablay et al. (1995) determined a pyroclastic column height of about 15 km, a wind speed of 10 m/s and a minimal area of 40 km^2 covered by pyroclastic fall deposits (see Fig. 14.18). Magma mixing (phonolitic and phonotephritic compositions), a shallow magma chamber and a relatively high percentage of volatiles are the main parameters invoked by Ablay and others to explain the most explosive phase of the Montaña Blanca events.

12.3.2 Gravitational Collapse of Phonolitic Domes and Lava Flow-Driven Explosive Eruptions

Small and disperse outcrops of phonolitic volcaniclastic deposits have been recently reported on the northern slopes of the TVC (del Potro et al. 2009; García et al. 2011). The eastern outcrops on that slope are located close to the western flank of Pico Cabras, in the area of Los Benjamines (del Potro et al. 2009), and the western outcrops close to El Boquerón and Roques Blancos (García et al. 2011). These volcaniclastic deposits are estimated to be of Holocene age by correlation with the general stratigraphy previously defined by the abundant radiometric ages found in Carracedo et al. (2007).

Volcaniclastic deposits of both these areas (the western and eastern outcrops of Teide's northern slope) present similar characteristics: poor lateral continuity, deposit slope angles of 20–30°, variable thickness (<1 to several metres), massive with no bedding or internal flow structures, matrix-supported and a consistently monogenetic (phonolitic) nature of clasts and matrix (Fig. 12.6b). According to del Potro et al. (2009) these features point to a block-and-ash flow deposit derived from gravitational collapses of incandescent asymmetrical domes and/or fronts of lavas flowing over break-in-slope areas.

García et al. (2011) described another type of volcaniclastic deposit found near the Abrunco volcano, which they interpreted as pumice-rich ignimbrites (Fig. 12.6c). Due to the presence of these pyroclastic deposits, the authors argue for potentially higher explosivity in the volcanic history of the TVC than generally assumed: "*The fact that the volume of pyroclastic deposits visible today is small compared to that of lavas does not necessarily imply that explosive activity has been insignificant in the recent evolution of Teide–Pico Viejo. On the contrary, we claim that phonolitic explosive activity has been more significant than previously thought in Teide–Pico Viejo during the Holocene. The evidence we have presented for the syn-depositional erosion of ignimbrites suggests that heavy rainfalls may have occurred in the area during these eruptions, which could explain the rapid disappearance of a significant part of these deposits*".

This type of pyroclastic deposit, however, has not been observed in abundance in the older TVC stratigraphic sequences, either along surficial outcrops or along the *galerías* (water tunnels) crossing the north flank of Teide at different locations and depths (e.g. Carracedo et al. 2007; Márquez et al. 2008; Boulesteix et al. 2012). Moreover, the rapid erosion, suggested by these authors as the rationale to explain the scant volume of these deposits, should have affected the Las Cañadas Volcano succession in the same way, and yet, pyroclastic deposits related to magmatic explosive eruptions are highly abundant there (e.g. Huertas et al. 2002; Edgar et al. 2007).

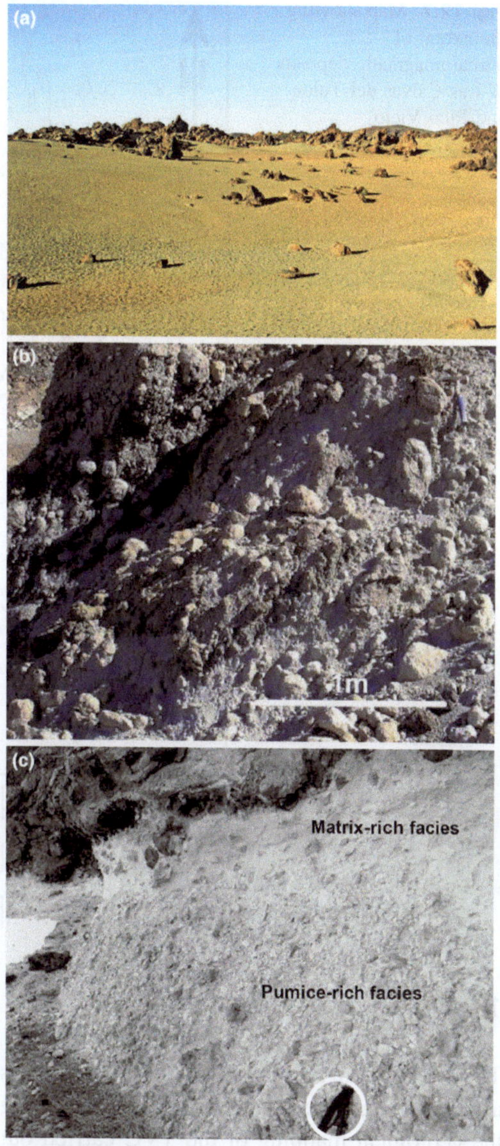

Fig. 12.6 Close-up view of some pyroclastic deposits in the TVC. **a** Pumice pyroclastic fall deposits from Montaña Blanca eruption mantling previous topography. **b** Block-and-ash deposit at Los Benjamines outcrop formed by gravitational collapse of a phonolitic dome (from del Potro et al. 2009). **c** Close-up photograph of El Abrunco ignimbrite showing matrix-rich and pumice-rich facies at Abrunco outcrop. Note the encircled walking stick handle–15 cm for scale) (from Garcia et al. 2011)

12.4 Phreatomagmatic Explosive Eruptions in the TVC

Phreatomagmatic explosive eruptions involving both mafic and felsic magmas, are relatively frequent in the geological evolution of the TVC, but voluminous deposits are scarce. The main outcrops are located at the northern flank of Teide (Calvas del Teide) and encircle the summit of Pico Viejo (Fig. 12.7). There, the water involved in the magma-water interaction must have been derived from snow and ice that accumulate during winter periods.

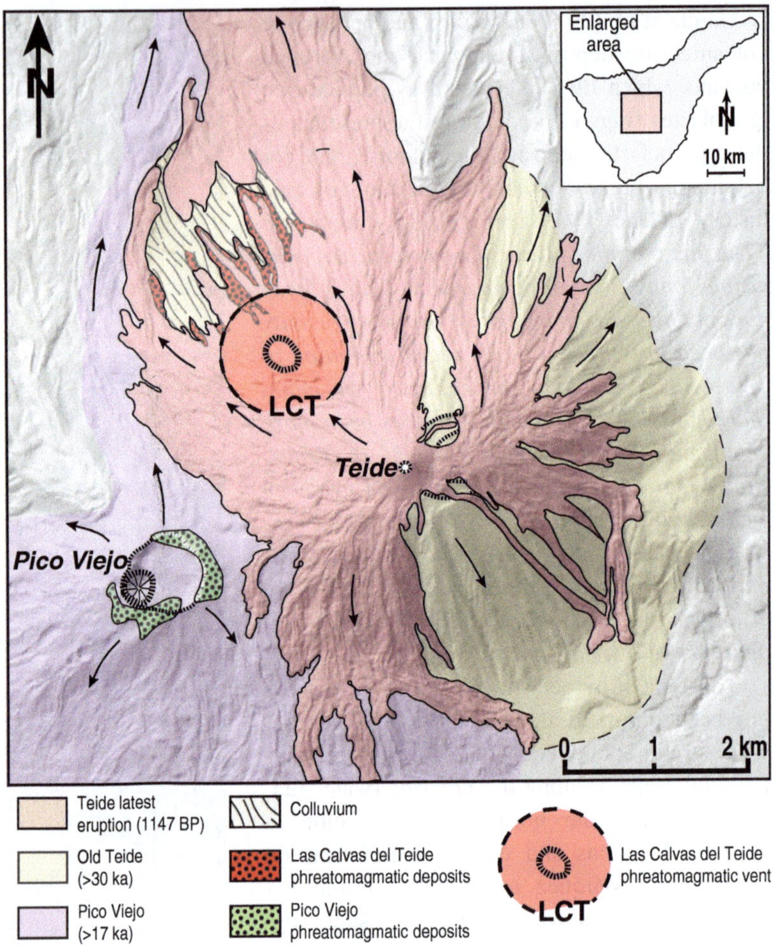

Fig. 12.7 Map showing the extent of phreatomagmatic deposits of Las Calvas del Teide and Pico Viejo

12.4.1 Las Calvas del Teide

This very visible formation was first identified as a phreatomagmatic eruption from Teide during the geological mapping of the TVC carried out between 2003 and 2006 (Pérez Torrado et al. 2004; Carracedo et al. 2007). The deposits appear as thick and off-white, indurated volcaniclastic slabs, devoid of any vegetation (locally known as Teide's bald patches) (Fig. 12.8a and b). Pérez Torrado et al. (2004) described them as being interbedded with thin "Old Teide" phonolitic lava flows (>30 ka), and capped by flows of Pico Viejo volcano (>17 ka). The entire sequence is partially covered by the latest (1150 ± 140 yr BP) phonolitic Teide event (see Fig. 12.8a and b). The scattered outcrops appear along the gullies that cut into the NW flank of the Teide stratocone (Fig. 12.9) and give a sense of the depth of erosion and the subsequent infill of younger lavas. The full stratigraphic column has been resolved by correlation of partial exposures at different localities, which all have slopes that are consistently ~35° (see Fig. 12.9).

The stratigraphic column is composed of six distinct units (Fig. 12.10) (Pérez Torrado et al. 2004, 2006). Thin (<1 m) phonolite flows form the bottom and top (Fig. 12.10a and f) and appear interlayered between the pyroclastic deposits (Fig. 12.10c). Volcaniclastic units show two different lithofacies: laminated, fine grained (Fig. 12.10b and e), and coarse-grained, massive beds (Fig. 12.10d). The laminated deposits appear as alternating layers of coarse (5–10 cm) and fine (1–15 mm) clasts, embedded in an intensely indurated cineritic matrix (Fig. 12.8c and d).

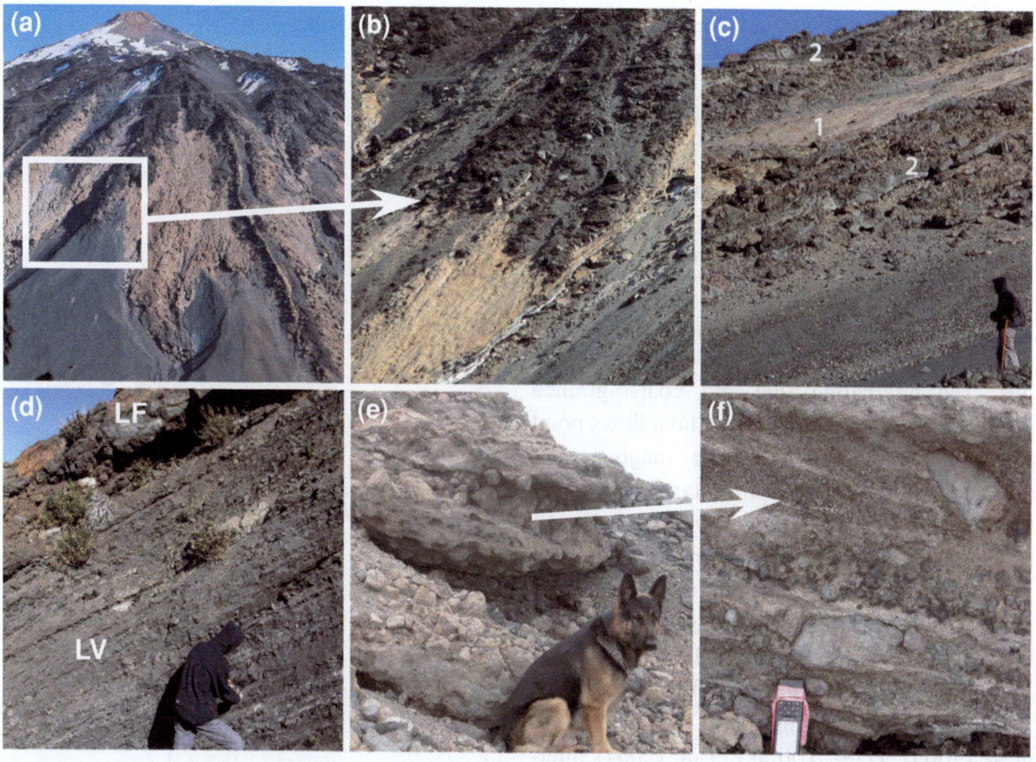

Fig. 12.8 Images of the Las Calvas del Teide phreatomagmatic deposits. **a** General view of the area. Note the indurated and thick white volcaniclastic slabs devoid of any vegetation (Teide's bald patches). **b** Close-up view of the volcaniclastic slabs covered by the Lavas Negras flows. **c** Phreatomagmatic volcaniclastic deposits of Las Calvas del Teide (*1*) alternating with phonolitic lava flows of "Old" Teide (*2*). **d** Close-up view of laminated volcaniclastics deposit (LV) topped by an "Old" Teide phonolitic lava flow (LF). **e** Volcaniclastic deposit outcropping down-slope at Las Calvas del Teide, interpreted as a debris flow deposit. **f** Detail of the debris flow deposit in (**e**)

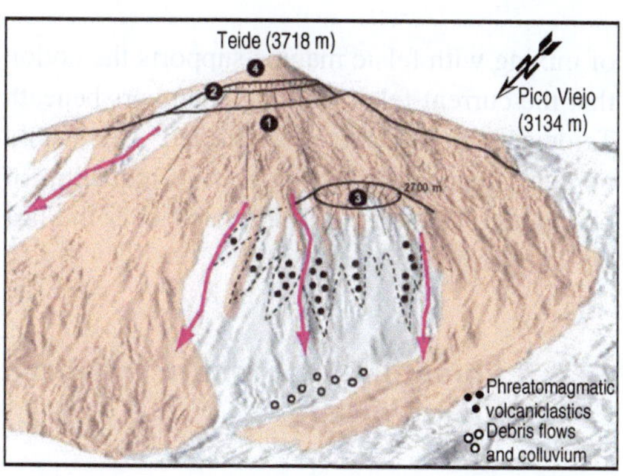

Fig. 12.9 Outcrops of the volcaniclastic and associated deposits from the Las Calvas del Teide phreatomagmatic eruption. *1* Lavas Negras (ca. 1.2 ka). *2* "Old" Teide (>30 ka). *3* Las Calvas del Teide inferred phreatomagmatic vent. *4* Summit cone of the Lavas Negras eruption

Internal depositional features including parallel and low-angle cross-lamination, scour and fills and bomb sags, have been identified in the fine grained layers. Coarse-grained beds show erosive bases, a clast-supported matrix, and a diffuse clast-size distribution with a tendency towards normal grading of the size and number of clasts. Both lithofacies show angular and poorly vesiculated clast types, some of them obsidian.

Of these clasts, del Potro et al. (2009) distinguished two types with similar characteristics to the interbedded lava flows, which they interpreted to be a co-magmatic feature. These clasts are affected by pervasive networks of microfractures, which are typical of magma–water interaction. Del Potro and co-workers suggested that the characteristic induration of the thin laminated matrix-supported units could have been derived from near-syn-depositional alteration of the vitric components by means of a geo-autoclave-type mechanism (Gottardi 1989; Pérez-Torrado et al. 1995). Alteration to zeolite facies would have occurred during the initial cooling of the pyroclastic deposit, implying the involvement of a significant hydrous component during the eruption.

The observed features suggest that these deposits have formed as phreatomagmatic surges. The alternating fine and coarse-grained laminations and the interlayered lava flows point to changing conditions in the magma-water interaction at different eruptive stages, and imply the proximity of the emission vent (c.f. Clarke et al. 2009). The most likely eruptive scenario involves the opening of a lateral vent during a period when the stratocone was covered by a thick cap of snow or ice (Fig. 12.11), a setting first proposed by Pérez Torrado et al. (2004, 2006) and later confirmed by del Potro et al. (2009). Pérez Torrado et al. (2004) interpreted the bulge observed on the mid-northern flank of Teide as being a reflection of the phreatomagmatic vent. An earlier interpretation by Ablay and Martí (2000), however, related the bulge to the scarp of the Icod lateral collapse.

The Calvas del Teide phreatomagmatic deposits change down-slope into debris flow and colluvium deposits (see Fig. 12.9), although the actual transition is covered by later lavas and scree deposits. Therefore, it is currently impossible to determine if a continuous lateral change from pyroclastic to debris flow facies exists. Alternatively, the latter may have formed from erosion of the former (see Fig. 12.8e and f).

Finally, an aspect to analyse is the presence of a N–S trending, sub-vertical dyke intruding the basal flows (Fig. 12.12a) of the stratigraphic sequence. The dyke feeds a thin, scoriaceous flow with abundant olivine phenocrysts (Fig. 12.12b). This mafic intrusion in a highly differentiated stratocone supports the idea that magmas constructing Teide volcano derive from a common deep source that also feeds rift zone volcanism (see Chap. 7). On the other hand, the arrival of mafic magma to the flank of Teide without signs

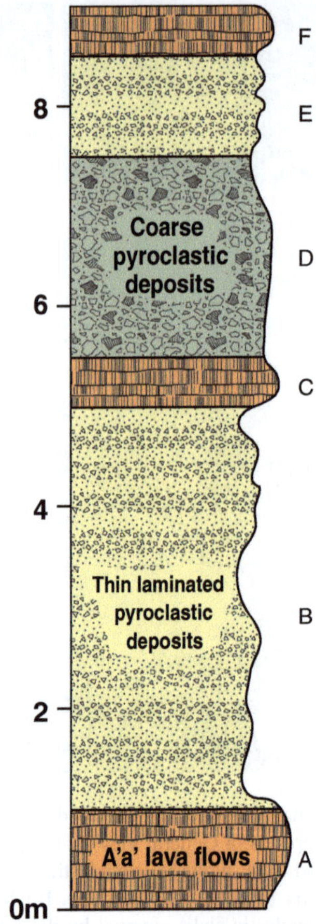

Fig. 12.10 Schematic stratigraphic column of the Las Calvas del Teide area: volcaniclastic deposits (*B*, *D*, *E*) are intercalated with phonolitic lava flows of "Old" Teide (*A*, *C*, *F*) (modified from Pérez-Torrado et al. 2004)

of mixing with felsic magma supports the notion that the current felsic holding chambers beneath Teide are small or largely solidified. This would be consistent with the progressive reduction of phonolitic eruptions of Teide itself over the last 30 ka (Carracedo et al. 2007).

12.4.2 Phreatomagmatism in the Pico Viejo Volcano

Deposits derived from a phreatomagmatic eruption, similar in many aspects to those of the Calvas del Teide, outcrop inside the Pico Viejo crater and mantle the summit slopes of the volcano (Fig. 12.13). The area covered by this Pico

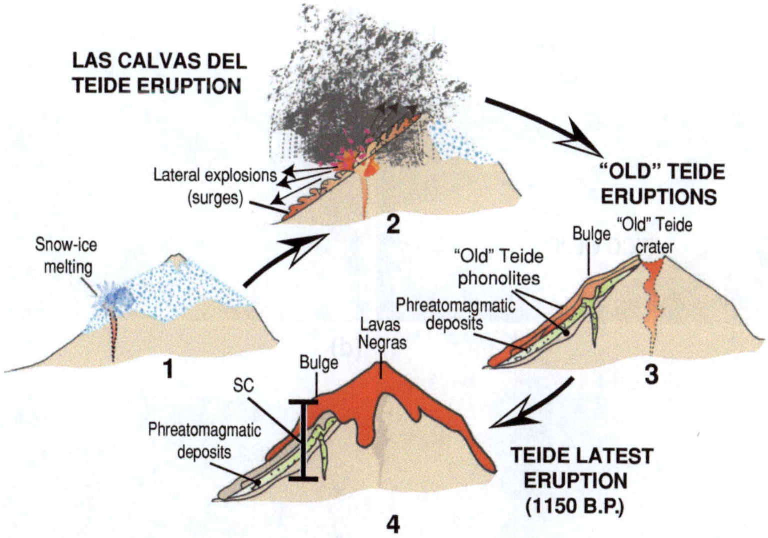

Fig. 12.11 Sketch depicting the probable cause of the Las Calvas del Teide phreatomagmatic event—by the interaction of magma with snow-ice meltwater. SC, stratigraphic column in Fig. 12.10 (modified from Carracedo et al. 2008)

Fig. 12.12 Basaltic dyke cutting a phonolitic flow of "Old" Teide (>30 ka), a relationship that implies the ultimate feeder system to Teide is the same as that of the rift zones

Viejo phreatomagmatic eruption is notably smaller than that of Calvas del Teide (see Fig. 12.7). The relative stratigraphic position of the Pico Viejo deposits indicates a younger age than that of the Calvas del Teide deposits. Ablay and Martí (2000) relate part of this formation to the eruption of Roques Blancos, dated at 1790 ± 60 yr BP (Carracedo et al. 2007).

A detailed study of these deposits was carried out by Ablay and Martí (2000). They describe a 40–80 cm thick layer of weakly indurated, cross-laminated, red-grey surge deposits (Fig. 12.14), dominated by 1–10 mm sized juvenile fragments of plagioclase-basanite scoriae in a heterolithic matrix. Above this lies a scoriaceous plagioclase-basanite spatter and a short-flowing a'a lava originating from the Pico Viejo crater rim (L in Fig. 12.13c). A final layer of unconsolidated explosion breccia (Fig. 12.13d), composed of angular rock fragments up to 1 m across and without juvenile components, can be correlated with the formation of an explosion pit in the SW part of the crater floor (see Fig. 12.13a–c). The onset of this phreatomagmatic eruption may have been caused, as in the Calvas del Teide event, by the interaction of magma with snow and ice

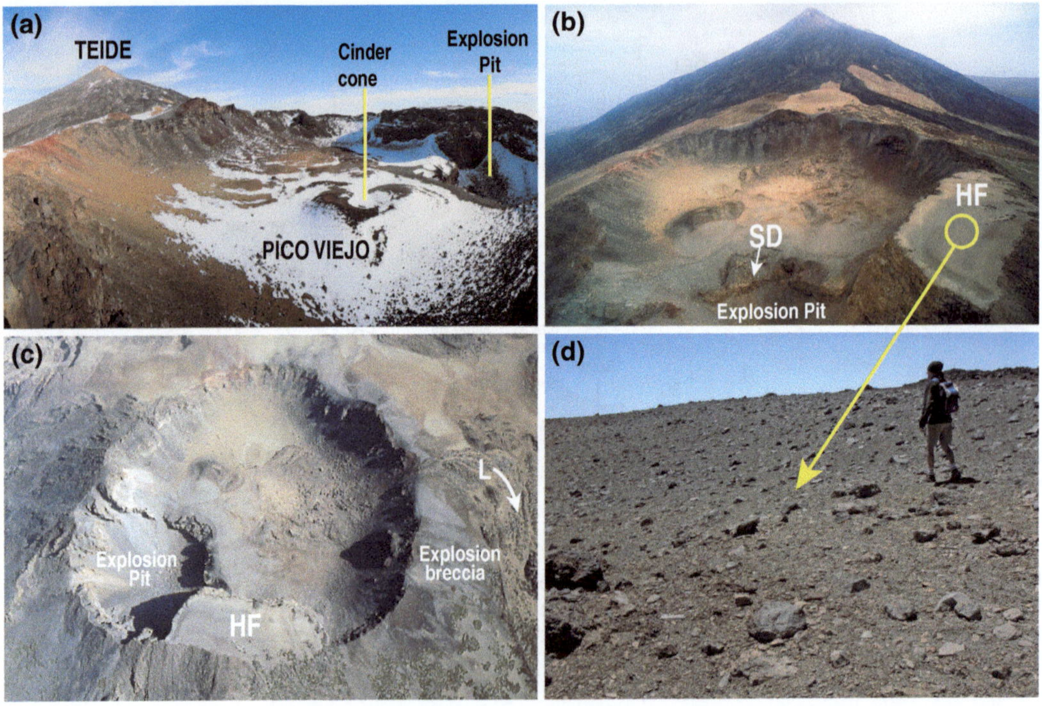

Fig. 12.13 Features of the phreatomagmatic eruption of Pico Viejo. **a** View from the western rim of Pico Viejo crater. **b** Southern part of the Pico Viejo crater wall showing the flat top of the crater-filling sequence. SD: surge deposits. Sub-horizontal flows (*HF*) are covered with explosion breccia deposits from the phreatomagmatic eruption. **c** Aerial view of the Pico Viejo crater showing the extensive explosion breccia deposits (grey colour). A lava flow (L) on the eastern side post-dates the phreatomagmatic event. **d** Close-up view of the explosion breccia deposit mantling the sub-horizontally bedded lava flows that partially fill the Pico Viejo crater (*HF* in **b** and **c**)

Fig. 12.14 Close-up view of pyroclastic surge deposits from the Pico Viejo phreatomagmatic eruption outcropping in the wall of the explosion pit (SD in Fig. 12.13b), which are similar to those mantling the sub-horizontally bedded lava flows (HF Fig. 12.13 **b** and **c**)

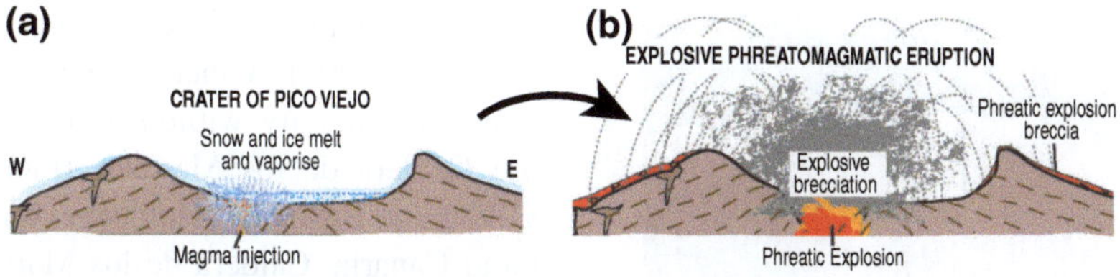

Fig. 12.15 Sketch illustrating the probable cause of the Pico Viejo phreatomagmatic event—by the interaction of magma with snow-ice meltwater. **a** Approaching magma heats the crater floor and melts snow producing considerable amounts of meltwater. **b** Water filters through the highly porous volcanic lapilli and fractured lavas, interacting with the shallow-level magma to cause phreatic explosions (modified from Carracedo et al. 2008)

accumulated inside the Pico Viejo crater (Fig. 12.15). The presence of a permanent aquifer or lake in this area is improbable due to the very high porosity of the volcanic summit deposits.

12.4.3 Phreatomagmatism in the Canary Islands

Besides the outlined examples at Teide and Pico Viejo, examples of phreatomagmatic eruptions exist on all the Canary Islands (Fig. 12.16), with felsic phreatomagmatic events being generally less frequent. Local place names in the archipelago frequently refer to the characteristic white and yellow tones of hydrothermally altered basaltic tuffs (e.g. Caldera Blanca, Lanzarote; Montaña Amarilla, Tenerife), and to wider craters and lower aspect ratio cones of such composition (e.g. Montaña Escachada, flattened mountain, Tenerife). The morphology and size of these phreatomagmatic cones is varied, with numerous examples of tuff-cones, maars, and tuff-rings, particularly on the littoral platforms and outcrops in marine cliffs (Carracedo et al. 2001). The majority of these eruptions are triggered by direct interaction with sea water (e.g. Montaña Amarilla, Tenerife; La Caldereta, La Palma; Montaña Escachada, Tenerife, etc.). In this type of eruption the source of water is unlimited, and the eruption remains phreatomagmatic throughout (e.g. Montaña Amarilla). However, the transition from phreatomagmatic to purely volcanic mechanisms during a single eruption is also observed in some examples [e.g. Montaña Los Erales, (Clarke et al. 2005, 2009)]. The eruption of La Caldereta, a large tuff cone near Santa Cruz on La Palma, is an example of this, as during the final stages, a small Strombolian vent formed with lava flows nested in the centre of the volcano. Another instance is the El Golfo vent on Lanzarote, where magma encountered a finite amount of water at its initial stage of eruption and variable additions of water during subsequent eruptive styles.

Further recorded cases of volcanic activity that have been influenced by magma-water interaction are summarised in Table 12.1. The majority of these phreatomagmatic eruptions are basaltic, but some spectacular felsic events have also occurred in Tenerife [e.g. the phonolitic-trachytic Caldera del Rey twin caldera (De la Nuez et al. 1993)]. Sea water is assumed to be the primary hydrological source, e.g. Caldera Blanca and El Golfo (Lanzarote), Montaña Goteras and La Caldereta (La Palma), La Isleta (Gran Canaria) and Montaña Amarilla and Montaña Escachada (Tenerife). Yet, phreatomagmatic activity is not explained by seawater interaction in all cases, as examples exist far inland as well, such as the Hoyo Negro eruption in La Palma in 1949 (Klügel et al. 1999; White and Schmincke 1999). The Hoyo Negro vent complex on La Palma is located significantly above sea level (at 1,880 m elevation) and extends 400 m along the north–south trending rift that runs along the centre of the southern half of the island. Here, the onset of the 1949 San Juan eruption was characterised by phreatomagmatic activity emanating from a series of vents around the Duraznero crater (Klügel et al. 1999; White and Schmincke 1999). At such altitude, a direct or indirect seawater

Fig. 12.16 Examples of phreatomagmatic eruptions in the Canary Islands. **a** Caldera Blanca (Lanzarote). Seawater is assumed to be the primary water source for this eruption, which occurred on a shallow coastal platform. **b** The phreatomagmatic eruption of Hoyo Negro (1949, La Palma). The vent is located far inland, at the summit of the Cumbre Vieja rift, thus groundwater or surface water was most probably involved in the eruption. *Note* the height of the phreatomagmatic column can be estimated by comparison to the brave person venturing dangerously close to the eruptive vent. **c** Los Erales, a strombolian vent that began with a hydrovolcanic explosive eruption (*1* in the figure), and ended with an entirely dry solely magmatic strombolian eruptive style (*3*), after a transitional stage during which the water source became progressively exhausted (*2*)

influence appears most unlikely. In this scenario, groundwater or surface water may have been present due to ephemeral pools that formed on the surface of older impermeable ash deposits, hence the name of this area 'Llanos del Agua', referring to the presence of water (Carracedo and Day 2002). Other examples of phreatomagmatic activity without seawater interaction are Caldera de los Marteles (Pleistocene) and Caldera de Bandama (Holocene) on the island of Gran Canaria. Caldera de los Marteles, with its crater base located at 1,458 m above sea level, formed within a steep-sided valley (Barranco de Guayadeque), which suggests that phreatomagmatism was induced when rifting along fissures commenced and surface water gained access to the magmatic heat source (Schmincke et al. 1974). On the other hand, Caldera de Bandama, located along a watershed boundary area with its crater base at 217 m above sea level and about 150–200 m depth to the pre-eruption surface, is inferred to have formed by magmatic interaction with shallow groundwater sourced in a detritic layer interbedded in the volcanic substratum (Rodriguez-Gonzalez et al. 2012).

Despite a high abundance of felsic rocks in the TVC, explosive magmatic eruptions are scarce and produce mainly low-volume deposits. One exception is the Montaña Blanca subplinian event, however. Extensive volcaniclastic deposits in the TPV derive from explosive events related to phreatomagmatic eruptions involving both felsic (Las Calvas del Teide) and mafic (Pico Viejo) magmas. Effusive eruptions, mainly of strombolian type, are the most frequent eruptive style observed in the TVC, again with mafic and felsic magmas displaying this mode of eruption.

Araña et al. (2000), in proposing a surveillance network for the island of Tenerife, consider the main volcanic hazards to be lava flows and ash fallout, with little reference to the potential of phreatomagmatic eruptions. A clear record of recent and historic eruptions that display a hydrous influence exists both at Teide and on the other Canary Islands (Table 12.1). The realisation of phreatomagmatism as an uncertain variable in an otherwise low-explosivity eruptive regime (e.g. on the rifts), increases the need for improved understanding of this eruptive style in the Canary Islands and needs to be taken into account for the evaluation of societal vulnerability and risk assessment (see Chap. 14).

Table 12.1 Examples of historic and recent phreatomagmatic eruptions in the Canary Islands

Eruption	Island	Date/age	Duration (days)	Composition	Eruptive style	Proximal to shoreline
Phreatomagmatic eruptions						
Escachada	Tenerife	Pleistocene	n.k.	Basaltic	Phreatomagmatic	Yes
Montaña Amarilla	Tenerife	Pleistocene	n.k.	Basaltic	Phreatomagmatic	Yes
Caldera del Rey	Tenerife	Quaternary	n.k.	Phonolitic/ trachytic	Phreatomagmatic	Yes
Teide (Calvas del Teide)	Tenerife	Pleistocene	n.k.	Phonolitic	Phreatomagmatic	No
Pico Viejo crater	Tenerife	Holocene	n.k.	Basanitic	Phreatomagmatic	No
La Caldereta	La Palma	Pleistocene	n.k.	Basaltic	Phreatomagmatic	Yes
Montaña Goteras	La Palma	Holocene	n.k.	Basaltic	Phreatomagmatic	No
Montaña Amarilla	La Graciosa-Lanzarote	Quaternary	n.k.	Basaltic	Phreatomagmatic	Yes
Caldera Blanca	Lanzarote	Quaternary	n.k.	Basaltic	Phreatomagmatic	Yes
Caldera de Los Marteles	Gran Canaria	Quaternary	n.k.	Basaltic	Phreatomagmatic	No
Caldera de Bandama	Gran Canaria	Holocene	n.k.	Basanitic	Phreatomagmatic	No
La Isleta	Gran Canaria	Quaternary	n.k.	Basaltic/ basanitic	Phreatomagmatic	Yes
Mixed eruptions						
El Golfo	Lanzarote	Quaternary	n.k.	Basaltic	Phreatomagmatic opening phase-Strombolian	Yes
Los Erales	Tenerife	Quaternary	n.k.	Basaltic	Phreatomagmatic opening phase-Strombolian	No
San Juan (Hoyo del Banco, Duraznero, Hoyo Negro)	La Palma	1949	38	Basanitic/ tephritic	Phreatomagmatic opening phase-Strombolian	No
Fuencaliente	La Palma	1677	66	Basanitic	Strombolian with phreatomagmatic phase	No
El Charco	La Palma	1712	56	Basanitic/ tephritic	Strombolian-phreatomagmatic	No
Tinguaton Tao	Lanzarote	1824	90	Basaltic	Strombolian with final phreatomagmatic phase	No

n.k. not known

Sources Klügel et al. (1999); Carracedo et al. (2001), (2007); Carracedo and Day (2002); Clarke et al. (2009); Rodriguez-Gonzalez et al. (2009)

References

Ablay GJ, Martí J (2000) Stratigraphy, structure and volcanic evolution of the Pico Teide-Pico Viejo formation, Tenerife, Canary Islands. J Volcanol Geotherm Res 103:175–208

Ablay GJ, Ernst GGJ, Martí J, Sparks RSJ (1995) The ~2 ka subplinian eruption of Montaña Blanca, Tenerife. Bull Volcanol 57:337–355

Ablay GJ, Carroll MR, Palmer MR, Martí J, Sparks RSJ (1998) Basanite-Phonolite lineages of the Teide-Pico Viejo volcanic complex, Tenerife, Canary Islands. J Petrol 39:905–936

Albert-Beltran JF, Araña V, Diez JL, Valentin A (1990) Physical-chemical conditions of the Teide volcanic system (Tenerife, Canary-Islands). J Volcanol Geotherm Res 43:321–332

Araña V, Barberi F, Ferrara G (1989a) El Complejo volcanico del Teide-Pico Viejo. In: Araña V, Coello J (eds) Los Volcanes y la Caldera del Parque Nacional del Teide (Tenerife, Islas Canarias), ICONA, Madrid, pp 101–126

Araña V, Aparicio A, Garcia Cacho L, Garcia Garcia R (1989b) Mezcla de magmas en la región central de Tenerife. In: Araña V, Coello J (eds) Los Volcanes y La Caldera del Parque Nacional del Teide (Tenerife, Islas Canarias). Serie Técnica, ICONA, pp 269–298

Araña V, Felpeto A, Astiz M, Garcia A, Ortiz R, Abella R (2000) Zonation of the main volcanic hazards (lava flows and ash fall) in Tenerife, Canary Islands. A proposal for a surveillance network. J Volcanol Geotherm Res 103:377–391

Boulesteix T, Hildenbrand A, Gillot P-Y, Soler V (2012) Eruptive response of oceanic islands to giant landslides: new insights from the geomorphologic evolution of the Teide-Pico Viejo volcanic complex (Tenerife, Canary Islands). Geomorphology 138:61–73

Büttner R, Dellino P, Zimanowski B (1999) Identifying modes of magma/water interaction from the surface features of ash particles. Nature 401:688–690

Carracedo JC, Rodríguez Badiola E, Guillou H, Paterne M, Scaillet S, Pérez Torrado FJ, Paris R, Rodríguez González A, Socorro S (2008) El Volcán Teide—Volcanología, Interpretación de Pasajes e Iterinarios Comentados. Caja Generál de Ahorros de Canarias, Santa Cruz de Tenerife

Carracedo JC, Day S (2002) Canary Islands. Terra Publishing, Hertfordshire, England

Carracedo JC, Rodríguez Badiola E, Guillou H, De La Nuez J, Pérez Torrado FJ (2001) Geology and volcanology of La Palma and El Hierro (Canary Islands). Estud Geol 57:175–273

Carracedo JC, Rodríguez Badiola E, Guillou H, Paterne M, Scaillet S, Pérez Torrado FJ, Paris R, Fra-Paleo U, Hansen A (2007) Eruptive and structural history of Teide volcano and rift zones of Tenerife, Canary Islands. Geol Soc Am Bull 119:1027–1051

Cas RAF, Wright JV (1987) Volcanic successions—modern and ancient. Allen and Unwin Ltd, London

Clarke H, Troll VR, Carracedo JC, Byrne K, Gould R (2005) Changing eruptive styles and textural features from phreatomagmatic to strombolian activity of basaltic littoral cones: Los Erales cinder cone, Tenerife, Canary Islands. Estud Geol 61:121–134

Clarke H, Troll VR, Carracedo JC (2009) Phreatomagmatic to strombolian eruptive activity of basaltic cinder cones: Montana Los Erales, Tenerife, Canary Islands. J Volcanol Geotherm Res 180:225–245

De la Nuez J, Alonso JJ, Quesada ML, Macau MD (1993) Edificios hidromagmáticos costeros de Tenerife (Islas Canarias). Rev Soc Geol Esp 6:47–59

del Potro R, Pinkerton H, Hüerlimann M (2009) An analysis of the morphological, geological and structural features of Teide stratovolcano, Tenerife. J Volcanol Geotherm Res 181:89–105

Edgar CJ, Wolff JA, Olin PH, Nichols HJ, Pittari A, Cas RAF, Reiners PW, Spell TL, Martí J (2007) The late Quaternary Diego Hernandez formation, Tenerife: volcanology of a complex cycle of voluminous explosive phonolitic eruptions. J Volcanol Geotherm Res 160:59–85

Fisher RV, Schmincke H-U (1984) Pyroclastic rocks. Springer-Verlag, Berlin

García O, Martí J, Aguirre G, Geyer A, Iribarren I (2011) Pyroclastic density currents from Teide-Pico Viejo (Tenerife, Canary Islands): implications for hazard assessment. Terra Nova 23:220–224

Gottardi G (1989) The genesis of zeolites. Eur J Mineral 1:479–487

Huertas MJ, Arnaud NO, Ancochea E, Cantagrel JM, Fúster JM (2002) Ar-40/Ar-39 stratigraphy of pyroclastic units from the Canadas volcanic edifice (Tenerife, Canary Islands) and their bearing on the structural evolution. J Volcanol Geotherm Res 115:351–365

Klügel A, Schmincke H-U, White JDL, Hoernle KA (1999) Chronology and volcanology of the 1949 multi-vent rift-zone eruption on La Palma (Canary Islands). J Volcanol Geotherm Res 94:267–282

Lorenz V (1987) Phreatomagmatism and its relevance. Chem Geol 62(1–2):149–156

Márquez A, López I, Herrera R, Martín-González F, Izquierdo T, Carreño F (2008) Spreading and potential instability of Teide volcano, Tenerife, Canary Islands. Geophys Res Lett 35:L05305. doi: 10.1029/2007GL032625

Martí J, Geyer A, Andujar J, Teixido F, Costa F (2008) Assessing the potential for future explosive activity from Teide-Pico Viejo stratovolcanoes (Tenerife, Canary Islands). J Volcanol Geotherm Res 178:529–542

Morrissey M, Zimanowski B, Wohletz K, Büttner R (2000) Phreatomagmatic fragmentation. In: Sigurdsson H (ed) Encyclopedia of Volcanoes. Academic Press, San Diego, pp 431–445

Pérez-Torrado FJ, Carracedo JC, Mangas J (1995) Geochronology and stratigraphy of the Roque Nublo cycle, Gran Canaria, Canary Islands. J Geol Soc (London, UK) 152:807–818

Pérez Torrado FJ, Carracedo JC, Paris R, Hansen A (2004) Descubrimiento de depósitos freatomagmáticos en las laderas septentrionales del estratovolcán Teide (Tenerife, Islas Canarias): relaciones estratigráficas e implicaciones volcánicas. Geotemas 6:163–166

Pérez Torrado FJ, Carracedo JC, Paris R, Rodríguez Badiola E, Hansen A (2006) Erupciones freatomagmáticas del complejo volcánico Teide-Pico Viejo. In: Carracedo JC (ed) Los Volcanes del Parque Nacional del Teide. Serie Técnica, Organismo Autónomo Parques Nacionales, Ministerio de Medio Ambiente, Madrid, pp 345–357

Rodríguez-Badiola E, Pérez-Torrado FJ, Carracedo JC, Guillou H (2006) Petrografía y Geoquímica del edificio volcánico Teide-Pico Viejo y las dorsales noreste y noroeste de Tenerife. In: Carracedo JC (ed) Los volcanes del Parque Nacional del Teide/El Teide. Pico Viejo y las dorsales activas de Tenerife. Naturaleza y Parques Nacionales-Serie Técnica, Organismo Autónomo Parques Nacionales Ministerio de Medio Ambiente, Madrid, pp 129–186

Rodriguez-Gonzalez A, Fernandez-Turiel JL, Perez-Torrado FJ, Hansen A, Aulinas M, Carracedo JC, Gimeno D, Guillou H, Paris R, Paterne M (2009) The Holocene volcanic history of Gran Canaria island: implications for volcanic hazards. J Quater Sci 24:697–709

Rodriguez-Gonzalez A, Fernandez-Turiel JL, Perez-Torrado FJ, Paris R, Gimeno D, Carracedo JC, Aulinas M (2012) Factors controlling the morphology of monogenetic basaltic volcanoes: the Holocene volcanism of Gran Canaria (Canary Islands, Spain). Geomorphology 136:31–44

Schmincke H-U, Brey G, Staudigel H (1974) Craters of phreatomagmatic origin on Gran Canaria, Canary Islands. Naturwissenschaften 61:125–127

Walker GPL (1973) Lengths of lava flows. Philos Trans R Soc London, Ser A 274:107–118

White JDL, Schmincke H-U (1999) Phreatomagmatic eruptive and depositional processes during the 1949 eruption on La Palma (Canary Islands). J Volcanol Geotherm Res 94:283–304

Wiesmaier S, Deegan F, Troll V, Carracedo J, Chadwick J, Chew D (2011) Magma mixing in the 1100 AD Montaña Reventada composite lava flow, Tenerife, Canary Islands: interaction between rift zone and central volcano plumbing systems. Contrib Mineraol Petrol 162:651–669

Wohletz KH (1983) Mechanisms of hydrovolcanic pyroclast formation—grain-size, scanning electron-microscopy and experimental studies. J Volcanol Geotherm Res 17:31–63

Wolff JA, Storey M (1983) The volatile component of some pumice-forming alkaline magmas from the Azores and Canary-Islands. Contrib Mineraol Petrol 82:66–74

Geophysical Investigations of the Teide Volcanic Complex 13

Vicente Soler-Javaloyes and Juan Carlos Carracedo

Abstract

While our geological understanding of Tenerife and that of the TVC has experienced significant advances in the last decades, comprehensive geophysical investigations into Teide's volcanism are still in their infancy. Geophysical data, however, are essential to the identification of deep features and processes relevant to the current state of activity of the TVC, such as the presence or absence of active magma chamber(s), their precise location, and their potential eruptible volume(s). This chapter gives an up-to-date summary of the currently ongoing geophysical studies concerning the TVC complex, and addresses the P–T conditions of the current Teide magma chamber system derived from gas emission data, gravimetry, aeromagnetics, seismicity, and from ground deformation studies.

13.1 Introduction

The geological understanding of the Teide Volcanic Complex, the most recent phase of volcanic activity of Tenerife, has significantly advanced over the last decade (e.g., Ablay et al. 1998; Carracedo et al. 2007; Wiesmaier et al. 2011). An even more complete understanding of this volcanic complex requires more precise knowledge of its structure, and crucially, establishing the presence or absence and exact location of any active magmatic chamber(s), thus making geophysical investigations essential. Low eruptive and seismic activity (Fig. 13.1 and Chaps. 7 and 8) has, however, hindered geophysical interest in Teide for a considerable time.

The analysis of published geophysical data shows that the main research efforts have traditionally been focussed on a few specific topics. The first modern geophysical studies in the Canary region gave priority to the analysis of the continental vs oceanic nature of the crust underlying the multiple islands composing the archipelago (MacFarlane and Ridley 1968; Rothe and Schmincke 1968; Dash and Bosshard 1969; Bosshard and MacFarlane 1970), a topic still of interest today (e.g., Geyer and Martí 2010, 2011; Carracedo et al. 2011). Later, Banda et al. (1981)

V. Soler-Javaloyes (✉)
Estación Volcanológica de Canarias, IPNA-CSIC, La Laguna, Tenerife, Spain

J. C. Carracedo
Departamento de Física (GEOVOL), Universidad de Las Palmas de Gran Canaria, Las Palmas de Gran Canaria, Canary Islands, Spain

J. C. Carracedo and V. R. Troll (eds.), *Teide Volcano*, Active Volcanoes of the World, DOI: 10.1007/978-3-642-25893-0_13, © Springer-Verlag Berlin Heidelberg 2013

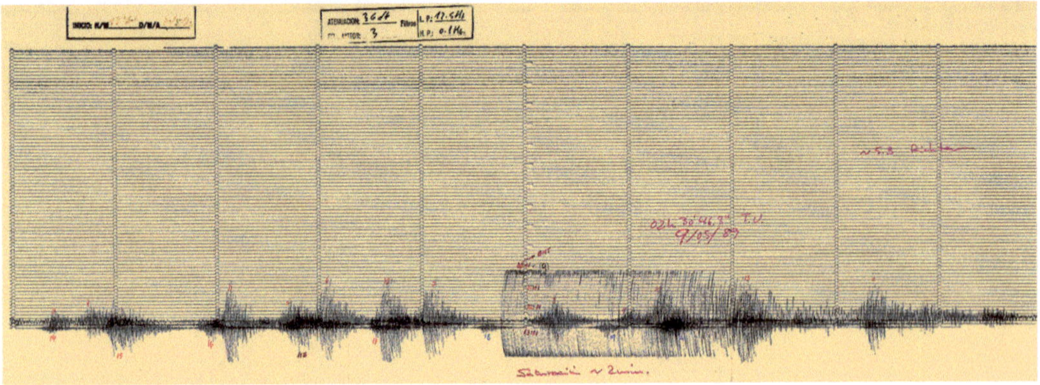

Fig. 13.1 The May 1989 earthquake, located off the south coast of Tenerife, with a magnitude of 5.3 on the Richter scale, is the biggest earthquake recorded in the Canary Islands to date

provided compelling evidence in favour of an oceanic crust underlying the entire Canarian archipelago. In the following years marine geophysical studies explored the ocean floor around the Canaries, identifying magnetic anomaly lineations and other crustal structures. (Uchupi et al. 1976; Roeser 1982; Dañobeitia and Colette 1989; Holik and Rabinowitz 1991; Roest et al. 1992; Watts 1994; Watts et al. 1997; Canas et al. 1998; Urgelés et al. 1998; Martínez and Buitrago 2002).

In the wake of the Mount St. Helens eruption in 1980, many of the marine geophysical and geological surveys in the Canaries began to focus on gathering seismic and sonar data to evaluate the number and characteristics of giant landslide deposits present on the flanks of the islands (Watts and Masson 1995; Masson 1996; Urgelés et al. 1997, 1999; Gee et al. 2001; Masson et al. 2001; Watts and Masson 2001).

Onshore surveys on Tenerife were largely centred on the task of gathering information to support or contrast the different models for the origin of the Las Cañadas Caldera, the two main schools of thought being on the one side those that favour a "Krakatau style" vertical collapse caldera (Araña 1971), with those on the other supporting a lateral gravitational collapse mechanism (MacFarlane and Ridley 1968). García et al. (1989) and Pous et al. (2002) gave priority in their studies to the morphology of the basal structure of the Las Cañadas Caldera, which they used as the best approach to understand its origin. They analysed the internal structure of the area by means of magnetotelluric (MT) profiles measured on the northern and western flanks of Las Cañadas Volcano.

For the remainder of the chapter, however, we shall summarise the work pertaining to the Teide-Pico Viejo (TPV) complex. We will address, in fumarolic sequence, the fumarolic activity, gravimetry, aeromagnetics, seismicity and ground deformation at the complex to gain a better understanding of the current state of activity.

13.2 Resolving the Current P–T Conditions of the Teide Magma Chamber Using Gas Emission Data

The island of Tenerife shows spectacular evidence of fumarolic activity and hydrothermal alteration related to the past emission of volcanic gas (i.e., Los Azulejos, Fig. 13.2a). Volcanic gas emission and elevated temperatures are frequently found inside *galerías*, posing a significant hazard to miners and scientists alike. (Fig. 13.2b). The analysis of fumarolic activity at the summit crater of Teide (Fig. 13.2c–f) is used by several authors to obtain information on the characteristics of the magma chamber system beneath Teide Volcano. This involves determination of the

Fig. 13.2 Different expressions of fumarolic activity in the TVC. **a** Los Azulejos, pre-Teide (Las Cañadas Volcano) hydrothermal fluid alteration is clearly visible in the field displaying *bluish*, *greenish*, and *yellowish* colours. Analcime and clay minerals from the smectite and illite groups, together with iron and manganese oxides are widely found in this area. **b** *Galerías* in Tenerife and **c** Fumaroles in the summit crater of Teide Volcano. **d** Hot steam and SO_2 producing yellow sulphur deposits at the summit crater of Teide. **e** Aerial view of the medieval summit crater of Teide Volcano showing hydrothermal alteration (*white*). **f** Close-up view of the rim of the "Old Teide" volcano with clear signs of hydrothermal alteration and sulphur sublimation (*yellow patches*)

temperature and pressure at which the gas fractions formed, thus allowing assessment of the current activity of the magmatic system.

According to Albert-Beltrán et al. (1989), the gas geochemistry and isotopic ratios of Teide fumaroles correspond to temperature and

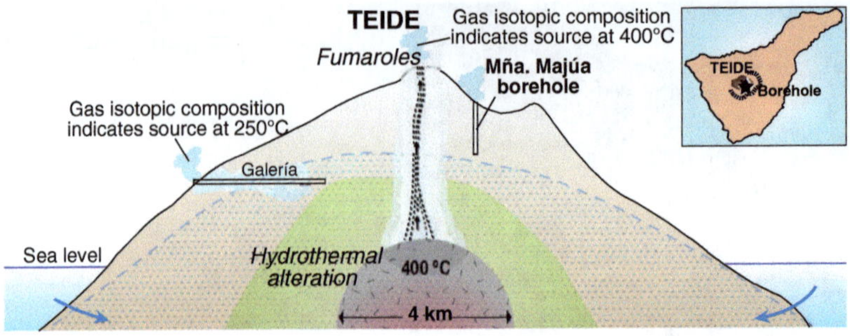

Fig. 13.3 Size, depth and temperature defined for the magma chamber of Teide. Thermodynamic model modified after Diez and Albert-Beltrán (1989)

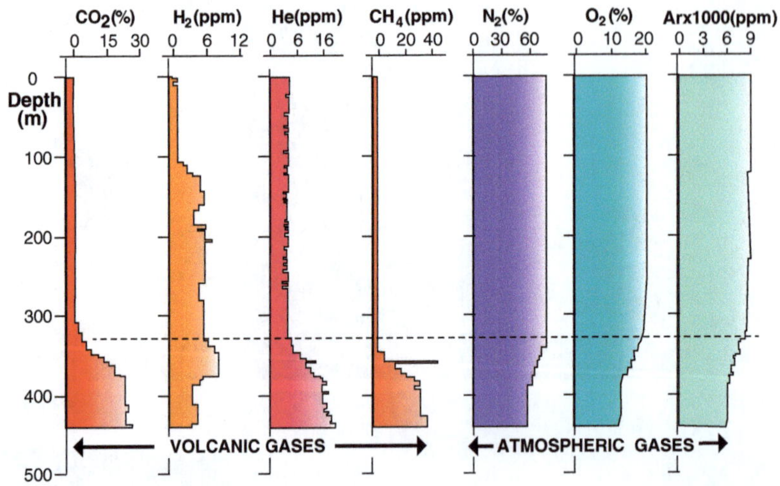

Fig. 13.4 Changes in gas composition with depth, measured in 2002 in the 500 m deep Mña. Majúa borehole, at the southern foot of Teide Volcano using a portable mass spectrometer (unpublished data, courtesy of M. Zimmer and J. Erzinger, GFZ Potsdam, Germany)

pressure conditions of 400 °C and 200–300 bar. The isotopic composition of gases measured within *galerías* (Albert-Beltrán et al. 1990), however, indicates a significantly lower temperature (250 °C), probably due to hydrothermal alteration in the underlying hot rock (Fig. 13.3).

Diez and Albert-Beltrán (1989) proposed a thermodynamic model that advocated a 400 ky-old, 4 km-diameter (30 km^3) spherical magma chamber at 400 °C, with its top located approximately at sea level. These studies emphasise the fact that 400 °C is the temperature of solid rock. The analysis of gas emissions therefore points to the absence of a shallow magma body capable of erupting or even causing microseismicity.

More recently, a deep borehole drilled inside the Las Cañadas Caldera to monitor water table changes allowed analysis of volcanic and atmospheric gases to a depth of 450 m into the volcanic sequence (Fig. 13.4). Through the analysis of CO_2 degassing at Teide Volcano crater, Melián et al. (2012) reported an increase in magmatic gases and total diffuse CO_2 associated with the main peak of seismic activity during 2004. They argued that "*associated temporal changes in seismic activity and magmatic degassing indicate that geophysical and fluid geochemistry signals are related in this system*".

13.3 Gravity Modelling

The first, and probably the most influential, gravity survey of Tenerife was carried out by MacFarlane and Ridley (1968). The Bouguer anomaly map they compiled shows a three-pointed star shape with a well-marked positive

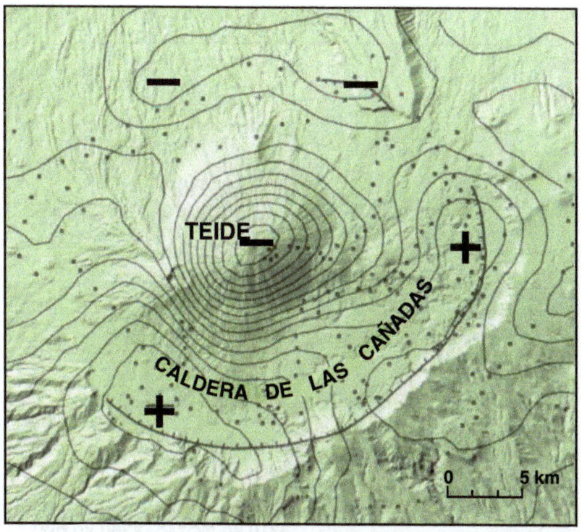

Fig. 13.5 Bouguer anomaly map of the Teide Volcano and the Las Cañadas Caldera. Note the presence of positive anomalies in the Las Cañadas Caldera rim, particularly high (over 60 mGal) in its western part (Boca de Tauce), interpreted as connected to a 10 km deep plutonic body. Surficial (3–4 km deep), low-density negative anomalies appear associated with Teide Volcano and the northern wall of the Caldera (after Vieira et al. 1989)

gravity ridge trending northeast and less-pronounced gravity ridges trending northwest and south (see Fig. 4.3). A marked gravity maximum of approximately 100 mgal magnitude is located slightly west of the Boca de Tauce escarpment. Comparing the Bouguer anomaly of Tenerife with those observed on the Hawaiian Islands, MacFarlane and Ridley (1968) interpreted the gravity ridges as reflecting a high concentration of dykes (i.e. volcanic rift zones). In their model, the island body is heavily intruded to form a high density (3.2 g/cm^3) conical core of mantle-like material, extending from the Moho to within ~4 km of the surface. The overlying, 2.9 g/cm^3, 1.5 km thick layer would represent derivative rocks produced by fractionation.

It is worth noting that in their gravimetric analysis of Tenerife, MacFarlane and Ridley (1968) were the first to point out that the Portillo and Boca de Tauce escarpments may "*not represent the bounding walls of a Krakatoan caldera but the head region of a massive landslide area, partially filled with the later volcanics from Pico Viejo and Teide*". Disregarding the ideas of MacFarlane and Ridley (1968), Araña (1971) concluded his study of the pre-Teide Las Cañadas Volcano proposing a vertical collapse for the origin of the Las Cañadas Caldera, created in a massive caldera-forming volcanic eruption. This idea started a long-lasting controversy, until Watts and Masson (1995) identified the deposits from a giant landslide north of Tenerife using swath bathymetry side-scan sonar. This they suggested resulted from the formation of the Las Cañadas Caldera (the Icod giant landslide).

In contrast, the first detailed gravity study to focus on the central part of Tenerife defined a negative Bouguer anomaly coinciding with the Teide-Pico Viejo volcanic complex itself (Fig. 13.5; Vieira et al. 1989). Araña et al. (2000) interpreted their own gravity data by defining three zones at different depths, approximately coinciding with inversion sections of the gravity model: a deep zone (below 12 km b.s.l.), corresponding to the basement of the island; an intermediate zone (12 km b.s.l. to 1 km a.s.l.), forming the submarine shield edifice; and the surficial zone (above 1 km a.s.l.), matching the Las Cañadas Volcano and the Teide Volcanic Complex.

More recently, Ablay and Kearey (2000) carried out a detailed gravity study of Tenerife to analyse the internal structure of the central part of the island and its active magma system. The maximum Bouguer anomaly is located near the rim of the western part of the Las Cañadas Caldera, and extends along the NW and NE rift zones. These authors attribute the main gravity high (asterisks in Fig. 13.6) to a large buried mafic volcano (Boca de Tauce volcano), which hosts a very dense plutonic complex. This "high-density body" might approximate the core of the Central Miocene shield, the first constructional stage of Tenerife as proposed by e.g. Guillou et al. (2004) and Carracedo et al. (2007). This core has been observed in *galerías* to form the central part of Tenerife to an elevation of about 1500 m (Coello 1973; Carracedo 1979), although this shield only outcrops at present in the Roque del Conde massif.

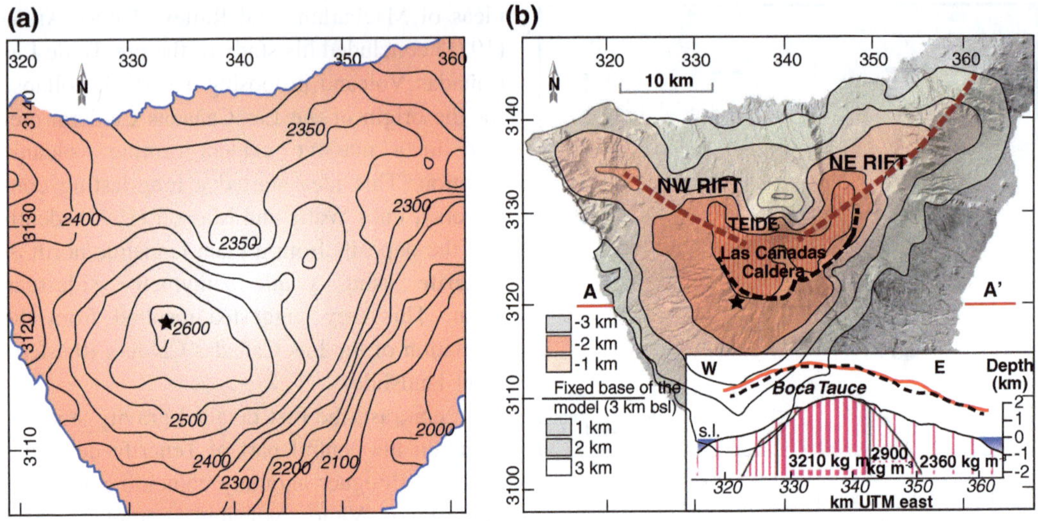

Fig. 13.6 **a** Bouguer anomaly contour map for the central part of Tenerife. Contour interval is 50 gu (1 gu = 0.1 mGal). Asterisk indicates the inferred location of the central vent systems of the Boca de Tauce volcano. **b** Map showing contours on the surface of a gravity model derived from the Bouguer anomaly map in **a.** Major surface structural features are shown for reference. Inset: W-E cross section through the gravity model, showing the dense (3.2 g/cm^3) body, probably corresponding to the plutonic core of the Central Miocene Shield of Guillou et al. (2004). Observed gravity anomalies (*red line*) and that calculated from the model (*dashed line*) are shown (modified from Ablay and Kearey 2000)

Gottsmann et al. (2008) analysed the "mass loss" identified by Vieira et al. (1989) in the Teide-Pico Viejo complex in detail, suggesting as a cause the low-density differentiated rocks that form these volcanoes, coupled with central hydrothermal alteration. The low-density anomaly associated with the central stratocones is, in their view, a shallow feature that extends from the surface down to about sea level. A deeper (>4 km b.s.l.) high-density body, however, forms the core of the central volcanic complex (Gottsmann et al. 2008). Their model favours a vertical collapse origin for the Las Cañadas Caldera but they point out that their gravity data are unsuitable to resolve present-day reservoirs that could contain eruptable material.

More recent research (Camacho et al. 2011) defines in detail the subsurface density structure of central Tenerife: a shallow (above sea level) low-density area matching the central stratocones and continuing northwards along the Icod Valley (Fig. 13.7a). The deeper structure (below ~2,000 m b.s.l.) appears as a diverse, high-density body (Fig. 13.7b), suggesting a complex composite plutonic structure, from Boca de Tauce to Montaña Blanca. Below about 6,000 m b.s.l., a positive anomaly appears, underlying the entire central part of the island (Fig. 13.7c). This is interpreted as a cumulate complex of mafic–ultramafic intrusions formed by fractionation of mantle-derived magma, similar to those observed in other oceanic islands (e.g., Hawaii; Borgia and Treves 1992). A W–E section of the model shows a low-density shallow feature (Fig. 13.7d), corresponding to the less dense volcanics of the central stratocones and the filling of the Las Cañadas Caldera. An alternative or complementary interpretation of the low density anomaly would be related to the presence of hydrothermally altered products.

13.4 Aeromagnetic Surveys

Aeromagnetic studies carried out in Tenerife provided new insights into the magnetic properties of subsurface rocks and the depth of the main magnetic layers, valuable information for

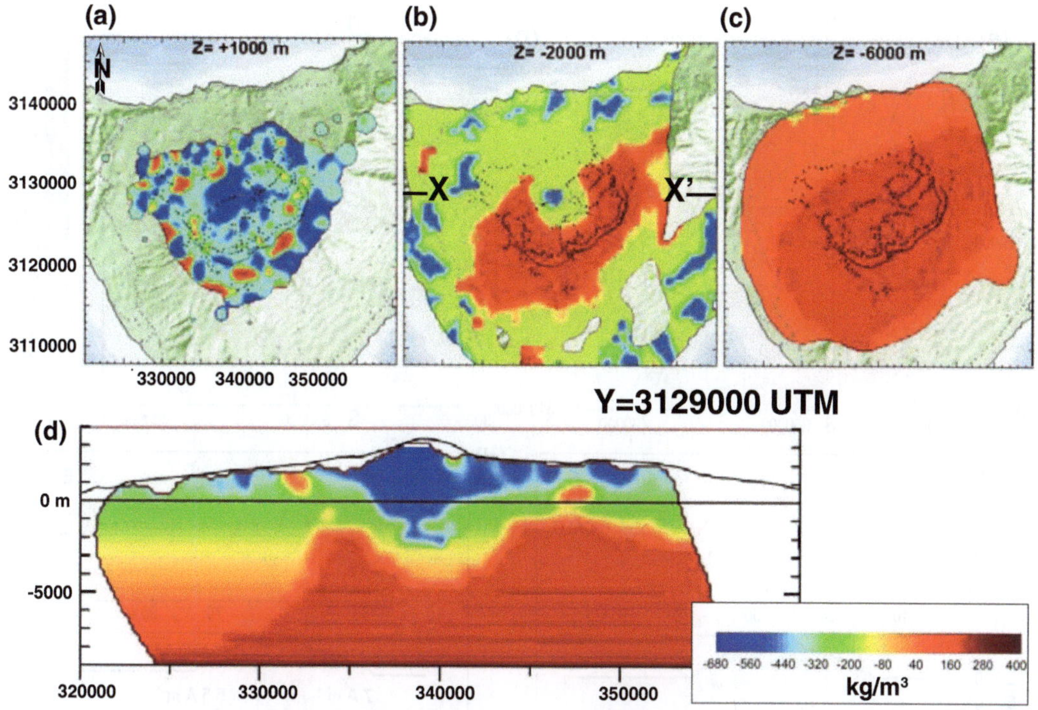

Fig. 13.7 Horizontal cross sections at various depths (+1,000, −2,000 and −6,000 m) in the 3D model for anomalous density (**a**, **b** and **c** in the figure). **a** WE vertical section (X-X' in **b**) is shown in (**d**). The main body of positive density (*dark red*) dominates the deep structure of the central part of Tenerife. However, in contrast, a low-density shallow structure corresponds to the Central Volcanic Complex and the less dense volcanics filling the Las Cañadas Caldera (modified from Camacho et al. 2011)

the definition of the internal geological structure of the island.

Araña et al. (2000) compiled a magnetic anomaly map of Tenerife before and after removing the effect of above-sea-level sources. The magnetic anomaly map was low-pass filtered to remove the subaerial magnetic sources, leaving only those originating from features located below sea level, thus distinguishing anomalies created by bodies deeper than 5 km b.s.l. from those shallower than 5 km b.s.l. (Fig. 13.8a, b respectively). The intense linear positive anomaly of Fig. 13.8a is interpreted by these authors as an uplifted block of oceanic crust. Forward modelling resulted in a 40 × 8 km strongly magnetised prismatic source, located between 5 and 12 km depth (Fig. 13.8c), probably formed by ultramafic cumulates. Superimposed on the deep linear anomaly, the model also showed two less intense positive anomalies, located below the Teide-Pico Viejo stratocones and the NE Rift Zone (Fig. 13.8d). A later high-resolution aeromagnetic anomaly map of Tenerife (García et al. 2007) confirmed the previous results, defining both the Teide-Pico Viejo and the anomaly corresponding to the NE Rift Zone in yet more detail (Fig. 13.9a, b).

13.5 Seismicity

13.5.1 Seismic Profiles

Seismic data have been used to determine the upper mantle and crustal structure of Tenerife. Bosshard and MacFarlane (1970) were the first to carry out marine refraction seismic investigations to define the depth of the mantle (15 km deep north of Tenerife) and confirm the oceanic

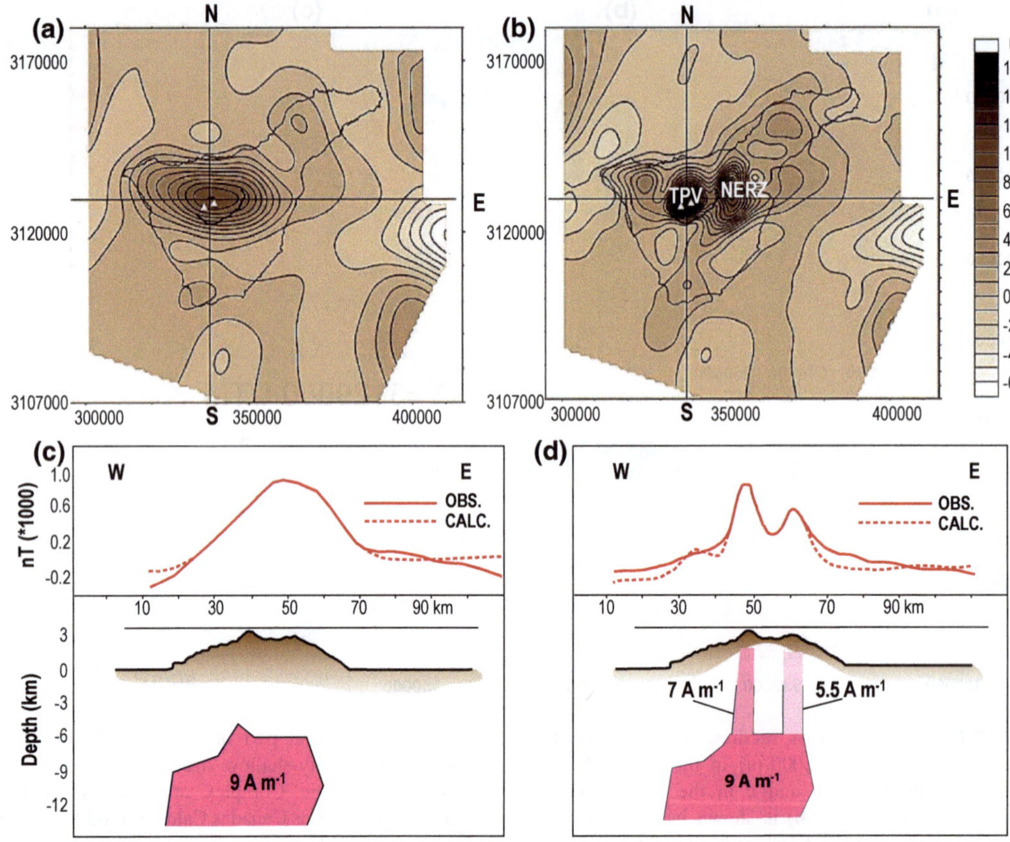

Fig. 13.8 **a, b** Regional magnetic anomaly maps, (**a** shows anomalies created by bodies deeper than 5 km b.s.l. and **b** those shallower than 5 km b.s.l.) and **c, d** models for a vertical source with an intensity of magnetisation of 9 A m−1. TPV: Teide-Pico Viejo volcanoes, NERZ: NE rift zone (modified from Araña et al. 2000)

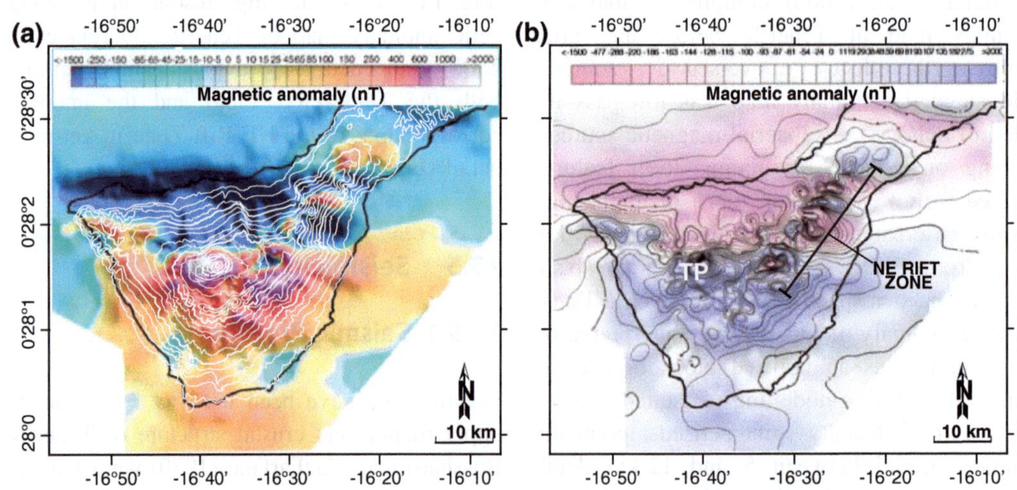

Fig. 13.9 **a** High-resolution aeromagnetic anomaly map (total intensity) of Tenerife highlighting anomaly intensity, **b** Contoured magnetic anomaly map (total intensity) of Tenerife Island plotted in two colour density interval to highlight anomaly frequency and wavelength. Both figures clearly highlight the higher intensity and frequency anomalies extending NE from Las Cañadas (the NE rift zone) (García et al. 2007)

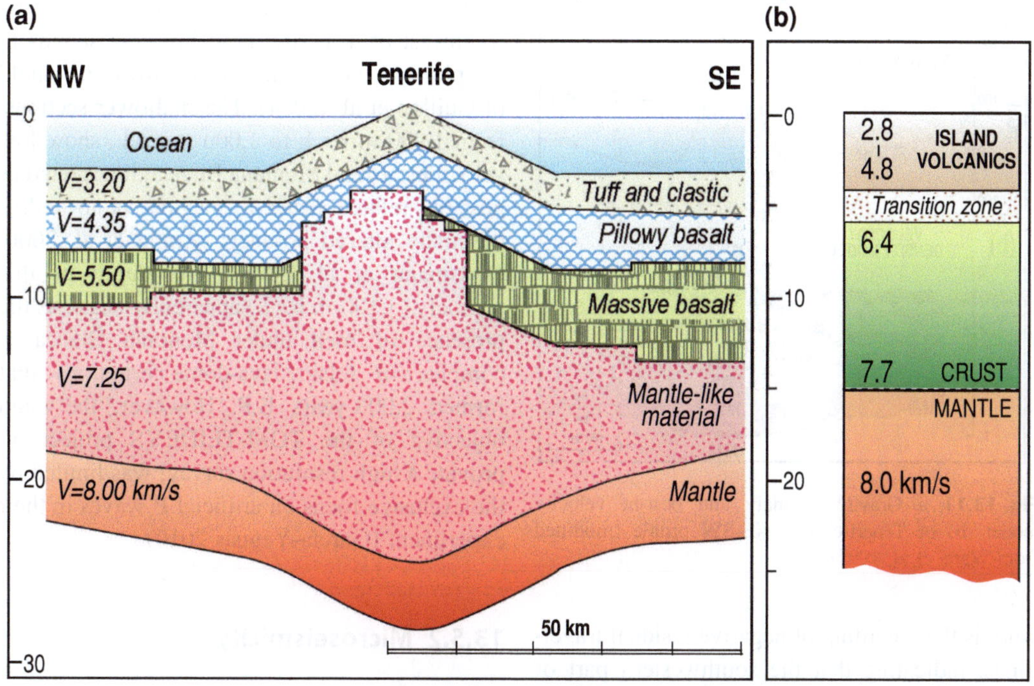

Fig. 13.10 **a** Crustal models for Tenerife constructed from gravity and seismic refraction data. A high-density, high-velocity body of "mantle-like material" underlies the central part of Tenerife, at about 5.5 km below the Teide-Pico Viejo central complex (modified from Bosshard and MacFarlane (1970). **b** Crustal model for Tenerife (from Banda et al. 1981)

character of the crust under the island (Fig. 13.10a). These early data and interpretations were later confirmed by abundant gravity and seismic investigations.

Banda et al. (1981) repeated the refraction seismic survey of the Canaries, confirming the depth of the Moho (15 km in Tenerife), reaching several interesting new conclusions, i.e., the oceanic character of all the islands, the absence of a common basement for the entire archipelago, the different islands constructed as independent volcanic edifices and, finally, the structural similarities between the island crust in the Canaries and in other oceanic islands (Fig. 13.10b).

Suriñac (1986) reported low velocities in the central part of Tenerife, suggesting the possibility of magma chamber(s) underneath the Teide volcanic complex. Watts et al. (1997) provided an interpretive two-dimensional model for Tenerife based on offshore seismic reflection and gravity profiles, seismic refraction data, and assumed relationships between seismic velocity and density. Their NE–SW cross section (Fig. 13.11) showed a volcanic core of 2.7 g/cm^3 overlain by a carapace of lower density material (2.5–2.3 g/cm^3).

However, wide-angle seismic investigations on the structure of Tenerife (Canales et al. 2000), which used a quasi-circular seismic profile around Tenerife, and airgun shots fired from the RRS Charles Darwin every 40 s, afforded a rather different conclusion. Seismic data revealed the absence of travel-time delays that could be associated with the presence of magma chambers. The authors, however, do not entirely exclude their existence, suggesting that either they are too small for the spatial resolution of the method applied, or the magma reservoirs feeding the recent volcanism of the Teide Volcanic Complex have cooled and largely crystallised. Another interesting conclusion of this

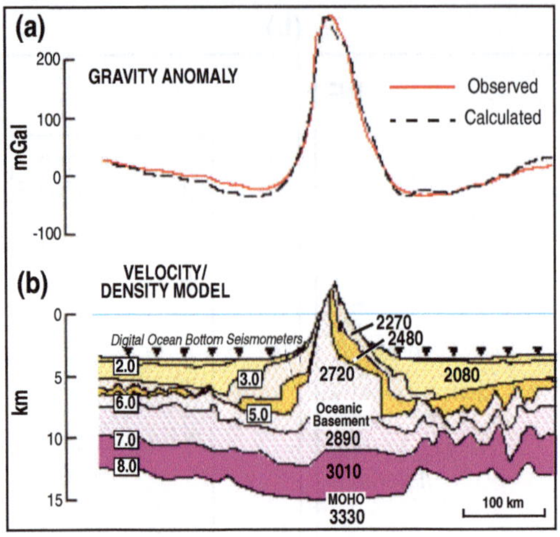

Fig. 13.11 a Gravity anomaly and best-fit velocity model (**b**) of Tenerife in a NE-SW profile (modified from Watts et al. 1997)

work is the recording of negative residual traveltimes, indicating that the southwestern part of Tenerife is characterised by a high P-wave velocity zone, coinciding with a gravity maximum previously modelled as a high-density body. In both cases, the data fit with the presence of the dense core of an old, large mafic pluton (and likely an associated volcano) north of the Miocene Roque del Conde massif. Once more, this high-density body is consistent with the plutonic core of the Central Miocene shield postulated by Guillou et al. (2004) and Carracedo et al. (2007), as mentioned above.

The latest seismic tomography survey, completed in 2007, focused on the Central Volcanic Complex of Tenerife (TOM-TEIDEVS), analysing the 3D structure of P-waves (García-Yegüas, 2010). This research, prompted by the increased seismicity in Tenerife in 2004, raised high expectations of solving the ambiguities still existent at the Teide Volcanic Complex (i.e., the presence, state of activity and eruptability of magmatic chambers). The results mainly confirmed the conclusions of the previous gravity and seismic studies outlined above, reaffirming the presence of a high-velocity central core of the island from 10,000 m b.s.l. and the high-velocity feature centred around Boca de Tauce (Fig. 13.12). The latter, comprising most of the southwest of Tenerife, probably corresponds to the plutonic core of the Central Miocene shield of Guillou et al. (2004). The shallower sections, from 1,900 m b.s.l. to 1,000 m a.s.l., show low velocities around Montaña Blanca, interpreted as possibly related either to magma, or to hydrothermally altered products (Araña et al. 2000; Gottsmann et al. 2008). The depth of this anomaly (3,100 m b.s.l.) apparently suggests the presence of melt under Montaña Blanca, a complex of felsic lava-dome eruptions that formed 2,000 years B.P. However, the main objective of the TOM-TEIDEVS project, to provide a high resolution survey, was limited by the exclusive focus on artificial P-waves in their experiment (García-Yeguas 2010).

13.5.2 Microseismicity

Seismicity associated with the Central Volcanic Complex of Tenerife is of low frequency and very low magnitude, particularly if compared with the island of Hawaii. The highest magnitude earthquake recorded in the Canaries (M 5.3) occurred in May 1989, and only a few events have since been noticed in Tenerife (i.e., a few in 2004). Hazards related to seismic activity at Teide Volcanic Complex are thus negligible (except during true volcanic unrest and eruptions).

Instrumental microseismicity, even of a very low rate (Mezcua et al. 1989), is an important tool in defining the internal structure of the volcanic system and as an early precursory signal to eruptions. However, a long recording period with a dense network of seismic stations around the volcano is needed to be able to analyse S-wave traveltimes and to define the presence of potential eruptable magma. This dense network and a long period of observation are essential in establishing a reliable baseline for forecasting eruptions, thus avoiding false alarms. Unfortunately, the approach has been the opposite, with some authors explicitly advising against the deployment of a dense network of

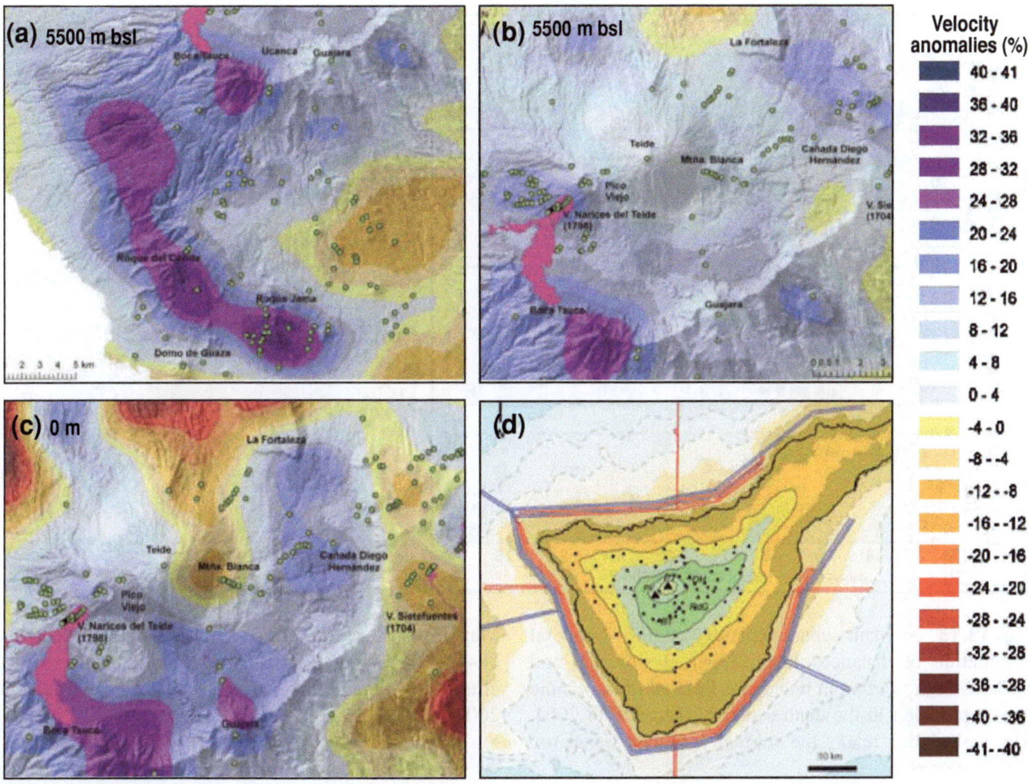

Fig. 13.12 Horizontal sections of the 3D model in different parts of Tenerife: **a** SW Tenerife section at 5,500 m.b.s.l. showing the high-velocity feature possibly related to the Central Miocene shield (Guillou et al. 2004); **b** 5500 m.b.s.l. section of Las Cañadas Caldera, comprising a high-velocity body without any indication of magma chambers beneath the central volcanoes; **c** Idem at sea level showing a low-velocity feature focused on Montaña Blanca. **d** Map of TOM-TEIDEVS 3D tomography area showing air-gun shots (*coloured lines*), and station locations (*black dots*) (from García-Yeguas 2010)

permanent seismic stations around the central volcanoes of Tenerife (Mezcua et al. 1989), and only after the impact of the 2004 seismic events, which received intense media coverage was the number of seismic stations installed on Tenerife increased (see Fig. 14.8).

One of the main problems encountered in the analysis of microseismicity in Tenerife is the presence of strong background noise at low frequencies, interpreted as caused by oceanic load. Noise levels are too high to detect other seismic waves of similar frequencies (Almendros et al. 2000). Further studies have shown the presence of background noise in the Teide Volcanic Complex at most operational frequencies (Almendros et al. 2004).

During one of the most active seismic phases (May 12–31, 2004) three seismic antennas were deployed in Las Cañadas Caldera, allowing for the first time the precise localisation of seismic sources (Almendros et al. 2006). The greater part of the earthquakes recorded before the tremor started occurred at depths between 15 and 20 km (Fig. 13.13a), and clustered north of Teide Volcano (Fig. 13.13b) after the tremor started, there was greater dispersion in the depth of events, with epicentres spreading under Las Cañadas Caldera (Fig. 13.13c). These data exclude any activity related to a shallow magma chamber under Teide Volcano, suggesting that, if it does exist, it is largely solidified and without large volumes of eruptible magma.

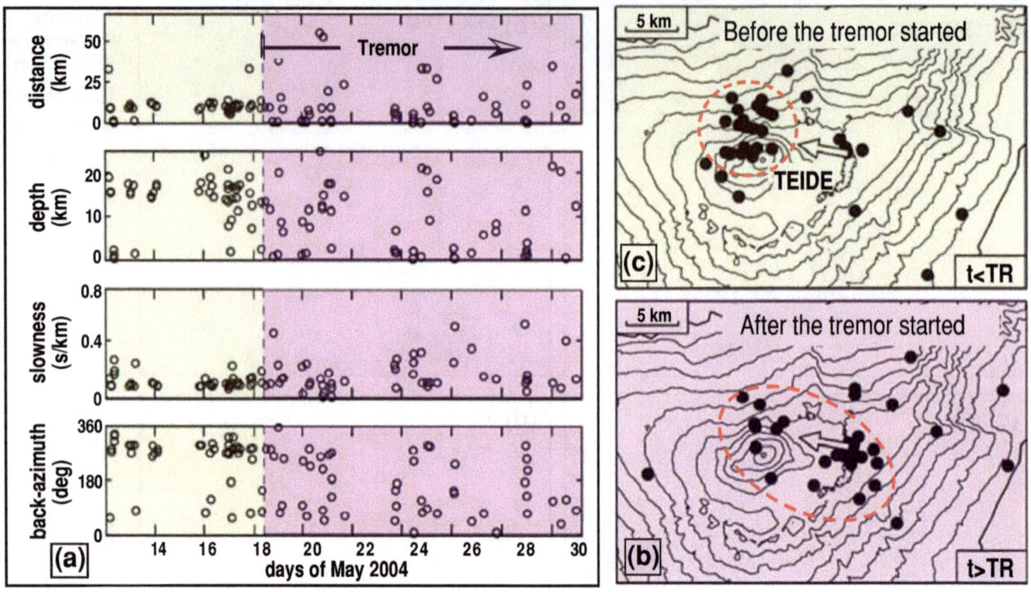

Fig. 13.13 **a** Main characteristics of the temporal distribution of distance, depth, apparent slowness and back-azimuth (*arrows* in **b** and **c**) estimated from seismic events recorded in the central part of Tenerife in 2004. The dashed line marks the start of a tremor signal (on May 18). **b** Epicentres of earthquakes recorded before the onset of the tremor. **c** Epicentres of earthquakes recorded after the beginning of the tremor (after Almendros et al. 2006)

13.6 Ground Deformation

Teide is frequently claimed by geoscientists and particularly the media to be structurally unstable. However, as yet no compelling evidence has been published to substantiate this view.

Early investigations using a geodetic network indicated the absence of significant deformation in the area of the Las Cañadas Caldera and Teide Volcano (Sevilla and Martín 1986; Vieira and Sevilla 1989). Additional work, applying Synthetic Aperture Radar Interferometry (InSAR) in Tenerife, in combination with Global Positioning System (GPS) data, confirmed the lack of significant deformation at Las Cañadas (Fernández et al. 2004, 2005). Later data, obtained using Differential Synthetic Aperture Radar Interferometry (DInSAR), detected a subsidence of 3–5 mm/yr at the NW, NE and S rift zones, which was particularly significant (15 mm/yr) at the location of the last historical eruptions on the island (Fernández et al. 2009) (see Fig. 14.8). This work was apparently capable of detecting a continuous long-term subsidence of the central area of Tenerife (3–4 mm/yr), in connection with the gravitational sinking of the dense core of the island. More localised deformation patterns were correlated by these authors with water table depletion caused by groundwater extraction.

Berrocoso et al. (2010) used data from a different GPS to point out that the main geological structures of the island (rift zones and the Las Cañadas Caldera) are the main cause of current deformation on Tenerife. These authors repeated the experiment after the period of increased seismicity in Tenerife in 2004 (using radar images acquired from 1992 to 2005 by the European Remote Sensing satellite (ERS), apparently observing variations in the deformation time series of the selected pixels associated with the beginning of the seismic crisis in 2004.

These observations, however, show a general SW–NE displacement of Tenerife and the Canaries coherent with the movement of the African plate, with a mean velocity between 20 and 25 mm/yr (Fig. 13.14a). When the effect of

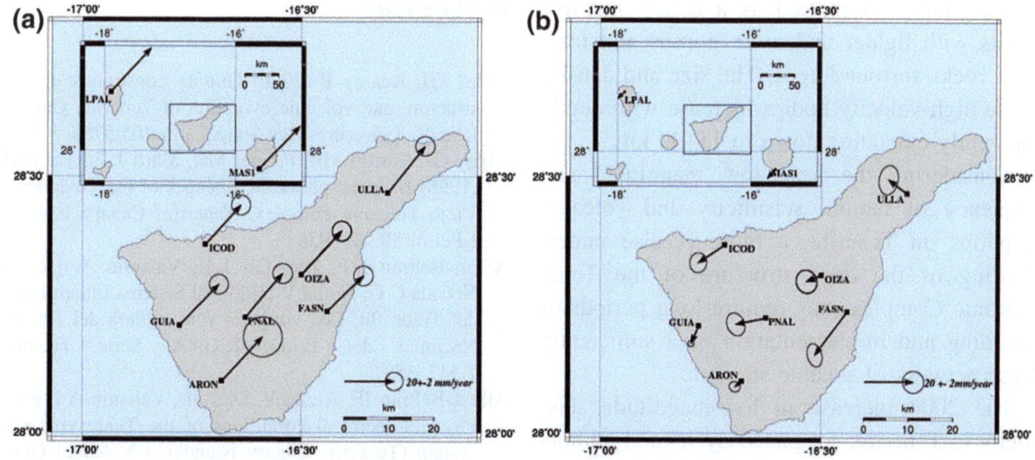

Fig. 13.14 a GPS velocities showing displacement of Tenerife in the global context of the African plate. b After the mean velocity of the plate is removed (from Berrocoso et al. 2010)

the plate displacement is removed, the data are scattered and difficult to interpret (Fig. 13.14b).

13.7 The Broader Picture

The study of the main characteristics of the possible magma chamber(s) of Teide and the assessment of its potential eruptibility has been approached using different geochemical and geophysical methodologies.

Several gas geochemistry, isotope geochemistry and modelling studies concur in relating the gas system of the TVC (both in *galerías* and at the summit of Teide) to an essentially magmatic source located approximately at sea level beneath the volcano. The temperature of this spherical, 4 km diameter post-caldera magma chamber has been estimated at about 400 °C. Independent and isolated smaller (about 200 m in diameter) magmatic bodies associated with the NW and NE rift zones show lower temperatures (250 °C).

Gravity (Bouguer anomalies) surveys have revealed important structural features in Tenerife: (1) A three-pointed star gravity ridge system, interpreted as being associated with the Tenerife rift zones, (2) A gravity minimum in the Teide area, attributed to the Teide-Pico Viejo stratocones, and probably related to the low-density differentiated rocks forming these volcanoes, increased by hydrothermal alteration, and (3) A marked gravity maximum in the western rim of Las Cañadas Caldera, interpreted as a buried old volcano (Boca de Tauce volcano), with a very dense plutonic complex, that likely represents the core of the Central Miocene shield, the first constructional shield of Tenerife.

Several seismic tomography studies carried out on Tenerife have been effective in obtaining three-dimensional images of the island. Active seismic surveys using air guns and a dense network of seismic stations, either onshore or on the ocean floor around the island, were intended primarily to define low-velocity anomalies that could be related to the postulated Teide magma chamber, since low velocities are related to partially molten rocks. However, active seismic experiments have a fundamental shortcoming because they can only generate P-waves, the S-waves being associated only with natural seismic activity that is insufficient on the island for this purpose. Nevertheless, the various tomographic surveys agree in the lack of any evidence of extensive magma reservoirs underneath the central volcanoes. Tenerife shows a high-velocity central core and high-velocity anomalies parallel to the NW and NE rift zones, encircled by low-velocity anomalies at the flanks of the island, interpreted as a dense intrusive plexus with lighter and more porous and fractured rocks surrounding it. The size and density of the high-velocity bodies increase with depth, apparently extending down to 15–20 km.

Considering the very low magnitude and frequency of natural seismicity and volcanic eruptions on Tenerife, a more precise understanding of the deep structure of the Teide Volcanic Complex may require long periods of recording and implementation of a sufficiently dense network of seismic stations.

The 2004 increase in low-magnitude seismicity in Tenerife, suggested by several authors to be "volcanic unrest", was most probably related to a deep injection of magma that was arrested at an early stage (Gottsmann et al. 2008). This seems coherent with a similar process that took place on the island of El Hierro in 2011, which culminated in a submarine eruption (Carracedo et al. 2012a, b). Whilst in 2004 seismicity on Tenerife was of low rate and magnitude (only a few earthquakes of M 2.0–M 2.6), on El Hierro in 2011 more than 10,000 events were recorded, 95 of them with magnitudes between M 3.0 and M 4.6. A particularly strong (M 4.4) seismic event probably coincided with the opening of the eruptive conduit and the onset of the submarine eruption there.

Differences in the pattern and intensity of seismicity in 2004 on Tenerife and 2011 on El Hierro suggest that an injection of deep magma was in both cases the source of seismicity. As may be expected, these deep intrusions seem to be a relatively frequent process and a common prelude to many Canarian submarine and subaerial eruptions. However, only a small fraction of such intrusions are capable of perforating the crust to produce a volcanic eruption. The majority of them probably become arrested before boring through the crust, and may be the source of several recorded seismic crises that ended without any surficial volcanic manifestation in the Canaries, as for example occurred in 1793 on El Hierro (Darias Padrón 1929; Hernández Pacheco 1982) and in 1936 on La Palma (Martel San Gil 1960).

References

Ablay GJ, Kearey P (2000) Gravity constraints on the structure and volcanic evolution of Tenerife, Canary Islands. J Geophys Res Solid Earth 105:5783–5796

Ablay GJ, Carroll MR, Palmer MR, Marti J, Sparks RSJ (1998) Basanite-phonolite lineages of the Teide-Pico Viejo volcanic complex, Tenerife, Canary Islands. J Petrol 39:905–936

Albert-Beltrán J F, Diez-Gil J L, Valentin A, De La Noceda C G, Araña V (1989) El Sistema fumaroliano del Teide. In: Los volcanes y la caldera del Parque Nacional del Teide, ICONA, Serie Técnica 7:347–358

Albert-Beltrán JF, Araña V, Diez JL, Valentin A (1990) Physical-chemical conditions of the Teide volcanic system (Tenerife, Canary Islands). J Volcanol Geotherm Res 43:321–332

Almendros J, Ibáñez JM, Alguacil G, Morales J, Del Pezzo E, La Rocca M, Ortiz R, Araña V, Blanco MJ (2000) A double seismic antenna experiment at Teide Volcano: existence of local seismicity and lack of evidences of volcanic tremor. J Volcan Geotherm Res 103:439–462

Almendros J, Luzon F, Posadas A (2004) Microtremor analyses at Teide Volcano (Canary Islands, Spain): assessment of natural frequencies of vibration using time-dependent horizontal-to-vertical spectral ratios. Pure Appl Geophys 161:1579–1596

Almendros J, Ibáñez M, Carmona E, Zandomeneghi D (2006) Array analyses of volcanic earthquakes and tremor recorded at Las Cañadas Caldera (Tenerife Island, Spain) during the 2004 seismic activation of Teide Volcano. J Volcanol Geotherm Res 160:285–299

Araña V (1971) Litología y estructura del edificio Cañadas, Tenerife (Islas Canarias). Estud Geol 27: 95–135

Araña V, Camacho AG, García A, Montesinos FG, Blanco I, Vieira R, Felpeto A (2000) Internal structure of Tenerife (Canary Islands) based on gravity, aeromagnetic and volcanological data. J Volcanol Geotherm Res 103:43–64

Banda E, Dañobeitia JJ, Surinach E, Ansorge J (1981) Features of crustal structure under the Canary Islands. Earth Planet Sci Lett 55:11–24

Berrocoso M, Carmona J, Fernández-Ros A, Pérez-Peña A, Ortiz R, García A (2010) Kinematic model for Tenerife Island (Canary Islands, Spain): geodynamic interpretation in the Nubian plate context. J Afr Earth Sci 58:721–733

Borgia A, Benedetta Treves B (1992) Volcanic plates overriding the ocean crust: structure and dynamics of Hawaiian volcanoes. J Geol Soc (London UK) 60:277–299

Bosshard E, MacFarlane DJ (1970) Crustal structure of the Western Canary Islands from seismic refraction and gravity data. J Geophys Res 75:4901–4918

Camacho AG, Fernandez J, Gottsman J (2011) The 3-D gravity inversion package GROWTH2.0 and its application to Tenerife Island. Spain. Comput Geos 37:621–633

Canales JP, Dañobeitia JJ, Watts AB (2000) Wide-angle seismic constraints on the internal structure of Tenerife, Canary Islands. J Volcanol Geotherm Res 103:65–81

Canas JA, Ugalde A, Pujades LG, Carracedo JC, Soler V, Blanco MJ (1998) Intrinsic and scattering seismic wave attenuation in the Canary Islands. J Geophys Res 103:37–50

Carracedo JC (1979) Paleomagnetismo e historia volcánica de Tenerife. Aula Cultura Cabildo Insular de Tenerife, Santa Cruz de Tenerife, p 81

Carracedo JC, Rodríguez Badiola E, Guillou H, Paterne M, Scaillet S, Pérez Torrado FJ, Paris R, Fra-Paleo U, Hansen A (2007) Eruptive and structural history of Teide volcano and rift zones of Tenerife, Canary Islands. Geol Soc Am Bull 19:1027–1051

Carracedo JC, Fernandez-Turiel JL, Gimeno D, Guillou H, Klügel A, Krastel S, Paris R, Perez-Torrado FJ, Rodriguez-Badiola E, Rodriguez-Gonzalez A, Troll VR, Walter TR, Wiesmaier S (2011) Comment on the distribution of basaltic volcanism on Tenerife, Canary Islands: implications on the origin and dynamics of the rift systems by Geyer A, Martí J (2010) Tectonophysics 483:310–326. Tectonophysics 503:239–241

Carracedo JC, Pérez-Torrado FJ, Rodriguez-Gonzalez A, Fernandez-Turiel JL, Klügel A, Troll VR, Wiesmaier S (2012a) The ongoing volcanic eruption of El Hierro, Canary Islands. Eos Trans AGU 93:89–90

Carracedo JC, Pérez-Torrado FJ, Rodríguez-González A, Soler V, Fernández-Turiel JL, Troll VR, Wiesmaier S (2012b) The 2011 submarine eruption in El Hierro (Canary Islands). Geol Today 28:53–58

Coello J (1973) Las series volcánicas en subsuelos de Tenerife. Estud Geol 29:491–512

Dañobeitia JJ, Collette BJ (1989) Estudio mediante sísmica de reflexión de un grupo de estructuras submarinas situadas al Norte y Sur del Archipiélago Canario. Acta Geol Hisp 24:147–163

Darias Padrón DV (1929) Noticias generales sobre la historia de la isla de El Hierro. Imprenta de M. Curbelo, La Laguna (Tenerife), p 407

Dash BP, Bosshard E (1969) Seismic and gravity investigations around the western Canary Islands. Earth Planet Sci Lett 7:169–177

Diez JL, Albert- Beltrán JF (1989) Modelo termodinámico de la cámara magmática del Teide. In: Los volcanes y la caldera del Parque Nacional del Teide, ICONA, Serie. Técnica 7:343–346

Fernández J, González-Matesanz FJ, Prieto JF, Staller A, Rodríguez-Velasco G, Alonso-Medina A, Charco M (2004) GPS Monitoring in the N–W part of the volcanic island of Tenerife, Canaries, Spain: strategy and results. Pure Appl Geophys 161:1359–1377

Fernández J, Romero R, Carrasco D, Tiampo KF, Rodríguez-Velasco G, Aparicio A, Araña V, González-Matesanz FJ (2005) Detection of displacements in Tenerife Island, Canaries, using radar interferometry. Geophys J Int 160:33–45

Fernández J, Tizzani P, Manzo M, Borgia A, González PJ, Martí J, Pepe A, Camacho AG, Casu F, Berardino P, Prieto JF, Lanari R (2009) Gravity-driven deformation of Tenerife measured by InSAR time series analysis. Geophys Res Lett 36:L04306. doi: 10.1029/2008GL036920

García A, Araña V, Astiz M, Ortiz R (1989) Modelo magneto-telúrico de la región central de Tenerife. In: Los volcanes y la caldera del Parque Nacional del Teide, ICONA, Ser. Técn 7:335–341

García A, Chiappini M, Blanco-Montenegro I, Carluccio R, D'Ajello Caracciolo F, De Ritis R, Nicolosi I, Pignatelli A, Sánchez N, Boschi E (2007) High resolution magnetic anomaly map of Tenerife, Canary Islands. Ann Geophys 50:689–697

García Yeguas MA (2010) Estudio de heterogeneidades laterales de volcanes activos: Tomografia sismica de alta resolucion de la isla de Tenerife, anomalias de propagacion de ondas sísmicas de la isla Decepcion y otros efectos. Tesis doctoral, Universidad de Granada

Gee MJR, Watts AB, Masson DG, Mitchell NC (2001) Landslides and the evolution of El Hierro in the Canary Islands. Mar Geol 177:271–293

Geyer A, Martí J (2010) The distribution of basaltic volcanism on Tenerife, Canary Islands: implications on the origin and dynamics of the rift systems. Tectonophysics 483:310–326

Geyer A, Martí J (2011) The distribution of basaltic volcanism on Tenerife, Canary Islands: Implications on the origin and dynamics of the rift system, reply to the comment by Carracedo et al. Tectonophysics 503:234–238

Gottsmann J, Camacho AG, Martí J, Wooller L, Fernandez J, Garcia A, Rymer H (2008) Shallow structure beneath the central volcanic complex of Tenerife from new gravity data: implications for its evolution and recent reactivation. Phys Earth Planet Inter 168:212–230

Guillou H, Carracedo JC, Paris R, Pérez Torrado FJ (2004) K/Ar ages and magnetic stratigraphy of the Miocene-Pliocene shield volcanoes of Tenerife, Canary Islands: implications for the early evolution of Tenerife and the Canarian hotspot age progression. Earth Planet Sci Lett 222:599–614

Hernández Pacheco A (1982) Sobre una posible erupción en 1793 en la isla de El Hierro (Canarias). Estudios Geológicos 38:15–25

Holik JS, Rabinowitz PD (1991) Effects of Canary hotspot volcanism on structure of oceanic crust off Morocco. J Geophys Res 96:12039–12067

MacFarlane DJ, Ridley WI (1968) An interpretation of gravity data from Tenerife, Canary Islands. Earth Planet Sci Lett 4:481–486

Martel San Gil M (1960) El volcán San Juan. La Palma (Canarias), Madrid, p 233

Martínez W, Buitrago J (2002) Sedimentación y volcanismo al este de las islas de Fuerteventura y Lanzarote (Surco de Fúster Casas). Geogaceta 32:51–54

Masson DG (1996) Catastrophic collapse of the volcanic island of Hierro 15 ka ago and the history of landslides in the Canary Islands. Geology 24:231–234
Masson DG, Watts AB, Gee MJR, Urgeles R, Mitchell NC, Le Bas TP, Canals M (2001) Slope failures on the flanks of the western Canary Islands. Earth Sci Rev 57:1–35
Melián G, Tassi F, Pérez N, Hernández P, Sortino F, Vaselli O, Padrón E, Nolasco D, Barrancos J, Padilla G, Rodríguez F, Dionis S, Calvo D, Notsu K, Sumino H (2012) A magmatic source for fumaroles and diffuse degassing from the summit crater of Teide Volcano (Tenerife, Canary Islands): a geochemical evidence for the 2004–2005 seismic–volcanic crisis. Bull Volcanol 74:1465–1483
Mezcua J, Ortiz R, Buforn E, Galán J, Herraiz M, Martínez Solares JM, Rueda J, Sánchez Venero M (1989) Estudio de la actividad sísmica en la isla de Tenerife. In: Los volcanes y la caldera del Parque Nacional del Teide, ICONA, Serie Técnica 7:397–404
Pous J, Heise W, Schnegg PA, Muñoz G, Martí J, Soriano C (2002) Magnetotelluric study of the Las Cañadas Caldera (Tenerife, Canary Islands): structural and hydrogeological implications. Earth Planet Sci Lett 204:249–263
Roeser HA (1982) Magnetic anomalies in the magnetic quiet zone off Morocco. In: Seibold E (ed) Von Rad HK, Sarthein M. Geology of the northwest African continental margin. Springer, Berlin, pp 61–68
Roest WR, Danobeitia JJ, Verhoef J, Collette BJ (1992) Magnetic anomalies in the Canary basin and the mesosoic evolution of the central North Atlantic. Marine Geophys Res 14:1–24
Rothe P, Schmincke H-U (1968) Contrasting origins of the eastern and western islands of the Canarian Archipelago. Nature 218:1152–1154
Sevilla M, Martín M (1986) Geodetic network design for crustal deformation studies in the Caldera of Teide area. Tectonophysics 130:235–248
Suriñac E (1986) La estructura cortical del Archipiélago Canario. Resultados de la interpretación de perfiles sísmicos profundos. An Fis S B 82:62–77
Uchupi E, Emery KO, Bowin CO, Phillips JD (1976) Continental margin off western Africa: Senegal to Portugal. Am Assoc Pet Geol Bull 60:809–878
Urgelés R, Canals M, Baraza J, Alonso B, Masson DG (1997) The last major megalandslides in the Canary Islands: the El Golfo debris avalanche and the Canary debris flow, west Hierro Island. J Geophys Res 102:305–323
Urgelés R, Canals M, Baraza J, Alonso B (1998) Seismostratigraphy of the western flanks of El Hierro and La Palma (Canary Islands): a record of Canary Island volcanism. Mar Geol 146:225–241
Urgelés R, Masson DG, Canals M, Watts AB, Le Bas T (1999) Recurrent large-scale landsliding on the west flank of La Palma. Canary Islands: J Geophys Res 104:331–348
Vieira A, Sevilla M (1989) Red geodésica de La Caldera de Las Cañadas. In: Los Volcanes y La Caldera del Parque Nacional del Teide, ICONA, Serie Técnica 7:423–426
Vieira R, Gonzalez Camacho A, De Toro C (1989) Modelo gravimétrico de la región central de Tenerife. In: Los volcanes y la caldera del Parque Nacional del Teide, ICONA, Serie Técnica 7:329–333
Watts AB (1994) Crustal structure, gravity anomalies and flexure of the lithosphere in the vicinity of the Canary Islands:Geophys J 119:648–666
Watts AB, Masson DG (1995) A giant landslide on the north flank of Tenerife, Canary Islands. J Geophys Res 100:49–487
Watts AB, Peirce C, Collier J, Dalwood R, Canales JP, Henstock TJ (1997) A seismic study of lithospheric flexure in the vicinity of Tenerife, Canary Islands. Earth Planet Sci Lett 146:431–447
Watts AB, Masson DG (2001) New sonar evidence for recent catastrophic collapses of the north flank of Tenerife, Canary Islands. Bul Volcanol 63:8–19
Wiesmaier S, Deegan F, Troll V, Carracedo JC, Chadwick J, Chew D (2011) Magma mixing in the 1100 AD Montaña Reventada composite lava flow, Tenerife, Canary Islands: interaction between rift zone and central volcano plumbing systems. Contrib Mineral Petrol 162:651–669

Geological Hazards in the Teide Volcanic Complex

14

Juan Carlos Carracedo, Alejandro Rodriguez-Gonzalez, Francisco Jose Perez-Torrado, Jose-Luis Fernandez-Turiel, Raphaël Paris, Eduardo Rodríguez-Badiola, Gustavo Pestana-Pérez, Valentin R. Troll, and Sebastian Wiesmaier

Abstract

The island of Tenerife displays contrasted densities of population, from the densely occupied coastal zones (including tourist resorts, airport, energy facilities, etc.) to the sparsely populated forests and mountainous highlands, where most of the recent volcanic events are located. Considering the low frequency of historical eruptions (compared to Hawaii or Reunion Island for example), the assessment of geological hazards must also rely on the analysis and interpretation of prehistorical events, going back to at least the Late Quaternary. In this chapter, we review the hazards related to Teide's volcanism, but also those from increased seismicity and from slope instability. We discuss the origin of low magnitude earthquakes, and particularly the 2004 episode of unrest. New estimates on cumulative volumes for resurfacing by lava flows during the last few thousand years are provided to serve as a tool for building a lava flow hazard map of Tenerife. Hazards related to explosive

J. C. Carracedo (✉) · A. Rodriguez-Gonzalez
F. J. Perez-Torrado · V. R. Troll · S. Wiesmaier
Departamento de Física (GEOVOL),
Universidad de Las Palmas de Gran Canaria,
Las Palmas de Gran Canaria, Canary, Islands, Spain

J.-L. Fernandez-Turiel
Institute of Earth Sciences Jaume Almera,
ICTJA-CSIC, Barcelona, Spain

R. Paris
Université Blaise Pascal, UMR 6524,
Clermont-Ferrand & CNRS, France

E. Rodríguez-Badiola
Museo Nacional de Ciencias Naturales, CSIC,
Madrid, Spain

G. Pestana-Pérez
Consejería de Agricultura, Ganadería,
Pesca y Alimentación, Gobierno de Canarias, Spain

J. C. Carracedo and V. R. Troll (eds.), *Teide Volcano*, Active Volcanoes of the World,
DOI: 10.1007/978-3-642-25893-0_14, © Springer-Verlag Berlin Heidelberg 2013

activity are also considered and although possible, with phreatomagmatic eruptions being the most likely style anticipated, explosive events are of relatively low probability at Teide in the near future.

14.1 Introduction

Geological hazards are comparatively moderate in the Canary Islands in general, and in particular in the TVC. Rock falls, road blocks, coastal cliff slumps and *barranco* flash floods are, by far, the main cause of loss of life and property, mainly at the lower part on the northern, leeward side of the TVC (Fig. 14.1).

This chapter analyses the most characteristic and significant hazards associated with the TVC as an active volcanic complex, namely seismic hazards, eruptive hazards and those derived from ground deformation.

In an early and very influential geophysical study on the Canaries, Dash and Bosshard (1969) and Bosshard and MacFarlane (1970) postulated large, regional "crustal fractures" related to the African tectonics that apparently pre-date the formation of the Canaries. This seminal idea opened a long-lived debate not only on the origin of the Canaries, but on the potential risk of major "tectonic" earthquakes in the archipelago.

Geophysical and geological information increasingly favours the construction of the Canary Islands in relation to a mantle plume on a passive continental margin disassociating the Canaries from northern African seismicity and tectonism (Frizon de Lamotte et al. 2000; Martínez and Buitrago 2002; Faccenna et al. 2004) (Fig. 14.2). The intense and abundant seismicity concentrated in the plate boundary zone between the converging African and Eurasian plates fades in the oceanic prolongation of the Atlas fault system, creating a seismic gap between Africa and the Canaries (Fig. 14.3). This reflects the difficulties encountered by tectonic stresses to propagate from the weaker continental to the stronger oceanic crust (Steckler and Tenbrink 1986).

Fig. 14.1 Rockfall and associated road blockage on a road on the northern slopes of the TVC (July 5, 2011, www.laopinion.es)

Geological hazards in the Canaries are thus correlated with a separate geodynamic framework, characterised by low magnitude seismicity (only 1 event $M > 5$ in the recording period), and volcanic eruptions where the majority of Holocene eruptions have $VEI < 3$.

In addition to the favourable geological and geodynamic setting of the Canary Islands in relation to natural hazards, passive margins around the Atlantic Ocean are characteristically aseismic, which contrasts, e.g., with the Hawaiian Islands. Consequently, the Canaries are at low risk of significant tsunamis induced by large earthquakes, which in turn frequently reach the Hawaiian Islands (Wood et al. 2007).

By comparing the islands of Tenerife and Hawaii (the classical example of an age progressive island chain fed from a hotspot source fixed in the mantle), it becomes apparent that the larger volume and more active mantle plume of Hawaii explains the considerably greater severity of geological hazards there, at least in the historical period of both archipelagos.

14.2 Seismicity and Seismic Hazards in the TVC

Although intraplate earthquakes do occur in some continental lithospheric plates (e.g., North American, Eurasian), seismicity is generally

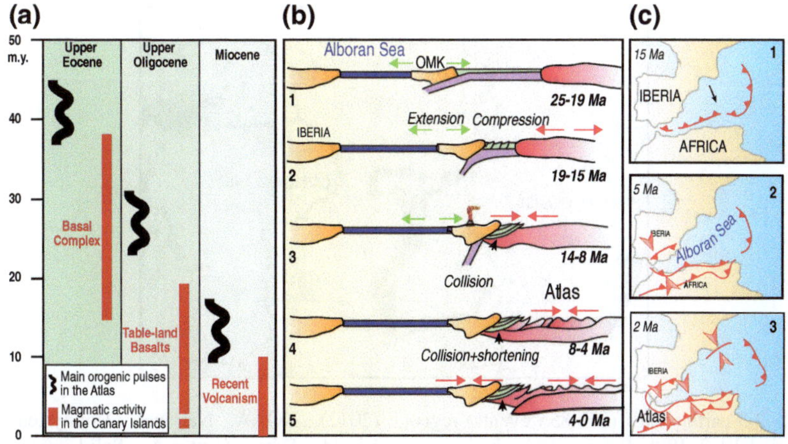

Fig. 14.2 **a** Correlation established by Anguita and Hernán (1975) between the main phases of Canarian volcanism and the three orogenic pulses involved in the genesis of the Western High Atlas structures (Ambroggi 1963). **b** Cenozoic evolution and active deformation in the North African margin (from Frizon de Lamotte et al. (2000). **c** Atlas folding apparently began in Early Pleistocene (from Faccenna et al. 2004)

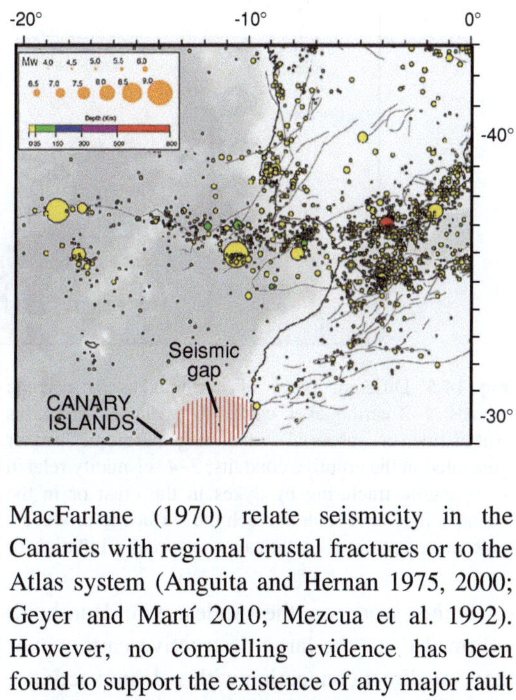

Fig. 14.3 The abundant seismicity in North Africa fades in the oceanic prolongation of the Atlas towards the Canary Islands, creating a seismic gap that apparently contradicts the alleged link of the archipelago with the Atlas tectonics. The figure shows earthquakes with M > 4 recorded from 1901 to 2004 (modified from Serpelloni et al. 2007)

negligible in the interiors of the lithospheric plates without the presence of oceanic islands (e.g. Hawaiian and Canary Islands in Fig. 14.4). Hawaii, for example, is far removed from geological fault lines, but the hot spot below the Hawaiian Islands causes sufficient stress in the crust to render Hawaii a very active seismic region (e.g., Klein et al. 2001). Earthquakes in these settings are thus mainly generated by plume related magmatism and the stresses caused by island growth and associated lithospheric loading.

Some authors, following the early ideas of Dash and Bosshard (1969) and Bosshard and MacFarlane (1970) relate seismicity in the Canaries with regional crustal fractures or to the Atlas system (Anguita and Hernan 1975, 2000; Geyer and Martí 2010; Mezcua et al. 1992). However, no compelling evidence has been found to support the existence of any major fault connecting the Atlas with the Canaries in detailed geophysical studies of the area (Martínez and Buitrago 2002) or around the Canarian archipelago in general (Watts 1994, 1997; Funck et al. 1996; Urgeles et al. 1998; Krastel et al. 2001; Krastel and Schmincke 2002; Carracedo et al. 2011).

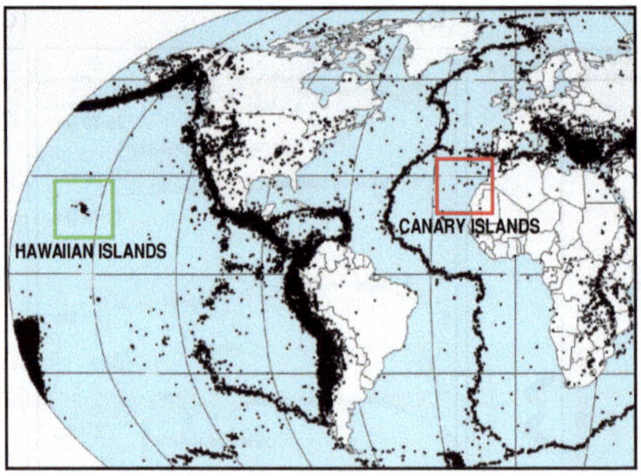

Fig. 14.4 M > 3.5 earthquakes (200,855 events) registered in the period 1963–1998. Seismicity in the interior of the plates is mainly related to the construction of the oceanic islands (from Earthquake Seismicity Catalog (2012), NOAA, National Geophysical Data Center, World Data Center-A Solid Earth Geophysics. http://www.ngdc.noaa.gov/hazard/earthqk.shtml)

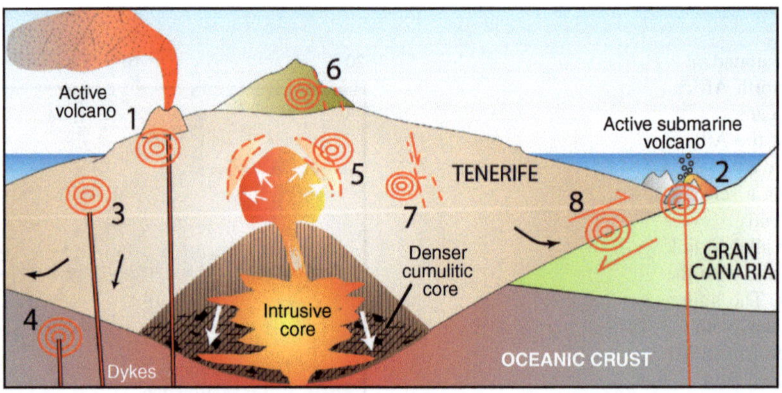

Fig. 14.5 Different types of earthquakes in oceanic islands: *1–2* earthquakes caused by volcanic eruptions (submarine or subaerial), including harmonic tremor generated in the eruptive conduits; *3–4* seismicity related to hydraulic fracturing by dykes in the crust or in the island edifice, that seldom reach the surface to produce a volcanic eruption; *5* earthquakes caused by inflation of shallow magmatic chambers; *6–7* earthquakes due to gravitational normal faulting; *8.* earthquakes associated with low-angle, inverse faults that originate by gravity-driven lateral escape of island sectors in the denser island basement, e.g. Volcano spreading (Carracedo et al. 2005). See text for explanation

In his work on the Andes, von Humboldt affirmed that "the large destructive earthquakes have no direct connection with volcanic activity, that only cause small local seismicity…" (von Humboldt and Bonpland 1805). Klein et al. (1987) deduced from studies in Hawaii that "All earthquakes in Hawaii are ultimately attributable to volcanism…", defining those earthquakes as "volcanic" that are in or immediately adjacent to magma conduits, and "tectonic" earthquakes as those that are removed from a magma conduit but derive their driving stress from dyke intrusion or volcano growth and deformation. In agreement with these concepts, seismic hazards decrease in the Canarian Archipelago with island age, as do both magmatic activity and edifice instability. Earthquakes in oceanic islands thus have different causes that group in two main categories: those associated with magmatic processes, and those originated by instability in the islands (Fig. 14.5). Earthquakes caused by volcanic eruptions (submarine or

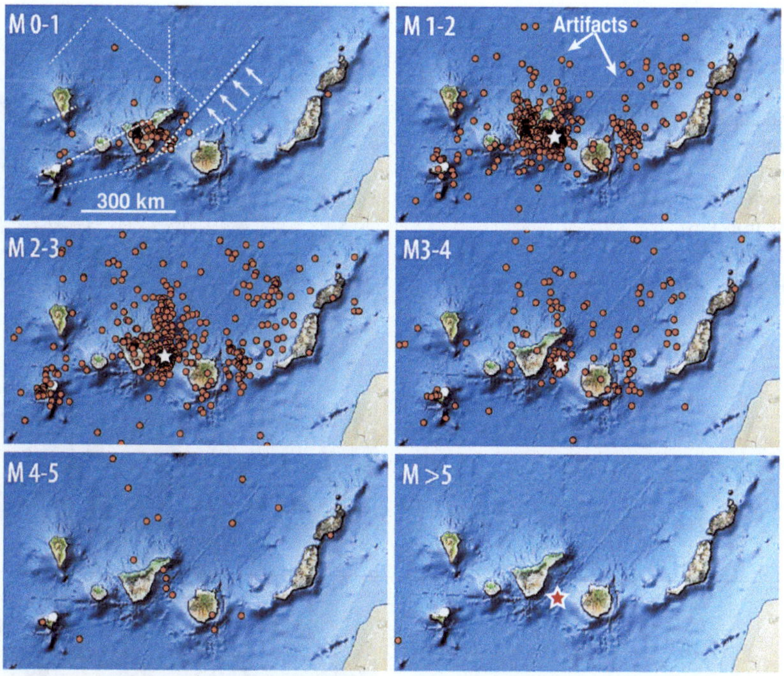

Fig. 14.6 Earthquakes recorded by the Spanish Geographical Institute (IGN) in the period 1980–2010, arranged as a function of their magnitude. The greater part of this seismicity has magnitudes below M 3.0. Only one earthquake of $M > 5$ has been recorded in the Canaries (M 5.2, May 1989). The epicentre distribution lacks any concordance with linear structures, such as the alleged faults of Geyer and Martí (2010) (*white dashed lines*), which are actually artifacts from *ship track lines* (Google Earth image). *The asterisk shows* the location of an active submarine volcano (Krastel and Schmincke 2002)

subaerial) belong to the first group and include harmonic tremor generated in the eruptive conduits (1 and 2 in Fig. 14.5). This category also includes the seismicity related to hydraulic fracturing by dykes in the crust or within the island edifice, that seldom reach the surface to produce a volcanic eruption (3 and 4 in Fig. 14.5); or that caused by inflation of shallow magmatic chambers (5 in Fig. 14.5).

The second category comprises earthquakes due to gravitational normal faulting (6 and 7 in Fig. 14.5), or associated to low-angle, inverse faults that originate by gravity-driven lateral escape of island sectors in the denser island basement, e.g. volcano spreading (8 in Fig. 14.5).

The majority of earthquakes in the Canaries in the period 1980–2010 (Fig. 14.6) have magnitudes between M1 and M3. The greater part of this seismicity is concentrated in the area of Tenerife, with two apparently different sources, one focused between Tenerife and Gran Canaria, and the other in the western part of Tenerife. The former is clearly defined by M 3–4 events forming a tight cluster, whereas the latter group is more disperse and shows only magnitudes $M < 3$.

The group of epicentres between Tenerife and Gran Canaria coincides with the alleged fault of Dash and Bosshard (1969) (dashed lines in Fig 14.6) and with a group of submarine volcanoes, some of them (e.g., Hijo de Tenerife) likely active (Schmincke and Rihm 1994; Schmincke and Graf 2000). As shown in Figs. 14.6 and 14.7, the epicentres between Tenerife and Gran Canaria present a Gaussian rather than the linear distribution which would be expected in case of fault-related seismicity (Carracedo et al. 2011).

The seismicity recorded for the island of Tenerife is comparatively low, particularly if compared with the island of Hawaii for a similar period of time (Fig. 14.7).

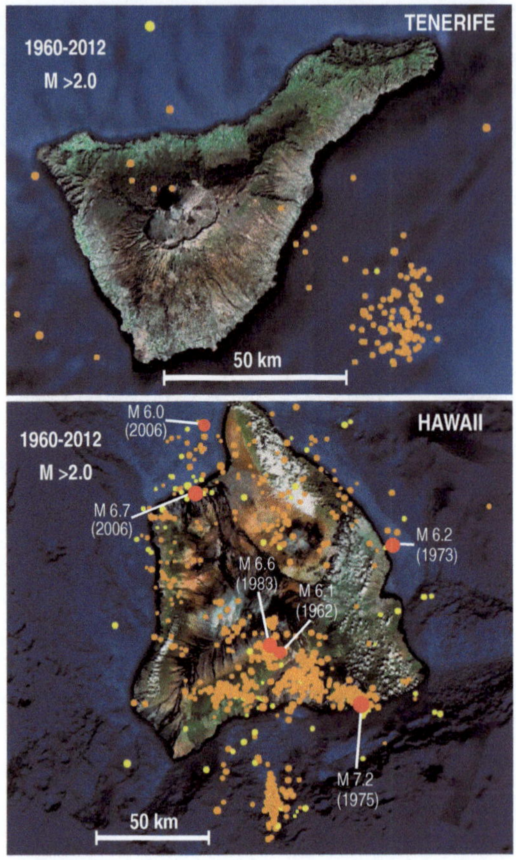

Fig. 14.7 Comparison of the M > 2.0 seismicity of Tenerife and Hawaii in the period 1960–2012 (IGN and USGS Earthquake Hazards Program, Rectangular Area Earthquake Search http://earthquake.usgs.gov/earthquakes/eqarchives/epic/epic_rect.php)

In the Hawaiian Islands, where high-quality earthquake hypocentre data have been collected by the Hawaiian Volcano Observatory since the late 1950s (the Hawaiian Volcano Observatory (HVO) has collected earthquake data since 1912, but its seismic network was modernised in the late 1950s). A wealth of information on seismic and magmatic processes is available from over 60 stations on the Big Island set up to monitor volcanic and tectonic features. In the Canaries, the period of adequate instrumental seismic data collection is very short and the number of seismic stations scant. The analysis of seismicity and assessment of seismic hazards associated with the TVC, in turn, has been hindered by the lack of instrumentation and, consequently, a database capable of establishing a baseline to define potential unrest episodes. The increase in the number of low magnitude events (M < 2.0) in May 2004, with several felt in the NW of the island according to IGN reports (M 2.0 to M 2.6, yellow dots in Fig. 14.8), was considered by some authors to be unusual "with much evidence pointing to a reawakening of volcanic activity" (García et al. 2006; Gottsmann et al. 2006; Martí et al. 2009; Dominguez-Cerdeña et al. 2011). However, other authors contested this alarming claim based on the lack of baseline data to detect and interpret which movements are normal for the area and which might indicate volcanic unrest (Carracedo et al. 2005; Carracedo and Troll 2006).

Notably, this apparently "anomalous" seismic activity coincided with the upgrade of the seismic network of the island (one station located in the northeast of the island and one inside the Las Cañadas Caldera). After the 2004 alarm—that ended without any observed volcanic manifestation but had a strong negative impact in the local and international media—the

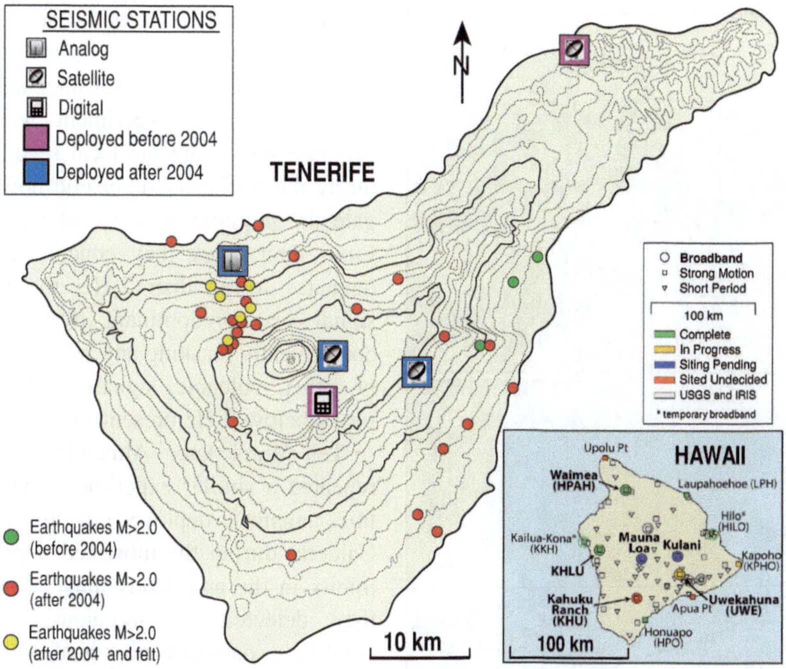

Fig. 14.8 Seismic network on Tenerife by the Spanish Geographical Institute (IGN) and $M > 2.0$ earthquakes recorded both before and after 2004. The seismic network deployed on the island of Hawaii is shown in the inset for comparison (from Shiro et al. 2009)

seismic network was upgraded to the present configuration of 5 stations, 4 of them deployed around the TVC (Fig. 14.8).

The origin of these earthquakes sparked much controversy within the scientific community. Although a deep magmatic intrusion was the favoured hypothesis, several different models were proposed. Using InSAR and GPS networks, Fernandez et al. (2003) detected deformation in the NW part of Tenerife, interpreted as subsidence caused by groundwater pumping operations performed on the island. These authors later associated the seismic crisis in 2004 to localized deformation correlated with the observed water table changes (Fernandez et al. 2005).

Gottsmann et al. (2006) observed gravity changes in the central part of Tenerife, relating the 2004 reactivation to a sub-surface mass addition without any significant widespread surface deformation, most likely caused by fluid migration at depth (arrival of new magma at depth, migration of hydrothermal fluids, or a combination of both).

Martí et al. (2009) regarded the unusual behaviour starting in spring 2004 as unequivocal evidence of volcanic unrest, based on the significant increase in seismicity onshore, the presence of volcanic tremor, the perturbation of the gravity field, the increase in fumarolic activity at Teide's crater and the opening of a new fracture with gas emission in the Orotava Valley.

Although a deep intrusion is a plausible explanation as to the origin of this perturbation, the alleged signs pointing to more specific volcanic unrest (e.g. harmonic tremor and anomalous gas emissions in the summit crater and inside the Orotava Valley) are far from evident.

The 2011/2012 volcanic eruption at El Hierro, with initially similar amplitudes, sheds new light on the 2004 Tenerife unrest. Numerous earthquakes M < 3.0 were recorded north of El Hierro by the Spanish Instituto Geográfico Nacional (IGN) since July 2011 related with intrusion of magma within the lower oceanic crust. However, conversely to what occurred in

Fig. 14.9 **a** Vent with vapour emissions with temperatures of 15–30 °C in the Orotava Valley, interpreted as a new fumarole in 2004 related to a N–S trending fracture. **b** Geological cross-section illustrating another explanation—the residual vapour emissions of hot water used for cleaning the equipment of a cheese factory located 80 m distant (after Carracedo and Troll 2006)

2004 in Tenerife, shifting seismic foci suggested that magma progressively accumulated and expanded laterally in a southward direction, causing a vertical surface deformation of about 40 mm (data from IGN). In October and November earthquake magnitude increased significantly (M > 3.0), with 5 events of M > 4.3 (IGN). The scenario changed dramatically at about 4 am on October 10, 2011, when the now frequent and strong seismicity ceased and was rather abruptly replaced by a continuous harmonic tremor, indicating the opening of a vent and thus the onset of the submarine eruption (Carracedo et al. 2012a, 2012b). A similar process of deep magma intrusion probably started in 2004 in Tenerife, but halted before approaching shallow levels to induce strong seismicity, ground deformation and eruption.

Almendros et al. (2000) carried out a field experiment in the TVC in 1994 with two short-period small-aperture seismic antennas. The objective of the experiment was to detect, evaluate and locate the local seismicity, particularly the background seismic noise to investigate the possible presence of volcanic tremor. From their work they concluded that at frequencies lower than 2 Hz, the oceanic load signal was predominant over other signals, consequently masking any weak volcanic tremor or other volcanic signals with predominant peaks below 2 Hz present in the Teide area.

Ten years later Almendros et al. (2007) repeated a similar experiment (with three seismic antennas deployed in the Las Cañadas Caldera) during the most active period of seismicity in the area (May–July 2004). This time they detected a continuous tremor (small amplitude, narrow-band signal with central frequency in the range 1–6 Hz) that started on May 18 and lasted for several weeks. This was the first time that volcanic tremor had been reported at Teide volcano. The model these authors outlined to explain the origin of the volcanic tremor is based on a deep magma injection under the northwest flank of Teide volcano. The primary process driving the tremor would be related to interactions between a shallow aquifer and an increased discharge of magmatic gases supplied by deep magma injection.

Finally, the new fracture with gas emission claimed to have opened in December in the Orotava Valley (Martí et al. 2009) is not a fracture but a lava flow cut by a *barranco* (Fig. 14.9). On cold days residual vapour emissions of a cheese factory located 80 m distant can be observed flowing through the porous basal scoria (Carracedo and Troll 2006). The cheese factory is operative and vapour emission is still active, despite the lack of continuous strong seismicity.

Low-magnitude seismicity has been continuously recorded on the island of Tenerife. Jimenez and Garcia-Fernandez (2000) carried out an experiment from December 1987 until October 1992 using a vertical-component seismic station from the network deployed by the Estación

Volcanológica de Canarias (CSIC) at the foot of Teide volcano (3 km from the summit). Routine analysis of this station's analog records during that period shows the occurrence of microseismic activity (only recorded at this station, and thus impossible to locate), that these authors correlated with intense precipitation periods and fluid circulation through fractures. Tarraga et al. (2006) and Carniel et al. (2008) interpret the seismic noise recorded in the island of Tenerife either as anthropogenic, or due to tectonic events.

In any case, the existence of anthropogenic and meteorological noise should be considered to allow better identification and discrimination of local seismicity associated with volcanic activity, to improve eruptive surveillance and prevent false alarms.

14.3 Volcanic Hazards in the TVC

A significant difference between Hawaiian and Canarian volcanism is the abundant felsic and potentially more hazardous eruptions in the latter (Clague 1987; Ablay et al. 1998; Rodríguez-Badiola et al. 2006; and Chaps. 9 and 10 in this volume). As illustrated in Fig. 1.16, the magmatic series of the TVC includes evolved rocks (phonolites, trachytes) and, therefore, eruptive mechanisms may be explosive, as described in Chap. 12.

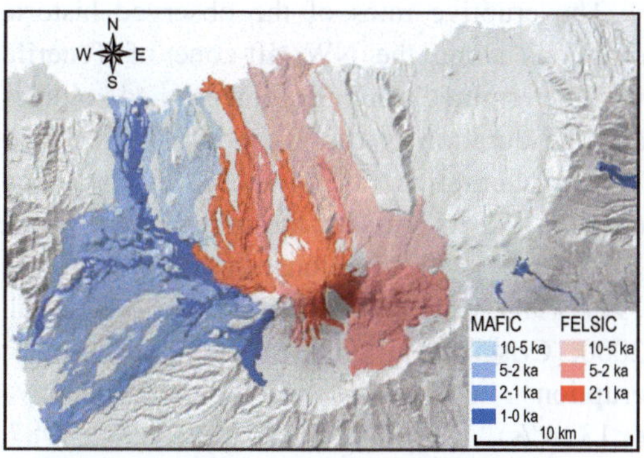

Fig. 14.10 Holocene distribution of *mafic* and *felsic* eruptions in the Teide Volcanic Complex (from Carracedo et al. 2007)

The interaction of the NW and NE rift zones of Tenerife in the construction of the post-Las Cañadas Caldera central Teide-Pico Viejo complex explains the coeval but distinctly separate mafic and felsic eruptions. The former relate to the rifts, particularly the NW rift zone, whereas the felsic volcanism is only associated to the central volcanoes, implying that explosive hazards are geographically confined (Fig. 14.10).

14.4 Lava Flow Hazards

Eruptive activity is relatively intense in Tenerife, especially in the rifts, even though the island is at present in the post-erosional rejuvenation stage (Carracedo 1994; Navarro and Farrujia 1989). During the Holocene, lava flows from the TVC have resurfaced a great part of the centre of Tenerife (Fig. 14.11). However, the potential for resurfacing large areas is considerably higher in the shield-stage islands, La Palma and El Hierro. In comparison, the greatest rate of resurfacing is found in the island of Hawaii, where historical eruptions (since 1778 A.D.) have covered most of Mauna Loa and Kilauea volcanoes (Fig. 14.12). The percentage area covered by lava in the last 750 years is >65 in the rift zones of both volcanoes (Mullineaux and Peterson 1974; Mullineaux et al. 1987).

The age and the main petrological and morphological features (Table 14.1), as well as the evaluation of cumulative volumes of lava flows (Fig. 14.13) of the recent volcanism are essential parameters to assess the eruptive hazards posed by the TVC. An important tool is a normalised lava flow resurfacing map that counts the number of lava flows on each given area of the volcano.

With this basic information, hazards related to lava flows can be summarised by answering some of the following questions: Which areas are most affected? What distance can lava flows travel considering their composition and viscosity? What volumes can be emitted and over what duration?

Calculations of the percentage area covered by lava flows in the TVC during the last 12 ky (Fig. 14.14) show that the NW rift-zone has

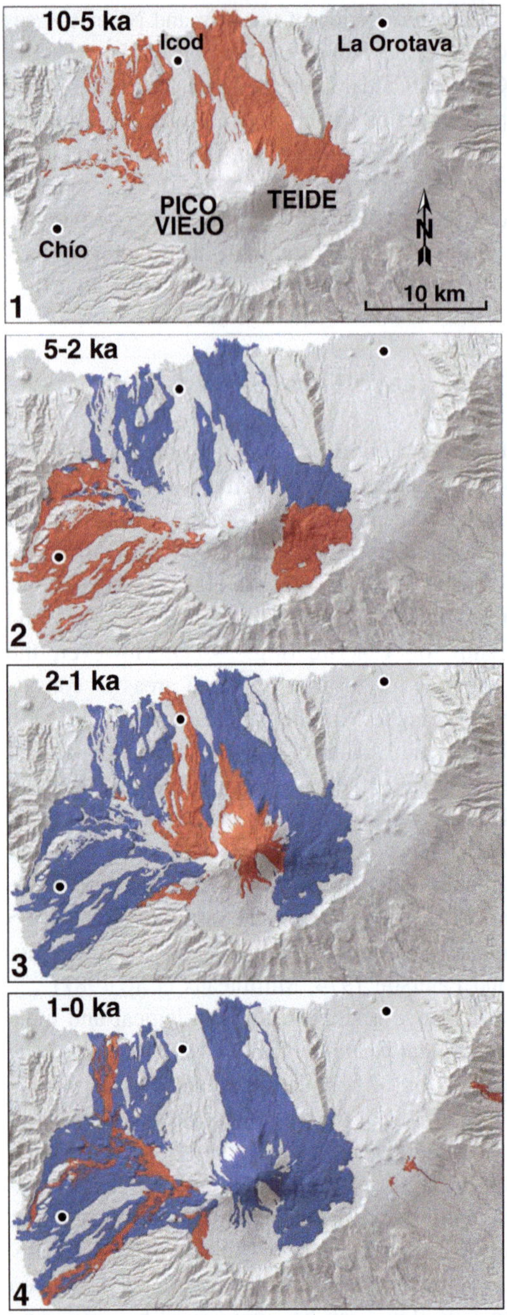

Fig. 14.11 Cumulative resurfacing of the Teide Volcanic Complex by lava flows in the last 10 ky (from Carracedo et al. 2007)

always been the most active with 95 % of its areal extent being renewed (covered) by lava flows. Despite minor historical eruptions (1704–1705), the NE rift-zone appears to be declining (resurfaced area <10 %). Almost 5.5 km^3 of lava flows were emplaced during the last 14 ky in Tenerife (Table 14.1), 1 km^3 from the NW rift-zone and 4.4 km^3 from the central edifices (Teide and peripheral domes). The thick felsic (phonolitic) flows from peripheral domes have greater volumes (e.g. 1.4 km^3 for Roques Blancos), whereas the basic lava flows of the NW rift-zone have volumes rarely exceeding 0.1 km^3 (e.g. 110×10^6 m^3 for Montañas de Chío).

Both felsic and basic flows are able to reach the coast, thus covering distances of 10–15 km. Lava flows in Hawaii reach exceptional lengths (>50 km) and eruption durations are sometimes very long (e.g., Mauna Ulu, Pu'u 'O'o), not only because of the greater rates and volumes and frequent formation of lava channels and tubes, but also because of the longer distances of eruptive centres (generally at the rift zones) from the coast in Hawaii. In both islands, the majority of lava flows reach the sea (see Figs. 14.11 and 14.12). Therefore, for hazard assessment, lava flow length is not a distinctive parameter, and hazard maps must consider that lava will flow over the "maximum" distance, i.e., all the way to the coast.

The cumulative volumes of basic and felsic lavas emitted during the last 15 ky in Tenerife is controlled by two pulses of felsic volcanism (~5000 BP and 1800–1200 BP), whereas the production of basic volcanism appears continuous through time at a rate of ca. 66×10^6 m^3/ky).

The eruptive rates of the observed historic eruptions along the NW rift-zone of Tenerife were determined using the total erupted volumes and the duration of the eruptions (Table 14.1). The high eruption rate of the 1706 Garachico eruption (71 m^3/s) is of the order of the 1669, 1980 and 1989 eruptions of Etna, and the 1977–1984 eruptions of Krafla (Harris et al. 2000; Crisci et al. 2003). The other historic eruptions in Tenerife were less productive (<13 m^3/s), similar to the 1983–1987, 1991–1993, 1999 and 2001 eruptions of Etna (Calvari et al. 1994). Considering the range of eruption rates and aspect ratios, the duration of

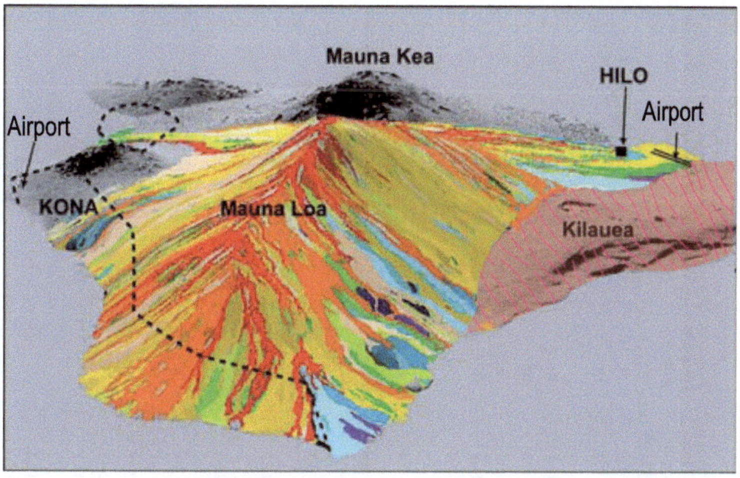

Fig. 14.12 Recent (<50 kyr) volcanism on the island of Hawaii. Note that historical eruptions (orange and red colours) have resurfaced the greater part of Kilauea Volcano and the rifts of Mauna Loa Volcano (US Geological Survey)

past eruptions of the NW rift-zone may have been typically in the order of 10 days to one month. With individual volumes greater than 500×10^6 m^3, the phonolitic eruptions of Roques Blancos, Pico Cabras, Abejera Alta and the latest eruption of Teide, the Lavas Negras, may have lasted for several months.

14.5 Hazard Maps

The main features of volcanism in the TVC served to compile a map of lava flow hazards and make comparisons with that of the island of Hawaii (Fig. 14.15). These maps need to be simple, with a limited number of zones, to capture the fundamental basis for models and risk mitigation efforts. The comparative analysis of lava flow hazard maps of both islands yields interesting conclusions. In both cases the areas with greater hazards are the rift zones. In Tenerife, significant hazards are also associated with the main central complex (Fig. 14.15a), whereas in Hawaii this is related to the main shield volcanoes, particularly the younger Mauna Loa and Kilauea (Fig. 14.15b).

In Tenerife, the greater part of the economic activity and the main settlements are concentrated in areas of old volcanism (the Mio-Pliocene shield volcanoes and the eastern part of the NE rift zone). This "spontaneous" land use is the result of agriculture having been the main economic resource of the island, with the sensible tradition of avoiding areas of recent volcanism, especially the rugged surfaces of 'a'ā flows, which are locally referred to as "*malpaíses*" (badlands).

Historically, therefore, the capital and the main populated areas, the airport, the harbour and the touristic areas were initially located in the NE of the island, on the eastern edge of the NE rift and the Orotava and Güímar Valleys.

Since 1960, massive new tourist developments have been established on the south flank of the island. This coincides with the advantage provided by the rim of the Las Cañadas Caldera, efficiently acting as an obstacle to the southward flow of lavas and small pyroclastic eruptions from the TVC, creating a topographically protected area (Fig. 14.14a).

In contrast, the economy and population of Hawaii is less protected from lava flow hazards. The capital, airports, harbours and the main tourist resorts have been built on the flanks of the very active Mauna Loa and Kilauea volcanoes (Fig. 14.15b).

In recent years there has been a growing tendency to "exaggerate" the geological hazards (seismic and eruptive) in the Canary Islands. Recent examples of these overstatements are the 2004 "reawakening of Canary Island's Teide volcano" (García et al. 2006), and the alleged potential collapse and tsunami at La Palma (Ward and Day 2001).

Table 14.1 Age and main characteristics of mafic and felsic eruptions of the Teide Volcanic Complex

Eruption	Age conventional	Age calibrated	Petrology	Location	Duration days	Area km^2	Δ alt. km	Runout length km	Aspect ratio	Slope angle degrees	Thickness (max) m	Thickness (mean) m	Volume 10^6 m^3	Eruption rate m^3/s
Chinyero	1909		Basanite	NW rift	10	2.7	0.46	4.6	2.2	5.7	6	4	10.8	12.5
Chahorra (Narices)	1798		Phonotephrite	Central	92	4.9	0.66	5.2	2.4	7.3	10	6	29.4	3.7
Garachico	1706		Basanite	NW rift	9	10.8	1.27	6.6	1.3	11.1	10	5	54.0	69.4
Arafo	1705		Basanite	NE rift	25	4.5	1.42	9.5	2.5	8.6	10	6	27.0	12.5
Siete Fuentes	1704–1705		Basanite	NE rift	6	0.3	0.16	1.7	6.5	5.4	6	4	1.2	2.3
Fasnia	1704–1705		Basanite	NE rift	21	0.9	1.12	5.5	6.5	11.8	10	7	6.3	3.5
Boca Cangrejo	350 ± 60 BP	1430–1660 AD	Tephrite	NW rift		5.8	1.34	11	1.8	7.0	7	5	29.0	
Reventada	990 ± 70 BP	900–1210 AD	Phonotephrite to phonolite	NW rift		21.8	0.93	9.9	1.1	5.4	8	6	130.8	
Volcán Negro			Tephrite to phonotephrite	NW rift		0.5	0.10	0.8	5.0	7.2		4	2.0	
Cuevas Negras			Phonoteph to phonolite	NW rift		10.8	2.45	17	1.3	8.3		5	54.0	
Teide (Lavas Negras)	1240 ± 120 BP	660–940 AD	Phonolite	Central		32.8	2.66	9.0	3.1	17.2	35	20	656.0	
Roques Blancos	1790 ± 120 BP	85–387 AD	Phonolite	Central		27.8	2.70	14.7	8.4	10.6	160	50	1390.0	
Volcán Los Hornitos	1930 ± 80 BP	39 BC–209 AD	Basanite	Central		3.2	1.05	7.7	2.5	7.8	6	5	16.0	
El Boquerón	2420 ± 140 BP	764–3942 BC	Phonolite	Central		2.1	1.24	3.5	18.3	20.8	60	30	63.0	
Los Gemelos			Phonolite	Central		0.2	0.18	0.7	39.6	14.9		20	4.0	
Mña. Mancha Ruana			Phonolite	Central		1.1	0.95	3.5	8.4	15.7		10	11	
Montañas Negras			Basanite	NW rift		4.8	1.30	7.3	2.0	10.3		5	24.0	
Volcán El Ciego	2660 ± 40 BP	900–790 BC	Phonotephrite	NW rift			1.05	11.3		5.3				
El Espárrago			Basanite	NW rift		9.2	1.36	9.7	1.5	8.1		5	46.0	
Montaña Cascajo			Basanite	NW rift		18.4	1.65	11.6	1.0	8.2		5	92.0	
Montaña Samara			Tephrite	NW rift		1.8	0.28	2.5	4.0	6.4		6	11.0	
Montaña la Botija			Basanite	NW rift		10.5	2.10	13.8	1.4	8.8		5	52.5	
Mñas. de Chío	3620 ± 140 BP	2197–1772 BC	Benmoreite	NW rift		22.0	2.06	15.6	0.9	7.6	6	5	110.0	
Montaña Bilma			Basanite	NW rift		5.8	1.20	9.6	1.8	7.2		5	29.0	

(continued)

Table 14.1 (continued)

Eruption	Age conventional	Age calibrated	Petrology	Location	Duration days	Area km^2	Δ alt. km	Runout length km	Aspect ratio	Slope	angledegrees	Thickness (max) m		
Thickness (mean) m	Volume 10^6 m^3	Eruption rate m^3/s	Montañetas Negras								Basanite	NW rift		0.2
0.06	0.8	5.9	4.3		3	0.6								
Mña. Cruz de Tea			Basanite	NW rift		3.5	0.46	5.2	2.4	5.1		5	18.0	
Mña. Majúa			Phonolite	Central		3.1	0.10	2.4	12.6	2.4		25	77.5	
Montaña Cruz Caldera			Trachyphonolite	Central		5.0	0.10	3.2	11.9	1.8		30	150.0	
Mña. Juan Evora			Basanite	Central		0.5	0.08		5.0	4.6		4	2.1	
Mña. Abejera Baja	4790 ± 140 BP		Phonolite	Central		3.4	1.06	5.6	19.2	10.9	100	40	136.0	
Mña. Abejera Alta	5170 ± 110 BP		Phonolite	Central		16.5	2.43	14.1	6.5	9.9		30	495.0	
Cueva del Ratón	5370 ± 50 BP		Phonotephrite	NW rift		15.0	1.60	9.9	1.1	9.3		5	75.0	
Pico Cabras			Phonolite	Central		15.5	2.30	12.5	11.3	10.6	110	50	775.0	
Mña. Los Conejos			Phonolite	Central		8.6	0.64	5.6	7.6	6.6		25	215.0	
Mña. Liferfe	7400 ± 40 BP		Phonotephrite	NW rift		9.1	1.50	9.1	1.5	9.5		5	45.5	
Mña. de Abeque			Mafic-basanite	NW rift			1.70	10.6		9.2				
Bocas de Dña María			Phonotephrite	Central		10.8	1.00	5.6	4.0	10.3		15	162.0	
Mña. Negra Tomillos	8220 ± 120 BP		Phonotephrite	Central		9.2	2.10	11.5	5.8	10.5		20	184.0	
Volcán del Portillo upper	11080 ± 160 BP		Phonotephrite	NE rift		4.7	1.40	8.3	4.1	9.7		10	47.0	
Volcán del Portillo lower	12020 ± 160 BP		Benmoreite	NE rift		5.3	0.90	6.2	3.8	8.3		10	53.0	
Mña. del Banco	12810 ± 60 BP		Basanite	NW rift		4.9	1.10	6.8	2.0	9.3		5	24.5	
Mña. Mostaza		15 ± 1 ky	Basanite	NE rift		2.9	0.05	4.2	2.1	0.7	6	4	11.6	
Volcán del Palmar		153 ± 6 ky	Basanite	NW rift		6.6	0.70	7	2.4	5.7		7	46.2	
Teno Alto		178 ± 6ky	Basanite	NW rift		4.4	0.70	4.5	3.0	8.9		7	30.8	
Mña. de los Silos		198 ± 8 ky	Basanite	NW rift		2	0.10	1.6	4.4	3.6		7	14.0	
Tierra del Trigo		261 ± 7 ky	Basanite	NW rift		3.0	0.70	3.9	3.6	10.3		7	21.0	
Mña. de Taco		706 ± 15 ky	Phonotephrite	NW rift		8.5	0.15	2	2.1	4.3		7	59.5	

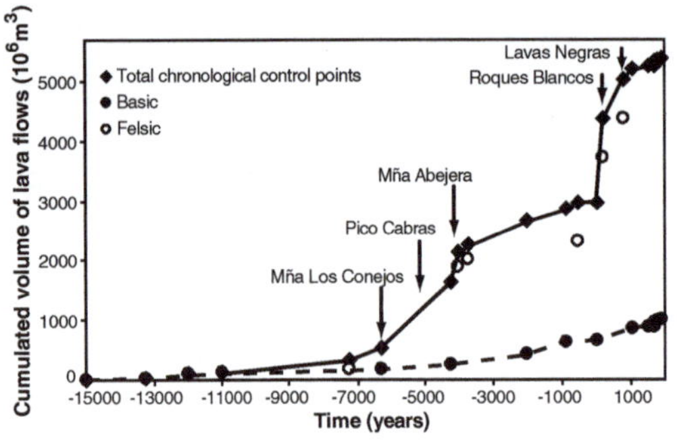

Fig. 14.13 Cumulative volumes of mafic and felsic volcanism during the last 15 ky in Tenerife. The most voluminous eruptions are indicated by arrows

Although the island of Tenerife (2,034 km^2, population 905,000 with 440 per km^2, >5 million annual visitors) is more densely populated than Hawaii (10,400 km^2, population 150.000 with 18 per km^2, 1.5 million annual visitors), the frequency and size of lava flows in Tenerife (volume, area covered, etc.) is lower than in Hawaii by about an order of magnitude. Furthermore, the historical distribution of population and economic infrastructure in Tenerife are in better "balance" with its volcanic history and eruptive hazards (Carracedo et al. 2006, 2007).

14.6 Topographic Control on Lava Flow Paths and Lava Inundation

The statement made by Klein (1982) that "Hawaiian eruptions are largely random phenomena displaying no periodicity; that is, future eruptions are relatively independent of the date of the last eruption" can be equally applied to eruptions in the Canaries, Tenerife and the TVC. In order to minimise the risk imposed by future eruptions the only means available are the detailed reconstruction of eruptive history to understand the scale of phenomena and adequate monitoring to give early warnings of imminent eruptive activity.

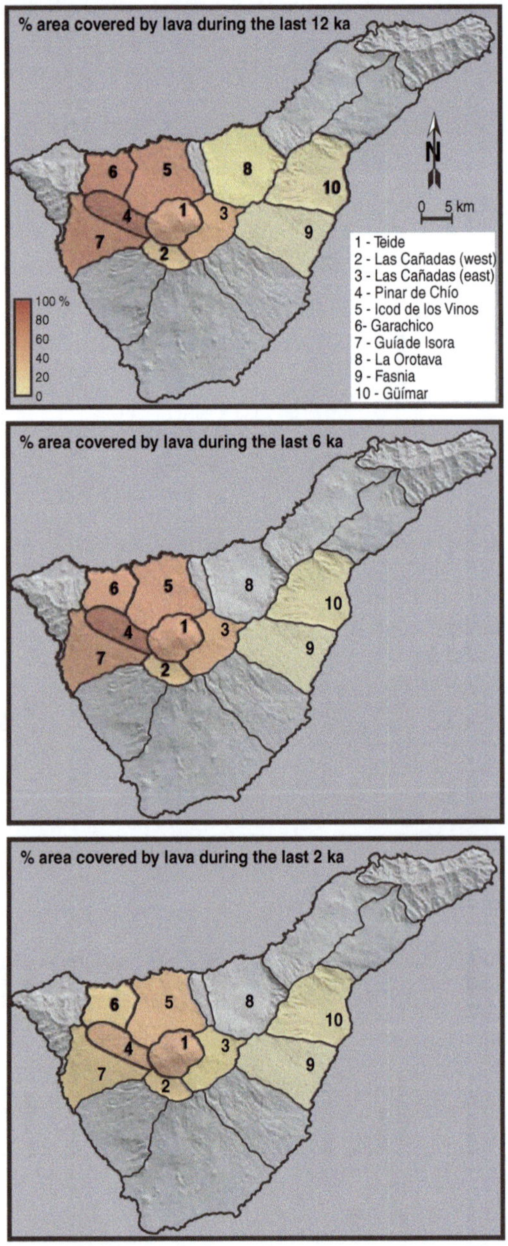

Fig. 14.14 Maps of percentage area covered by lava flows during the last 12, 6 and 2 ky, calculated after numerical mapping of the flows in a geographical information system (GIS). Main zones were defined using the topographical boundaries (hydrological basins) given and the pathways of recorded lava flows

Simulations of patterns of lava flow paths and lava inundation in different scenarios and comparison with the actual progress of well-known events (particularly historical eruptions), can provide useful information to better anticipate the behaviour of future events contrary to the

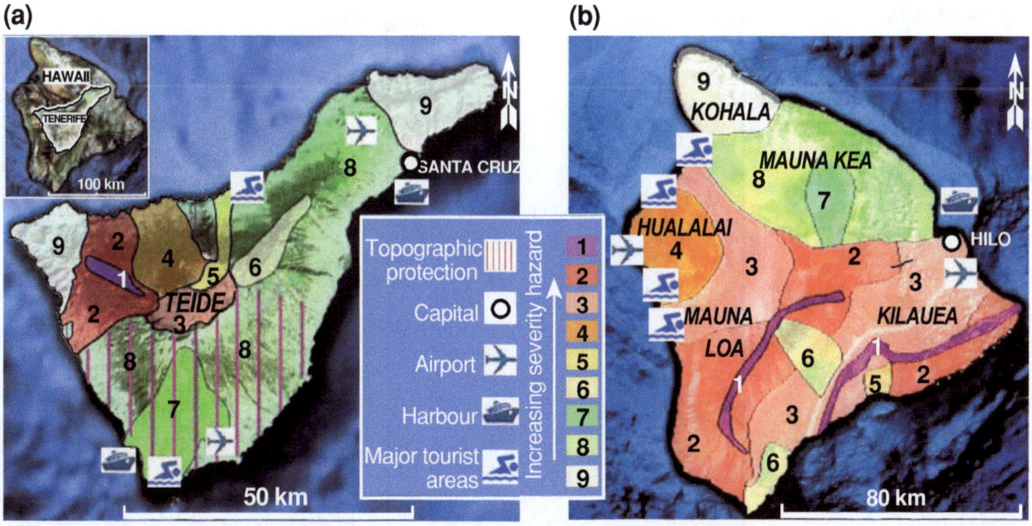

Fig. 14.15 **a** Lava flow hazard zone map of Tenerife. Hazard zones from lava flows are based chiefly on the location and frequency of Holocene eruptions (modified from Carracedo et al. 2007). **b** Current map showing volcanic hazard zones on the island of Hawaii, based on the probability of coverage by lava flows (USGS Hawaiian Volcano Observatory (HVO), http://pubs.usgs.gov/gip/hazards/maps.html)

statement by Klein (1982). Geographic Information System (GIS) and detailed geological mapping may provide valuable details to facilitate the design of strategic plans and to anticipate preventive measures.

The recently compiled GIS geological map of the TVC (Carracedo et al. 2007) provides a basis to assess the behaviour and effects of eruptions. A test of the "reliability" of these observations is to try to reconstruct historical eruptions by removing the lavas from the pre-eruption topography. Two examples are analysed here: the simulation of an eruption similar to the 1706 Garachico event (Fig. 14.16), and a simulation of the conditions required for an eruption to overflow the Las Cañadas Caldera and spill over the southern flank of the island (Fig. 14.17).

14.6.1 Inundation by a Potential Eruption Close to the 1706 Garachico Event

Figure 14.16 shows the location of a potential eruption that reproduces a situation similar to the 1706 event. Slope control will result in the partial inundation of the town of Garachico with an assumed lava thickness of 5 m reaching Garachico. Using a theoretical 10 m of lava to reach Garachico, most of the town will be flooded and the new harbour threatened. The harbour will be inundated and even buried if the eruption was to continue to the equivalent of 15–20 m thickness.

14.6.2 Overflow of the Las Cañadas Caldera

The topographic barrier formed by the wall of the Las Cañadas Caldera prevents lava flows from Teide and Pico Viejo volcanoes reaching the southern flank of the island. However, this topographic protection is incomplete because of several "*portillos*" (breaches), formed by backward erosion of the main *barrancos* (Fig. 14.17a). Some of these gaps, under 10 m high, can theoretically be overflown by volcanic eruptions similar to the 1798 Chahorra event.

A GIS simulation (Fig. 14.17b) shows that a 20 m thick layer of lava is required to allow the lava to enter Barranco de Erques directed towards the southern coast. However, even with

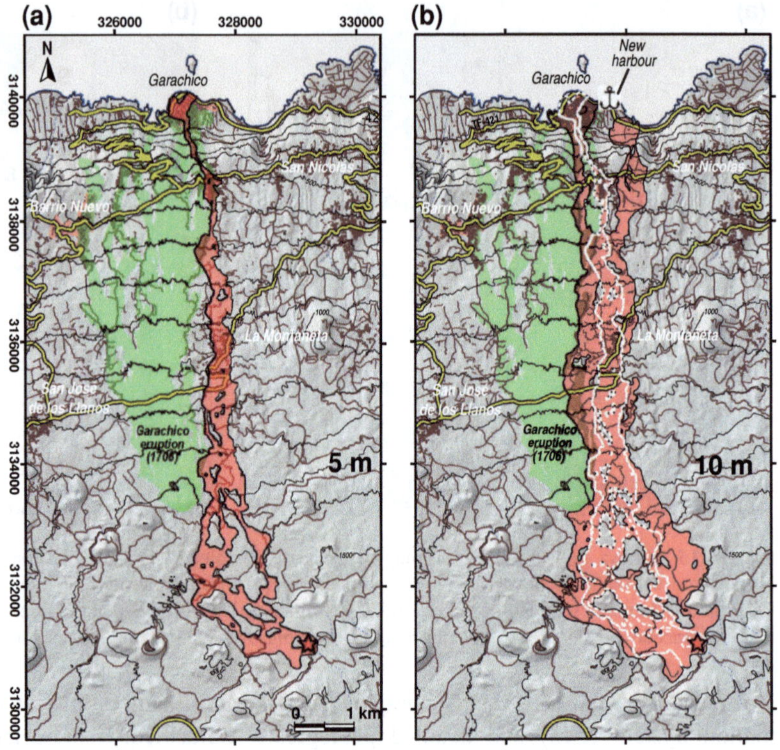

Fig. 14.16 GIS simulation of a potential eruption that could inundate the town of *Garachico*, to a similar exent as the 1706 event (**a**), and inundating it entirely (**b**)

larger volumes of lava, the hazard for the populated areas at the coast will be rather limited because the flow will still be confined inside the deep *barranco* and because the coast directly at the mouth of this *barranco* is uninhabited.

Eruptions with vents located close to the Chahorra volcano could take advantage of a similar gap at the head of Barranco del Fraile (Fig. 14.17a) to overflow the caldera rim and reach the coast. This scenario implies a more significant risk, since the village of Playa San Juan (5,000 inhabitants) is located at the mouth of this *barranco*, the foreseeable path of the lava flow. However, significant eruptive volumes would be required for this scenario.

14.7 Hazards Related to Felsic Volcanism in the TVC

Relatively frequent felsic explosive eruptions have taken place in the TVC during the Holocene (but none in the last thousand years), all of them located inside the Las Cañadas Caldera (see Fig. 14.10).

Contrasting the very explosive volcanism of the Las Cañadas Volcano (LCV) prior to the 200 ky Icod lateral collapse, the post-collapse nested construction of the TVC has overall been characterised by low-explosivity magmas. As described in Chaps. 10 and 12, despite the significant volume of felsic (phonolitic) volcanics (Ablay et al. 1998; Rodríguez-Badiola et al. 2006), there is no evidence in this latest volcanic cycle of the abundant plinian episodes of the pre-TVC activity of the Las Cañadas Volcano (>150 km^3 according to Edgar et al. 2007).

Low-explosivity eruptions, mainly mafic and felsic strombolian, characterise the TVC volcanism. However, there are some interesting examples of more violent episodes: the subplinian Montaña Blanca eruption (Ablay et al. 1995), and hydromagmatic eruptions of Pico Viejo and Calvas del Teide (Pérez Torrado et al. 2004; Pérez Torrado et al. 2006; del Potro et al. 2009). Garcia et al. (2011) reported the presence

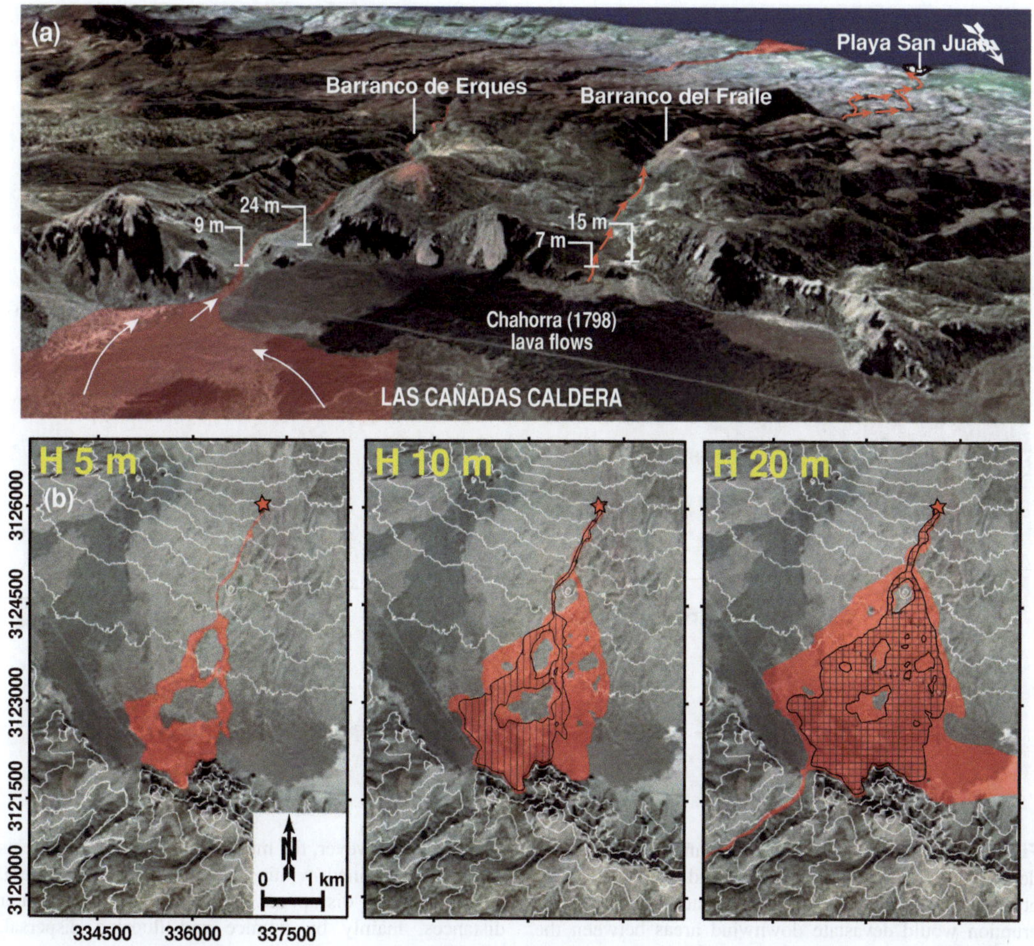

Fig. 14.17 **a** Oblique view from the NW of the southern wall of the Las Cañadas Caldera showing the breaches (portillos) carved by the backward erosion of the main *barrancos*. These gaps would be the pathways for a potential overflow of lava for future eruptions inside the Las Cañadas Caldera, threatening villages on the southern coast. **b** GIS simulation shows that for overflow to commence a 20 m thick lava flow is needed

of density current deposits, including ignimbrites and block and ash deposits, in the Holocene eruptive history of the Teide–Pico Viejo stratovolcanoes, probably associated with the gravitational collapse of domes or the front of phonolitic flows.

The Montaña Blanca eruption exemplifies the main explosive event in the TVC and the hazards that this type of volcanism presents. From this point of view, the most interesting episode of this complex (see Fig. 8.33) is the Montaña Blanca pumice fallout deposit, a single well-sorted massive bed extending northeast towards El Portillo in an elongated area reflecting southwest dominant winds during the pumice eruption (Ablay et al. 1995).

Ablay et al. (1995) carried out a detailed study of the pumice bed from sixty pits up to 3.5 m depth, constructing isopach and isopleth maps (Fig. 14.18). According to the criteria of Walker (1973) and Pyle (1989), these authors classify the deposit as subplinian, since the 0.01 T_{max} isopach greatly exceeds 40 km^2, and is probably of the order of 100–200 km^2.

According to Ablay et al. (1995), a repetition of this eruption would devastate downwind areas

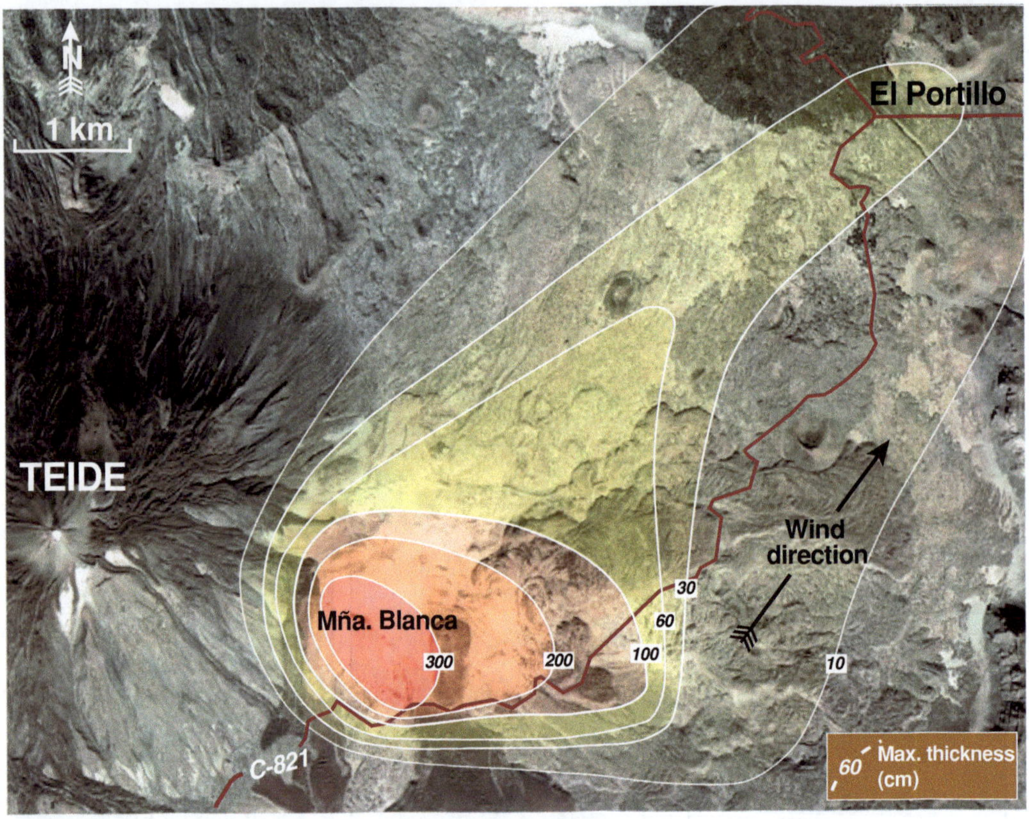

Fig. 14.18 Isopach map of the Montaña Blanca pumice deposit (contours in centimetres). Modified from Ablay et al. (1995). These authors claim that a repeat of this eruption would devastate downwind areas between the centre of Tenerife and the north-east coast within 20 km of the vent. However, the most significant adverse effects would be confined within the Las Cañadas Caldera, although some disruption may be caused at greater distances, mainly by pumice and fluorine dispersal, depending on wind direction and strength

between the centre of Tenerife and the north-east coast within about 20 km of the vent, and perhaps affect air traffic and even have widespread repercussions. However, the main impact of such an event would probably be confined inside the Las Cañadas Caldera and would depend on the wind direction and strength during the eruption.

Eruptions triggered by magma mixing have been relatively frequent in the TVC. This mechanism commonly produces very explosive events, but those recognised (e.g., Montaña Reventada) are consistently of relative low explosivity (Rodríguez-Badiola et al. 2006; Wiesmaier et al. 2011) and are restricted to the rift zone-central complex boundary (see Chap. 7).

14.8 Ground Deformation Hazards

Ground deformation is a main source of hazard in the Hawaiian Islands but is less significant in the Canaries. In Hawaii, vertical and horizontal motions of as much as 3–5 m occur in association with large earthquakes and the shallow migration and eruption of magma; slower motions of as much as 20 cm/yr are almost continuous, particularly near Kilauea's summit (Kauahikaua et al. 1994).

Ground deformation in Tenerife is very limited in comparison and related hazards appear negligible. Deformation is restricted to long-term gravitational sinking of the dense core of

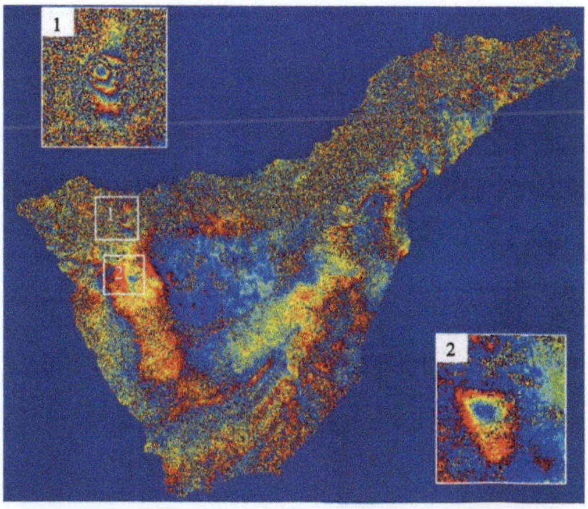

Fig. 14.19 Differential interferogram of Tenerife (1993–2000) showing two areas of subsidence interpreted to result from water table changes due to intensive extractions (from Fernandez et al. 2005)

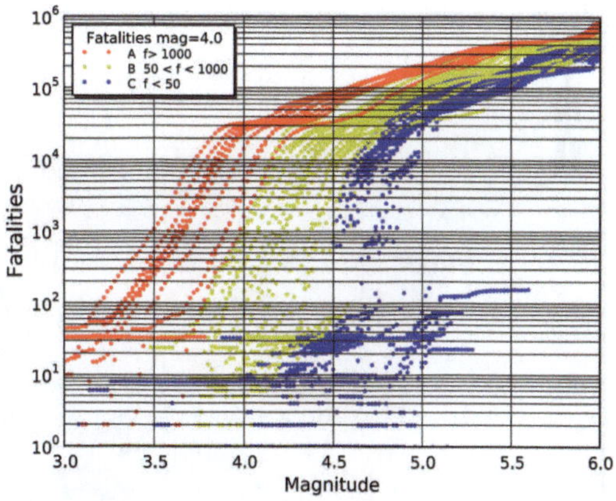

Fig. 14.20 Number of potential fatalities versus the magnitude of an eruption. The model shows that the major impacts (*red curves*) are produced by the Teide summit and north flank vents. According to this model "the emergency plan for Tenerife must provide for the evacuation of more than 100,000 people" (from Marrero et al. 2012)

the island. More localised deformation in the west of the island (1 and 2 in Fig. 14.19) has been associated with water table changes due to intensive extractions (Fernandez et al. 2005), and present a very isolated hazard only. These areas are well known and hence represent little danger to the population of the island.

14.9 The Present State of the TVC Plumbing System

In contrast to the relatively well-known volcanic plumbing system of Hawaiian volcanoes (e.g., Kilauea Volcano; Ryan 1987), very little information is available on the subcaldera magma reservoir of the TVC. Attempts to define the main features of the TVC plumbing system have led to contradictory interpretations. Albert-Beltran et al. (1986, 1990) and Diez Gil and Albert (1989) identified a central post-caldera magmatic chamber located at sea level, relatively small (~4 km in diameter) and at ~400 °C. Valentin et al. (1989) based their characterisation of this magmatic chamber on gas isotope geochemistry, concluding that the source of gases from Teide proceed from a 400 °C post-caldera magmatic chamber approximately at sea level depth underneath Teide.

In contrast, Ortiz et al. (1994) postulated a deeper magmatic chamber, almost completely solidified, sustaining that the Teide volcano is practically extinct: "The highest temperature of the 6 km deep Teide magmatic chamber is 400 °C, the final stage of cooling from the geological point of view". Araña et al. (1989) argued that this magmatic chamber is in a quasi-terminal state, probably formed by independent small magma bodies with temperatures as low as 250 °C. These authors claim that "the scientific thermodynamic models show that the present cycle of activity of Teide is in a terminal stage and, consequently, the explosivity of expected eruptions of Teide is similar to that of any other eruption in the Canaries". Araña et al. (1989) conclude that "We can assure that an eruption of Teide is virtually impossible, that it is an exhausted system". Ortiz et al. (1994 insisted in stating that "The work carried out by using petrology, fluid geochemistry, geophysics, etc., denotes the absence of magmatic chambers with the capability to produce explosive eruptions in the Teide central volcano. The hazards are magnified by the media and 'unscrupulous' research groups believing that overstating the risk will bring funds and promotion".

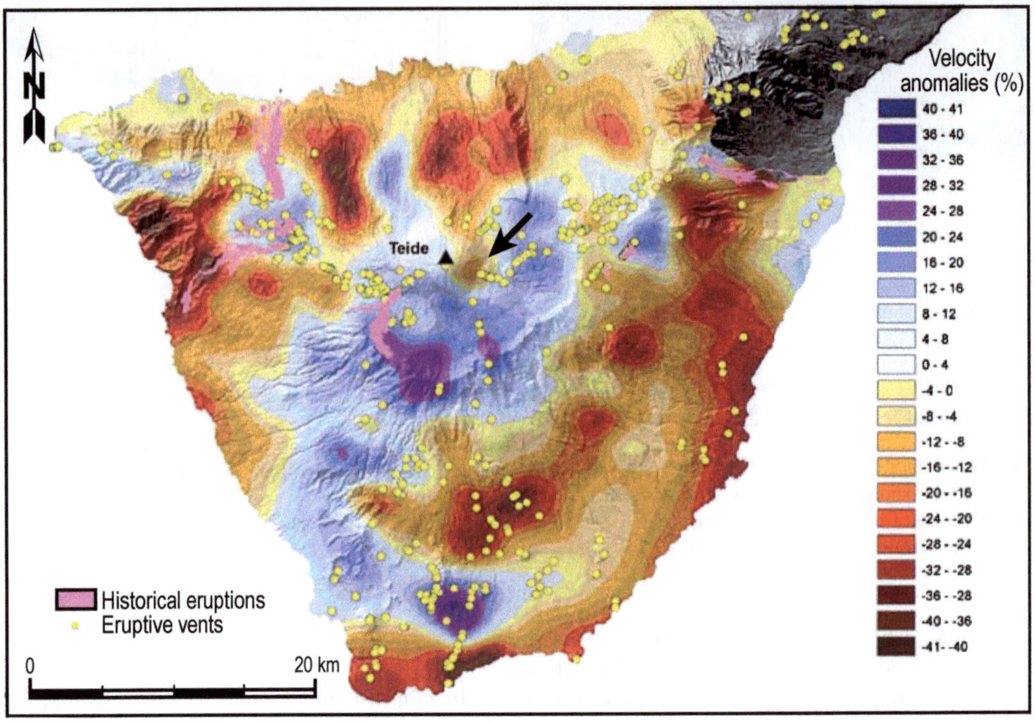

Fig. 14.21 Horizontal section at sea level of the seismic tomography model of Tenerife (from García Yeguas 2010)

This general notion changed with the increase in the number of low-magnitude events in May 2004 that was considered by some authors as unusual and pointed to a reawakening of Canary Islands' Teide Volcano (García et al. 2006; Gottsmann et al. 2006; Martí et al. 2009; Dominguez-Cerdeña et al. 2011). In a recent article, Marrero et al. (2012) assessed the impact that a possible eruption from the Tenerife Central Volcanic Complex would have on the population at risk, and urged the need to foresee "evacuating more than 100,000 persons in the case of an eruption warning in the Central Volcanic Complex of Tenerife Island. This situation would be similar to that of Vesuvius" (Fig. 14.20).

This assumption is at odds with the results obtained in a tomography study (performed using airgun shots) of Tenerife (García Yeguas 2010). In this model, the island of Tenerife shows a central nucleus of high velocities encircled by low velocity areas (Fig. 14.21). High velocity of seismic waves is commonly associated in oceanic islands to dyke swarms (rifts), plutonic bodies derived from the cooling of shallow magma reservoirs or dense cores of cumulates. The low velocity areas extending on the flanks of the island are probably associated with fractured and porous formations.

The only area of low velocity in Teide appears under Montaña Blanca in the section of the model corresponding to sea level and above (Fig. 14.22), but not in a deeper section (5,500 m b.s.l.). This has been interpreted by García Yeguas (2010) as associated with zones of hydrothermal alteration or with a shallow magma intrusion, although the latter interpretation is uncertain as only P-waves were used in this study.

14.10 The Present Risk Mitigation Challenge

Naturally, Teide poses hazards and we have now established that lava flow-, pyroclastic flow- and air-fall hazards are present at Teide (see also Martí et al., 2012). Of these, air-fall events

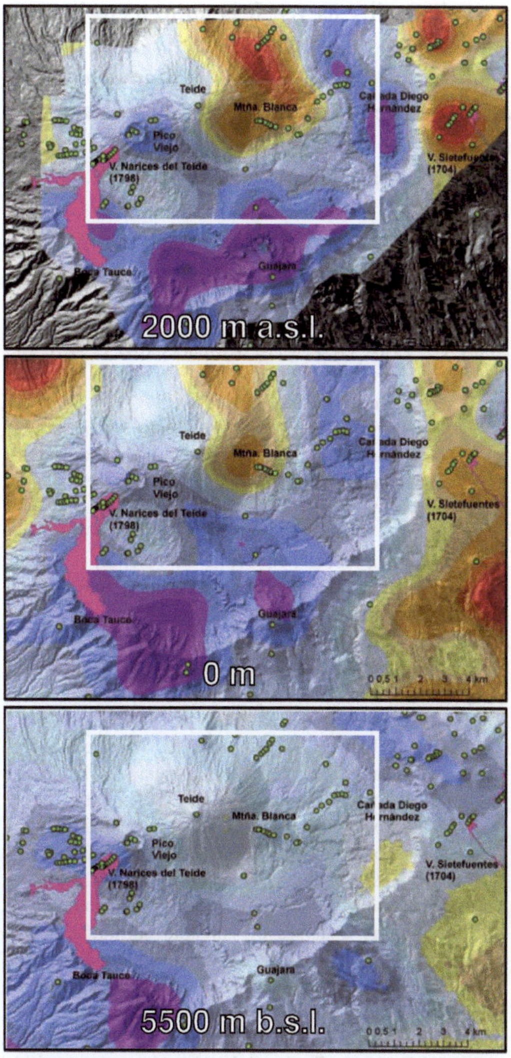

Fig. 14.22 Sections at different depths of the seismic tomography model of Tenerife. An area of low velocity appears at shallow depths in the zone of Montaña Blanca and the Abejeras-Pico Cabras lava domes (from García Yeguas 2010)

probably represent a more serious risk than for example lava flows, but the most serious of these is the potential threat of a large pyroclastic eruption (e.g., Martí et al., 2012; Marrero et al., 2012). Probabilities for such an event are low (e.g. with about 21 % likelihood during the next 100 years; Martí et al., 2012), but estimates of casualties of the order of 100,000 inhabitants are predicted for such a scenario if no precautions are taken (e.g. if no evacuation is performed).

On the other hand, large eruptions, for all we know, do not happen instantaneously, i.e. they do not occur without premonition. Volcanoes usually display periods of significant unrest prior to eruption (e.g., Carracedo et al. 2012a; Druitt et al. 2012) and large eruptions do signal their arrival. Teide, like all other volcanoes, is likely to signal frequently on its changing condition, allowing us to take precautions and move out of the way. This precursory behaviour should allow us to readily identify an impending eruptive scenario weeks to even months in advance, giving ample preparation and response time if thoroughly planned and managed well (cf. Carracedo et al. 2012a, b).

Given today's volcano-monitoring capabilities, future eruptions should cause limited damage if handled effectively and with foresight. It is noteworthy in this context that none of the historical eruptions in the Canaries have caused a single fatality as yet. It is up to us to learn to read Teide's messages correctly and prevent fatalities in the future as well to eventually achieve a life in symbiosis with the volcano, and move out of harm's way should Teide need to clear its throat again.

References

Ablay GJ, Ernst GGJ, Marti J, Sparks RSJ (1995) The ~2 ka subplinian eruption of Montaña Blanca, Tenerife. Bull Volcanol 57:337–355

Ablay GJ, Carroll MR, Palmer MR, Martí J, Sparks RSJ (1998) Basanite-phonolite lineages of the Teide-Pico Viejo volcanic complex, Tenerife, Canary Islands. J Petrol 39:905–936

Albert-Beltran JF, Arafia V, Diez JL, Filly A, Fontes JC, Garcia De la Noceda C, Ocafia L, Valentin A (1986) Modelo termodinámico de la actividad fumaroliana del Teide. An Fis 82:186–201

Albert-Beltran JF, Araña V, Diez JL, Valentin A (1990) Physical-chemical conditions of the Teide volcanic system (Tenerife, Canary-Islands). J Volcanol Geotherm Res 43:321–332

Almendros J, Ibanez JM, Alguacil G, Morales J, Del Pezzo E, La Rocca M, Ortiz R, Arana V, Blanco MJ (2000) A double seismic antenna experiment at Teide Volcano: existence of local seismicity and lack of evidences of volcanic tremor. J Volcanol Geotherm Res 103:439–462

Almendros J, Ibanez JM, Carmona E, Zandomeneghi D (2007) Array analyses of volcanic earthquakes and tremor recorded at Las Cañadas caldera (Tenerife Island, Spain) during the 2004 seismic activation of Teide volcano. J Volcanol Geotherm Res 160:285–299

Ambroggi R (1963) Etude géologique de versant méridional du Haut Atlas Occidental et de la plaine du Souss. Not Mém Géol Mar, 15:1–321

Anguita F, Hernan F (1975) Propagating fracture model versus a hot spot origin for Canary-Islands. Earth Planet Sci Lett 27:11–19

Anguita F, Hernan F (2000) The Canary Islands origin: a unifying model. J Volcanol Geotherm Res 103:1–26

Araña V, Barberi F, Ferrara G (1989) El Complejo volcanico del Teide-Pico Viejo. In: Araña V, Coello J (eds) Los Volcanes y la Caldera del Parque Nacional del Teide (Tenerife, Islas Canarias). ICONA, Madrid, pp 101–126

Bosshard E, MacFarlane DJ (1970) Crustal structure of western Canary-Islands from seismic refraction and gravity data. J Geophys Res 75:4901–4918

Calvari S, Coltelli M, Neri M, Pompilio M, Scribano V (1994) The 1991–1993 Etna eruption: chronology and lava flow-field evolution. Acta Vulcanol 4:1–14

Carniel R, Tarraga M, Jaquet O, Ortiz R, Garcia A (2008) The seismic noise at Las Cañadas volcanic caldera, Tenerife, Spain: Persistence characterization, and possible relationship with regional tectonic events. J Volcanol Geotherm Res 173:157–164

Carracedo JC (1994) The Canary Islands—an example of structural control on the growth of large oceanic-island volcanoes. J Volcanol Geotherm Res 60:225–241

Carracedo JC, Troll VR (2006) Seismicity and gas emissions on Tenerife: a real cause for alarm? Geol Today 22:138–141

Carracedo JC, Pérez Torrado FJ, Rodríguez Badiola E, Hansen Machín A, Paris R, Guillou H, Scaillet S (2005) Análisis de los riesgos geológicos en el Archipiélago Canario: origen, características, probabilidades y tratamiento. Anu Estud Atl 51:513–574

Carracedo JC, Rodríguez Badiola E, Guillou H, Paterne M, Scaillet S, Pérez Torrado FJ, Paris R, Criado C, Hansen A, Arnay M, González Reimers E, Fra- Paleo U, González Pérez R (2006) Los volcanes del Parque Nacional del Teide: El Teide, Pico Viejo y las dorsales activas de Tenerife. In: Carracedo JC (ed) Los Volcanes del Parque Nacional del Teide. Serie Técnica, Madrid, pp 175–199

Carracedo JC, Rodríguez Badiola E, Guillou H, Paterne M, Scaillet S, Pérez Torrado FJ, Paris R, Fra-Paleo U, Hansen A (2007) Eruptive and structural history of Teide Volcano and rift zones of Tenerife, Canary Islands. Geol Soc Am Bull 119:1027–1051

Carracedo JC, Guillou H, Nomade S, Rodriguez-Badiola E, Perez-Torrado FJ, Rodriguez-Gonzalez A, Paris R, Troll VR, Wiesmaier S, Delcamp A, Fernandez-Turiel JL (2011) Evolution of ocean-island rifts: the northeast rift zone of Tenerife, Canary Islands. Geol Soc Am Bull 123:562–584

Carracedo JC, Pérez Torrado FJ, Rodriguez-Gonzalez A, Fernandez-Turiel JL, Klügel A, Troll VR, Wiesmaier S (2012a) The ongoing volcanic eruption of El Hierro, Canary Islands. Eos Trans AGU 93:89–90

Carracedo JC, Pérez Torrado FJ, Rodriguez-Gonzalez A, Fernandez-Turiel JL, Troll VR, Wiesmaier S (2012b) The 2011 submarine volcanic eruption in El Hierro (Canary Islands). Geol Today 28:53–58

Clague DA (1987) Hawaiian xenolith populations, magma supply rates, and development of magma chambers. Bull Volcanol 49:577–587

Crisci GM, Gregorio S, Rongo R, Scarpelli M, Spataro W, Calvari S (2003) Revisiting the 1669 Etnean eruptive crisis using a cellular automata model and implications for volcanic hazard in the Catania area. J Volcanol Geotherm Res 123:211–230

Dash BP, Bosshard E (1969) Seismic and gravity investigations around the western Canary Islands. Earth Planet Sci Lett 7:169–177

del Potro R, Pinkerton H, Hurlimann M (2009) An analysis of the morphological, geological and structural features of Teide stratovolcano, Tenerife. J Volcanol Geotherm Res 181:89–105

Diez Gil JL, Albert JF (1989) Modelo termodinámico de la cámara magmática del Teide. In: Los volcanes y la caldera del Parque Nacional del Teide, vol 7. Serie Técnica. ICONA, pp 343–346
Dominguez-Cerdeña I, del Fresno C, Rivera L (2011) New insight on the increasing seismicity during Tenerife's 2004 volcanic reactivation. J Volcanol Geotherm Res 206:15–29
Druitt TH, Costa F, Deloule E, Dungan M, Scaillet B (2012) Decadal to monthly timescales of magma transfer and reservoir growth at a caldera volcano. Nature 482:77–97
Edgar CJ, Wolff JA, Olin PH, Nichols HJ, Pittari A, Cas RAF, Reiners PW, Spell TL, Martí J (2007) The late quaternary Diego Hernandez formation, Tenerife: Volcanology of a complex cycle of voluminous explosive phonolitic eruptions. J Volcanol Geotherm Res 160:59–85
Faccenna C, Piromallo C, Crespo-Blanc A, Jolivet L, Rossetti F (2004) Lateral slab deformation and the origin of the western Mediterranean arcs. Tectonics 23, TC1012, doi:10.1029/2002tc001488
Fernandez J, Yu TT, Rodriguez-Velasco G, Gonzalez-Matesanz J, Romero R, Rodriguez G, Quiros R, Dalda A, Aparicio A, Blanco MJ (2003) New geodetic monitoring system in the volcanic island of Tenerife, Canaries, Spain. Combination of InSAR and GPS techniques. J Volcanol Geotherm Res 124:241–253
Fernandez J, Romero R, Carrasco D, Tiampo KF, Rodriguez-Velasco G, Aparicio A, Arana V, Gonzalez-Matesanz FJ (2005) Detection of displacements on Tenerife Island, Canaries, using radar interferometry. Geophys J Int 160:33–45
Frizon de Lamotte D, Saint Bezar B, Bracène R, Mercier E (2000) The two main steps of the Atlas building and geodynamics of the western Mediterranean. Tectonics 19:740–761
Funck T, Dickmann T, Rihm R, Krastel S, Lykke-Andersen H, Schmincke H-U (1996) Reflection seismic investigations in the volcaniclastic apron of Gran Canaria and implications for its volcanic evolution. Geophys J Int 125:519–536
García A, Vila J, Ortiz R, Macià R, Sleeman R, Marrero JM, Sánchez N, Tárraga M, Correig AM (2006) Monitoring the reawakening of Canary Island's Teide volcano. Eos Trans AGU 87:61–72
Garcia O, Martí J, Aguirre G, Geyer A, Iribarren I (2011) Pyroclastic density currents from Teide-Pico Viejo (Tenerife, Canary Islands): implications for hazard assessment. Terra Nova 23:220–224
García Yeguas MA (2010) Estudio de heterogeneidades laterales de volcanes activos: tomografía sismica de alta resolución de la isla de Tenerife. Ph. D. Thesis, University of Granada, Granada
Geyer A, Martí J (2010) The distribution of basaltic volcanism on Tenerife, Canary Islands: Implications on the origin and dynamics of the rift systems. Tectonophysics 483:310–326
Gottsmann J, Wooller L, Martí J, Fernandez J, Camacho AG, Gonzalez PJ, Garcia A, Rymer H (2006) New evidence for the reawakening of Teide volcano. Geophys Res Lett 33:L20311. doi:10.1029/2006gl027523
Harris C, Smith HS, le Roex AP (2000) Oxygen isotope composition of phenocrysts from Tristan da Cunha and Gough Island lavas: variation with fractional crystallization and evidence for assimilation. Contrib Mineral Petrol 138:164–175
Jimenez MJ, Garcia-Fernandez M (2000) Occurrence of shallow earthquakes following periods of intense rainfall in Tenerife, Canary Islands. J Volcanol Geotherm Res 103:463–468
Kauahikaua JP, Moore RB, Delaney P (1994) Volcanic activity and ground deformation hazard analysis for the Hawaii geothermal project environmental impact statement. US Geol Surv Open-File Rep 94–553:44
Klein FW (1982) Patterns of historical eruptions at Hawaiian volcanoes. J Volcanol Geotherm Res 12:1–35
Klein FW, Koyanagi RY, Nakata JS, Tanigawa WR (1987) The seismicity of Kılauea's magma system. In: Decker RW, Wright RL, Stauffer PJ (eds) Volcanism in Hawaii, vol 2. US Geol Surv Prof Pap 1350, pp 1019–1185
Klein, FW, Frankel AD, Mueller CS, Wesson RL, Okubo PG (2001) Seismic Hazard in Hawaii: high rate of large earthquakes and probabilistic ground-motion maps. Bull Seismol Soc America 91(3):479–498
Krastel S, Schmincke H-U (2002) Crustal structure of northern Gran Canaria, Canary Islands, deduced from active seismic tomography. J Volcanol Geotherm Res 115:153–177
Krastel S, Schmincke H-U, Jacobs GL, Rihm R, Le Bas TP, Alibés B (2001) Submarine landslides around the Canary Islands. J Geophys Res 106:3977–3997
Marrero JM, Garcia A, Llinares A, Rodriguez-Losada JA, Ortiz R (2012) A direct approach to estimating the number of potential fatalities from an eruption: application to the central volcanic complex of Tenerife island. J Volcanol Geotherm Res 219:33–40
Martí J, Ortiz R, Gottsmann J, Garcia A, De La Cruz-Reyna S (2009) Characterising unrest during the reawakening of the central volcanic complex on Tenerife, Canary Islands, 2004–2005, and implications for assessing hazards and risk mitigation. J Volcanol Geotherm Res 182:23–33
Martí J, Sobradelo R, Felpeto A, Garcia A (2012) Eruptive scenarios of phonolitic volcanism at Teide-Pico Viejo volcanic complex (Tenerife, Canary Islands). Bull Volcanol 74:767–782
Martínez W, Buitrago J (2002) Sedimentación y volcanismo al este de las islas de Fuerteventura y Lanzarote (Surco de Fúster Casas). Geogaceta 32:51–54
Mezcua J, Buforn E, Udias A, Rueda J (1992) Seismotectonics of the Canary-Islands. Tectonophysics 208: 447–452
Mullineaux DR, Peterson DW (1974) Volcanic Hazards on the Island of Hawaii. U.S. Geol Surv Open-File Rep, pp 74–239
Mullineaux DR, Peterson DW, Crandell DR (1987) Volcanic hazards in the Hawaiian Islands. In: Decker

RW, Wright RL, Stauffer PJ (eds) Volcanism in Hawaii, vol 1. US Geol Surv Prof Pap 1350, pp 599–621

Navarro JM, Farrujia I (1989) Plan Hidrológico Insular de Tenerife. Zonificación hidrogeológica, aspectos geológicos e hidrogeológicos. Cabildo Insular de Tenerife

Ortiz R, Vieira V, Blanco MJ, Araña V (1994) Equipo de volcanología y geofísica volcánica. In: WOVO/IAVCEI/UNESCO (ed) Directory of Volcano Observatories, 1993-1994. WOVO/IAVCEI/UNESCO, Paris

Pérez Torrado FJ, Carracedo JC, Paris R, Hansen A (2004) Descubrimiento de depósitos freatomagmáticos en las laderas septentrionales del estratovolcán Teide (Tenerife, Islas Canarias): relaciones estratigráficas e implicaciones volcánicas. Geotemas 6:163–166

Pérez Torrado FJ, Carracedo JC, Paris R, Rodríguez Badiola E, Hansen A (2006) Erupciones freatomagmáticas del complejo volcánico Teide-Pico Viejo. In: Carracedo JC (ed) Los Volcanes del Parque Nacional del Teide. Serie Técnica, Madrid, pp 345–357

Pyle DM (1989) The thickness, volume and grainsize of tephra fall deposits. Bull Volcanol 51:1–15

Rodríguez-Badiola E, Pérez-Torrado FJ, Carracedo JC, Guillou H (2006) Petrografía y Geoquímica del edificio volcánico Teide-Pico Viejo y las dorsales noreste y noroeste de Tenerife. In: Carracedo JC (ed) Los volcanes del Parque Nacional del Teide/El Teide. Pico Viejo y las dorsales activas de Tenerife. Naturaleza y Parques Nacionales-Serie Técnica, Madrid, pp 129–186

Ryan MP (1987) Elasticity and contractancy of Hawaiian olivine tholeiite and its role in the stability and structural evolution of subcaldera magma reservoirs and rifts system. In: Decker RW, Wright RL, Stauffer PJ (eds) Volcanism in Hawaii, vol 2. US Geol Surv Prof Pap 1350, pp 1395–1447

Schmincke H-U, Graf G (2000) DECOS/OMEX II, Cruise no. 43, 25 Nov 1998–14 Jan 1999. METEOR-Berichte, vol 00-2. Universität Hamburg, Hamburg

Schmincke H-U, Rihm R (1994) Ozeanvulkan 1993, Cruise No. 24, 15 Apr–9 May 1993. METEOR-Berichte. University of Hamburg, Hamburg

Serpelloni E, Vannucci G, Pondrelli S, Argnani A, Casula G, Anzidei M, Baldi P, Gasperini P (2007) Kinematics of the Western Africa-Eurasia plate boundary from focal mechanisms and GPS data. Geophys J Int 169:1180–1200

Shiro B, Koyanagi K, Hirshorn B (2009) Status report: PTWC's seismic network

Steckler MS, Tenbrink US (1986) Lithospheric strength variations as a control on new plate boundaries—examples from the northern red-sea region. Earth Planet Sci Lett 79:120–132

Tarraga M, Carniel R, Ortiz R, Marrero JM, Garcia A (2006) On the predictability of volcano-tectonic events by low frequency seismic noise analysis at Teide-Pico Viejo volcanic complex, Canary Islands. Nat Hazards Earth Syst Sci 6:365–376

Urgeles R, Canals M, Baraza J, Alonso B (1998) Seismostratigraphy of the western flanks of El Hierro and La Palma (Canary Islands): A record of Canary Islands volcanism. Mar Geol 146:225–241

Valentin A, Albert JF, Diez JL, Garcia de la Noceda C (1989) Emanaciones magmáticas residuales en Tenerife. In: Los volcanes y la caldera del Parque Nacional del Teide, vol 7. Serie Técnica. ICONA, pp 299–310

von Humboldt A, Bonpland A (1805) Voyage aux régions équinoxiales du nouveau continent, fait en 1799–1804. Levrault, Paris

Walker GPL (1973) Explosive volcanic eruptions—a new classification scheme. Geol Rund 62:431–446

Ward SN, Day S (2001) Cumbre Vieja Volcano—Potential collapse and tsunami at La Palma, Canary Islands. Geophys Res Lett 28:3397–3400

Watts AB (1994) Crustal structure, gravity-anomalies and flexure of the lithosphere in the vicinity of the Canary-Islands. Geophys J Int 119:648–666

Watts AB, Peirce C, Collier J, Dalwood R, Canales JP, Henstock TJ (1997) A seismic study of lithospheric flexure in the vicinity of Tenerife, Canary Islands. Earth Planet Sci Lett 146:431–447

Wiesmaier S, Deegan F, Troll V, Carracedo JC, Chadwick J, Chew D (2011) Magma mixing in the 1100 AD Montaña Reventada composite lava flow, Tenerife, Canary Islands: interaction between rift zone and central volcano plumbing systems. Contrib Mineral Petrol 162:651–669

Wood N, Church A, Frazier T, Yarnal B (2007) Variations in community exposure and sensitivity to tsunami hazards in the state of Hawaii. US Geol Surv Sci Invest Rep 2007–5208:38

Author Biographies

Juan Carlos Carracedo: A prestigious research professor of the Spanish Research Council (CSIC) and director of the Estación Volcanológica de Canarias (EVC) until 2011 is now an emeritus research associate at the University of Las Palmas de Gran Canaria (GEOVOL GROUP). Prof. Carracedo has worked in the Canary Islands over 40 years and published over 200 scientific articles on the geology, palaeomagnetism and geomorphology of the Canaries. jcarracedo@proyinves.ulpgc.es.

Valentin R. Troll: Lectured in volcanology and petrology at Trinity College Dublin for seven years before taking up the Chair in Petrology and Geochemistry at Uppsala University, Sweden in 2008. Prof. Troll has worked on the volcanic phenomena of the Canary Islands since the late 1990s and has published over 25 scientific articles on the geology, petrology and geochemical characteristics of the archipelago. valentin.Troll@geo.uu.se.

J. C. Carracedo and V. R. Troll (eds.), *Teide Volcano*, Active Volcanoes of the World,
DOI: 10.1007/978-3-642-25893-0, © Springer-Verlag Berlin Heidelberg 2013

Index

J. C. Carracedo and V. R. Troll (eds.), *Teide Volcano*, Active Volcanoes of the World,
DOI: 10.1007/978-3-642-25893-0, © Springer-Verlag Berlin Heidelberg 2013

MIX
Papier aus verantwortungsvollen Quellen
Paper from responsible sources
FSC® C105338

If you have any concerns about our products,
you can contact us on
ProductSafety@springernature.com

In case Publisher is established outside the EU,
the EU authorized representative is:
Springer Nature Customer Service Center GmbH
Europaplatz 3, 69115 Heidelberg, Germany

Printed by Libri Plureos GmbH
in Hamburg, Germany